Management of Tropical Plantation-Forests and Their Soil Litter System

Litter, Biota and Soil-Nutrient Dynamics

Management of Tropical Plantation-Forests and Their Soil Litter System

Litter, Biota and Soil-Nutrient Dynamics

Editor
Mallapureddi Vikram Reddy

Science Publishers, Inc.
Enfield (NH), USA   Plymouth, UK

SCIENCE PUBLISHERS, Inc.
Post Office Box 699
Enfield, New Hampshire 03784
United States of America
Internet site: *http://www.scipub.net*

sales@scipub.net (marketing department)
editor@scipub.net (editorial department)
info@scipub.net (for all other enquiries)

Library of Congress Cataloging-in-Publication Data

Management of tropical plantation forests and their soil litter system: litter, biota and soil nutrient dynamics/editor, M. Vikram Reddy.
p.cm.
Includes bibliographical references (p.).
ISBN 1-57808-176-9
1. Forest soils--Tropics. 2. Forest litter--Biodegradation--Tropics. 3. Soil fertility--Tropics. 4. Biological diversity--Tropics. 5. Tree farms--Tropics--Management. I. Reddy, M.V.
SD390.3.T76 M36 2001
634.9--dc21

Published by Science Publishers, Inc., Enfield, NH, USA
Printed in India.

Dedicated
to
my Teachers

"Gurur Brahma Gurur Vishnu Gurur Devo Maheswarah,
Gurur Sakshath Para Brahma Tasmai Sri Guruve Namah"

FOREWORD

There is a global recognition of the need to promote and expand sustainable plantation forestry and farm forestry in the tropics. While areas under native forests are diminishing, those under planted forests are increasing. In a broad sense, the latter include industrial plantations designed for the production of pulp or solid timber for a variety of end products, farm forests and agro-forests, woodlots for fuel, and plantations established for environmental benefits such as land rehabilitation, bio-diversity rejuvenation or catchment protection. A relatively new goal of planted forests is to make use of their potential to sequester atmospheric carbon and thus assist the management of rising CO_2 levels and associated climatic changes.

In many cases, plantations are established and managed for multiple purposes. They are managed with high capital-intensive inputs to achieve high rates of production or may be grown under low intensity regimes. Similarly the scale of their establishment varies from large and contiguous planting to isolated, smallholdings in farms. They may be dominated by a species or mixture of species and may include a wide range of understory plants. The varying goals, purpose, structure and economic considerations influence the way they are managed. There is good basis to conclude that planted forests, if judiciously managed offer great opportunities for sustainable land use from which people can receive multiple benefits. Indeed, planted forest is a critical option in land use strategies in some circumstances, for example, large areas of farmland degraded by rising salinity in Australia, Pakistan and elsewhere.

Sensible management practices are essential if plantations are to deliver benefits regardless of the scale of the operation. The success of planted forests depends upon this sustained productivity and positive contributions to ecosystem functions. These outcomes are determined by a complex array of interactive processes related to soil, vegetation, biota and environment. Of these, the physical, chemical and biological processes in the soil-litter continuum have an overriding influence on nutrient and water availability. Management operations should ensure that the soil base (including litter) is protected and that disruptions to ecological processes—carbon, nutrient and water cycles—are managed within known boundaries of resiliencies and amelioration of the particular ecosystem.

Serious decline in productivity invariably follows the practice of litter raking and removal, still common in some countries. Similarly, long-term research and operational experience clearly show that retention of litter and slash (harvest residue), thus maintaining site organic matter and reserve nutrients is a prerequisite for sustaining and increasing productivity of successive crops of plantations. Soil and litter are the home of fundamentally important biota and biological activity (e.g., decomposition and mineralization) in the ecosystem. Litter layers are also among the most vulnerable components of the ecosystem; for example, they are readily lost in fire or by disturbance associated with site management practices.

The soil-litter biota system and the processes that govern nutrient dynamics have been a popular area of research in temperate, sub-tropical and tropical ecosystems. Because planted forests are expanding in the tropics and they are subject to a myriad of operations—small and large —it is important that the best use is made of existing knowledge, recognizing the critical importance of soil and litter to sustained productivity and ecosystem functions.

This book has brought together information on the interplay between soil, litter and biota that drives fundamental processes. Results from a wide range of plantations representing contrasting biophysical environments are discussed. In particular, the book presents the perspectives and experience of a broad spectrum of scientists with diverse expertise. I hope that this timely synthesis will advance our knowledge and pursuit of sustainable management of planted forests.

E.K. Sadanandan Nambiar
CSIRO Forestry and Forest Products
Kingston ACT 2604, Australia.

PREFACE

Tropical tree plantations, today are becoming increasingly a matter of global priority as they are not only the source of timber and pulp but also the sole agents to mitigate the environmental degradation effects of tropical deforestation, and help in carbon sequestration. Tree-plantation is one of the few human-induced activities that ameliorate the environmental quality particularly taking out CO_2 from the atmosphere. The present edition brings out a comprehensive state-of-the-art overview of the structure and function of soil-litter system in the tropical tree-plantation forests, which is the prime source of nutrients for trees that affects their growth and productivity. The first part of the book describes the litter dynamics in the soil-litter system—the litter production and decomposition and release of nutrients from the litter into the soil in tropical plantation-forests. The second part presents the dynamics of various beneficial edaphic organisms-microorganisms, and micro-, meso- or meio- and macro-fauna and their indispensable role in litter decomposition and nutrient cycling, and pedogenesis in the soil litter system. The final part deals with soil nutrient dynamics in the rhizosphere and their availability for tree growth and productivity in plantation forest ecosystems.

Since the last three decades while continuing my research on plantation-forest ecosystem structure and function with particular reference to soil-litter systems, I conceived the idea of bringing out this volume. Professor P.S. Ramakrishnan (now in Jawaharlal Nehru University, New Delhi) gave me an opportunity to carry out research on "Structure and function of pine-plantation forests" at North Eastern Hill University, Shillong in northeast Indian state of Meghalaya. Dr. J.R.B. Alfred and Professor R. George Michael were very generous in giving me a free hand in continuing my research on ecosystem-structure and –function of consumer and decomposed arthropod communities of these pine plantations. Late Professor D.N. Raychoudhury of Calcutta

University (Calcutta) was magnanimous showing me the path of research, which gave me confidence. I shall ever remain very much beholden to all of them, and to many others who have cooperated with me in plantation forestry research.

This book could not have been completed if the contributors of its various chapters would have not cooperated with me in submitting their articles timely, for which I am indeed thankful to all of them. I appreciate very much the willingness of the publishers, Springer-Verlag (Germany) and Elsevir Science (U.K.), and others permitting to reproduce and use the published data and figures, in various chapters of this book. Special thanks are due to my beloved wife for her cooperation and patience while I was working for this book.

Professor M. Vikram Reddy
Kakatiya University
Warangal-506 009, A.P., India

CONTENTS

Part—II: Biotic Dynamics

Part—III: Soil–Nutrient Dynamics

LIST OF CONTRIBUTORS

Airede, C.E.
Nigerian Institute for Oilpalm Research (NIFOR)
Benign-Akure Road
PMB 1030, Benin City,
Nigeria
Fax: 234-52-602486

Aisueni, N.O.
Nigerian Institute for Oilpalm Reserach (NIFOR)
Benign-Akure Road
PMB 1030, Benin City,
Nigeria
Fax: 234-52-602486
E-mail: nifor@infoweb.abs.net

Akinnifesi, F.K.
Laboratorio do Solos
Universidade Estadual do Maranhao (UEMA)
Caixa Postal 3004,
São Luis 65057-970 MA.
Brazil

Present Address:
SADC-ICRAF Agroforestry project,
Makoka Agricultural Research Station,
P.O. Box 134, Zomba, Malawi,
Fax: (265) 534298 or 534283.
E-mail: fakinnifesi@malawi.net.

Bernhard-Reversat, France
Lab. d'Ecol-des sols Tropicaux,
IRD (ORSTOM), Centre d'lle de France,
32 AVENUE Henri Varagnat,
93143 Bondy, France.
E-mail: France.Reversat@bondy.ird.fr

Costa, Enir Salazar
U.CI, Department of Earth System Science,
Rowland Hall 220,
92697-3100, Irvine CA,
U.S.A.
E-mail: salazar@essgrad.ps.uci.edu.

Fölster, H.
Institute of Soil Science and Forest Nutrition,
Büsginweg 2, D-37077 Göttingen,
Germany
Fax: 0551-39-3310
E-mail: Aspange@gwdg. de.

Gunadi, B.
Soil Ecology Lab
The Ohio State University,
Columbus, U.S.A.

Gunatilleke, I.A.U.N.
Department of Botany,
University of Peradeniya,
Peradeniya 20400
Sri Lanka
Fax: 948 388018 or 948 389026
E-mail: nimsav@botany.pdn.ac.lk.
savnim@slt.lk.

Kang, B.T.
15 Beechwood Drive,
Haverhill, MA 01832,
U.S.A.
E-mail: biauwkang@worldnet.att.net.

Dr. Lehmann, Johannes
Institute of Soil Science and Soil Geography
University of Bayreuth,
95440, Bayreuth,
Germany
Present address:
Department of Crops and Soil Sciences
909 Bradfield Hall
Cornell University, Ithaca
NY 14853, U.S.A.
Fax: (607) 255-8615
E-mail: CL273@cornell.edu

Loumeto, J.J.
Université de Brazzaville
BP 69, Brazzaville,
Congo.

Luizao, Flavio J.
INPA, Departmento de Ecologia
Caixa Postal 478,
CEP 69011-970, Manaus, A.M,
Brazil.

Luizao, Regina C.
INPA, Department de Ecologia,
Caixa Postal 478,
CEP 69011-970 Manaus A.M,
Brazil.

McMillan, A.
UCI, Department of Earth System Science,
Rowland Hall 220,
92697-3100, Irvine CA
USA.

Mailly, D. ing.f.
Ministere des Ressources Naturelles
Direction de la Recherche Forestiere,
2700, rue Einstein, Sainte-Foy, Que.,
Canada G1P 3W8
Fax: 418-643-2165
E-mail: daniel.mailly@mrn.gouv.qc.ca.

Margolis, H.A.
Centre de Recherche en Biologie Forestiere,
Faculté de Foresterie et Geomatique,
Université Laval, Sainte-Foy
Que., GIK 7P4, Canada.

Montagnini, Florencia
Management and Conservation of Forests and Biodiversity
Centro Agronomico Tropical de Investigacion y Enseñanza (CATIE)
7170 Turrialba,
Costa Rica
Present address:
School of Forestry and Environ. Studies, Yale University
310 Prospect St.,
New Haven ET 06511, U.S.A.
E-mail: florencia montagnini@yale.edu

Nambiar, E.K.S.
Plantations and Farm Forestry Program
CSIRO Forestry and Forest Products
P. O. Box E 4008, Kingston ACT 2604,
Australia
Fax: 61 2 6281 8313
E-mail: Sadanandan.nambiar@ ffp.csiro.au

Raizada, A.
Central Soil and Water Conservation Research and Training Institute (ICAR),
218 Kaulagarh Road,
Dehradun-248195, U.P., India

Rao, D.L.N.
Indian Institute of Soil Science,
Nabi Bagh,
Bhopal–462 038, M.P., India
Fax: 910755 733310
E-mail: dlnrao@iiss@ iiss.mp.nic.in

Reddy, M. Vikram
Department of Zoology and Environmental Science Program,
Applied Ecology Laboratory,
Kakatiya University,
Warangal–506 009, A.P. India
Fax: +91-8712-447188
E-mail: mallapureddymv@yahoo.com

Samra, J.S.
Central Soil and Water Conservation Research and Training Institute (ICAR)
218 Kaulagarh Road,
Dehradun-248195, U.P., India
Present address:
Natural Resource Management
Indian Council of Agricultural Research
Krishi Bhavan, New Delhi-110001,
India
Fax: 91 11 3382306, 3387298
E-mail: jssamra@icar.delhi.nic.in

Spangenberg, A.
Institute of Soil Science and Forest Nutrition
Büsginweg 2, D-37077, Göttingen,
Germany
Fax: 0551-39-3310
E-mail: aspange@ gwdg.de

Tian, G.
International Institute of Tropical Agriculture (IITA),
P.M.B. 5320, Ibadan,
Nigeria
E-mail: G.Tian@CGNET.com.

Verhoef, Herman A.
Institute of Ecological Science,
Vrije Universiteit,
de Boelelaan 1087,
1081 HV Amsterdam
the Netherlands
Fax: +31 2044 47123
E-mail: verhoef@bio.vu.nl

Xiaoming Zou,
Institute for Tropical Ecosystem Studies,
University of Puerto Rico,
P. O. Box 363682, San Juan,
Puerto Rico 00936-3682
U.S.A.
Fax: 787-758, 0815
E-mail: xzou@sunites.upr.clu.edn.

CHAPTER 1

INTRODUCTION: PERSPECTIVES OF SOIL-LITTER ENVIRONMENT IN TROPICAL PLANTATION–FORESTS IN RELATION TO LITTER, BIOTA AND NUTRIENT DYNAMICS

M. Vikram Reddy

Applied Ecology Laboratory, Department of Zoology and Environmental Science Prog.
Kakatiya University, Warangal–506009, A.P., India

ABSTRACT

A brief over-view of the causes and consequences of deforestation, and of the beneficial effects of afforestation in ameliorating environmental quality is presented, and the ecosystem functioning of the soil-litter system of tropical plantation-forests, in relation to litter, biota and soil-nutrient dynamics is emphasized.

Key Words: Tropical deforestation, afforestation, soil-litter system.

About a third of Earth's land surface is forested which covers ≈3.54 billion hectares (Noble and Dirzo, 1997). Tropical forests cover only six percent of the total area, and harbour more than 50 percent of the world's animal and plant species. It is a matter of grave concern that these forests are being lost at an alarming rate of about 14×10^6 million hectares a year worldwide (FAO, 1997), and there are similar losses in forest quality across the temperate and boreal forests of Canada, Europe, Russia and U.S.A. Of the 8,080 million hectares of forests in the world

8,000 years ago, about 62 per cent of the world's original forests have been lost. Such loss of forests is mainly due to clear-cutting and conversion of forests to other land-uses, known as 'deforestation'. According to a study of World Wide Fund for Nature, the Asia-Pacific region has already lost 88 per cent of its original forest cover, and only five percent of the remaining area is under protection. Global deforestation models link population, economic growth with deforestation (FAO, World Conservation Monitoring Centre, and World Wide Fund for Nature). According to FAO estimates, countries like India are losing their forest cover at a rate of 0.339 m hectares per year. Such unprecedented removal of forest cover, according to many experts, has two grave global environmental consequences—the emission of CO_2 and other green house gases, and loss of biodiversity (Menon and Bowa, 1998). Deforestation currently brings about 20 percent of CO_2 emission worldwide. Such tropical deforestation may adversely affect global climate, biodiversity conservation, watershed protection and the consequent reduction in soil and water quality (Nordin, 1991). Biological diversity in the tropical forests is also at stake as the forests are fast utilized, and are disappearing day-by-day. The deforested soils in many tropical countries, lose the top soil and the nutrients and as a consequence, the streams, rivers, reservoirs and impoundments are filled with sediment and silt, leading to siltation of these water bodies, which cause overflowing of water and flooding.

The chief causes of deforestation, according FAO, were unprecedented growth of population and rural poverty. In the past two and half decades, many tropical countries became more populated, and poorer particularly in rural areas, placing additional pressure on forests utilizing them for farming, cattle grazing and ranching, fuel wood gathering, making various wood products including structural-wood, construction of hydro-electric power projects and setting up of large industrial units particularly the paper and pulp industries. Nevertheless, one of the main causes of deforestation in these developing countries has been the illegal felling of the trees by forest contractors in league with the officials and even the State Government Ministers. In the tropical countries like India, the average forest cover even by the most optimistic estimates was only 23 per cent. However, in reality India has 11 per cent of its total area under good forest cover against 33 per cent as envisaged under its National Forest Policy 1952 and the revised policy of 1988. According to Kurt (1999) only nine per cent of India's land surface is forested today. In order to fullfil the national goal to bring one-third of the country's landmass under forest/tree cover, an additional area of 30 million hectares outside forest area is needed to be brought under forest cover, through innovative agroforestry programme. However, the most recent estimates stated that 76.52 million ha which is 23.28 per cent of the total geographical area is the recorded forest area of the country, and another 23.83 million ha (the 'culturable wastelands' –13.94 million ha + the 'fallow land and other than current fallows' –9.89 million ha) seem to be the potential areas on which afforestation can be extended (The Hindu Survey of the Environment 2001). An action programme to enlarge the area has been prepared which is to be implemented over a period of 20 years.

Tropical ecosystems as a whole, contain two-thirds of all the world's plant and animal species. The extinction of these species taking place today due to indiscriminate destruction of the tropical forests, is a world concern because of the daily losses of potential sources of new food, fibres and medicine (flowering plants from the world's tropical forests have produced 47 major pharmaceutical drugs and an estimated 328 more drugs are hidden somewhere in the forests' vast reaches). Thus, people worldwide are gradually recognizing the threat to themselves and their global neighbours, and are searching for alternatives to deforestation (e.g., Tropical Forestry Action Plan). Keeping in view the continued destruction of tropical forests it has been emphasized that there is an urgent need for greatly improved management of natural forests, the rehabilitation of degraded forests, and for planting trees in deforested (reforestation) and in marginal agricultural lands and wastelands (afforestation). "Reforestation" of land according to Evans (1983), is that which has carried forests within the last 50 years but where the previous crop is replaced by an essentially different one, or by renewal of essentially the same crop as before, whereas "Afforestation" of land is that where there has been no forest for at least 50 years, and the land is planted with a tree-crop.

Tree plantations in tropics, not only in poorly-wooded countries but also in countries with large natural reserves of timber useful for both domestic and industrial purposes, are therefore increasing rapidly. A 'Tree-plantation' is defined as a "forest crop" or stand raised artificially either by sowing or planting for production of wood, fuel wood and non-wood products (FAO, 1993). It is one of the major forms of land-use in tropics. Of the total forested lands, about 150 million ha are plantations (Noble and Dirzo, 1997). There were rather less than a million hectares of plantations in tropics by 1950, which increased to some 11.5 million by 1980 globally, and it was almost 44 million hectares by 1990 (FAO, 1995); by 1999, estimates reached to 60 million hectares. It clearly showed that the global coverage of tropical tree plantations has increased significantly in the last 20 year. According to the FAO study, in countries like India there was an annual afforestation of 1.44 million ha during 1981–90. Countries like Brazil with one of the largest areas of natural forests in the world—the Amazon rainforest, embarked on plantation programmes two decades back which reached a planting rate of half a million hectares per year. Keeping in view the future demand of wood, experts urged to continue investment in various types of tree-plantations. Even in countries with high population densities, such as India and Indonesia, millions of hectares of land has already been earmarked for plantation programmes. Because, plantations are not only sources of wood and relieve pressue on the natural forests but are also known to protect the environment, mitigating global warming by sequestering atmospheric carbon (Dixon *et al.*, 1991) and help in reducing the build-up of carbon dioxide which is the single most important "greenhouse" gas that pushes up the atmospheric temperature. However, according to the UN's Intergovernmental Panel on Climate Change (IPCC), the planned new forests which are called "Carbon Sinks" will swiftly become saturated with carbon and start returning most of their carbon to the atmosphere, temporarily accelerating global warming. Whether the tree-plantations absorb or release the gas, CO_2 these planted trees will always keep some CO_2 out of the atmosphere and provide other ecological benefits (Will Steffen of Sweden's Royal Academy of Sciences who chairs International Geosphere-Biosphere-Biosphere Programme-IGBP). A team of researchers, Jason Kaye and his colleagues from Colorado State University (USA) reported that carbon sequestration was significantly boosted when the composition of tree stand included nitrogen-fixing trees. They have discovered that the tree stands which were interplanted with albizia trees, *Albizia falcataria* were able to sequester more carbon than areas where only eucalypts, *Eucalyptus saligna* were planted. They have further reported that in tree-plantations where the two species were inter-planted, the stand contained twice as much as carbon in trees as monocrop stand. Besides, the stands of pure albizia sequestered about 20 precent more carbon in the soil system compared to the pure eucalypts stand.

According to Richard Birdsey of U.S. Department of Agriculture Forest Service Laboratory in Radnor Pennsylvania, there is enough land available for further forestry to increase carbon sequestration by perhaps up to 40 per cent at least until 2040. However, carbon sequestration would decrease rapidly after 2045 when land for tree-plantation would be used up (Moffat, 1997).

Tree plantations stabilize the soil and prevent its erosion, and soil degradation (Nambiar and Brown, 1997), control water run-off in catchment areas, provide shelter from wind and heat, and against sound and dust storms which are much needed particularly in tropical countries. For example, casuarina plantations have been used as very extensive shelterbelts along the east coastal dunes of the Indian

Peninsula. Moreover, one of the principal benefits of plantations is to raise agricultural yields by decreasing environmental hazards. According to experts, community forestry leading to rapid afforestation and reforestation targets would alleviate the indigence in about 30 heavily deforested countries of the world A study in the Orissa state of India recently showed that plantations in social forestry projects have the potential for substantial welfare improvements for the target rural population, especially women (Kohlin and Ostwald, 2001). Therefore, it has been estimated that at least 129.6 million hectares of tree plantations are needed by the year 2000 to meet the demand for fuel and industrial wood products, and to stabilize soil and water resources in the regions prone to water and wind erosion of soil (World Watch Analysis).

The plantations which are mostly monoculture ecosystems, however, have certain limitations particularly in species richness and diversity, and long-term maintenance of productivity compared to polyculture systems and natural forests. The organic matter and nutrient cycling processes at the decomposer sub-system are not that efficient in monoculture plantation forest ecosystems as in natural forests. Little organic matter accumulates beneath mature tropical rain forests while much may build up under plantation forests. Nutrient uptake in natural forests is relatively constant from year to year with nutrients being rapidly recycled in the ecosystem while in plantation forests the process of litter decomposition is relatively slow; and thus, the nutrients are recycled slowly, and the nutrient uptake varies with age of the stand.

Plantation forestry affects the soil quality-physico-chemical and biological parameters. Soil physical characteristics such as soil moisture, bulk density and compaction change. The inherent soil moisture reserves are used up in plantations, eventually resulting in dried soil conditions. The harvesting operations gradually deteriorate the soil by compacting it and alter its bulk density. The chemical properties of the soil are also affected. The nutrients are removed from the soil into the trees as they grow and from the ecosystem as they are harvested. The changes may occur in the chemical characteristics of surface soil layers as the litter layer and organic matter becomes dominated by one species particularly in case of monoculture systems, and the chemical composition and decay characteristics of the litter become uniform (Evans, 1983). In plantations where a litter layer of one species develops, and if other plants are suppressed or disappear from such plantations, the uniformity of dead organic matter results in development of humus from exclusively one litter type and the uniformity in the chemical composition of the lecheate water which percolates through the soil. Thus, a single type of food available for the decomposer biota, both microorganisms and fauna, may restrict their biodiversity (Ananthakrishnan, 1996). In polyculture tree-plantations, influence of litter on soil chemical status can be important because the leaves of different species do not decay and release their nutrients at the same time. The structure and composition of litter of a species, the associated edaphic biological conditions at the soil-litter system, and the presence of leaves of other vegetation including that of undergrowth in the plantation, affect the rate of litter decomposition, nutrient dynamics and soil health.

The accumulation of large quantities of decayed litter in broad leaved-species of tropical plantations is not a common feature. In some monoculture plantations like teak (*Tectona grandis* Linn. f.) all other forms of vegetation are usually suppressed and the large leaves are readily decayed. However, the light crowns of most of the *Eucalyptus* plantations lead to other ground flora species and there occurs faster litter breakdown and decomposition. Even heavy litter fall from leguminous trees does not lead to accumulation of litter on the soil surface, because the high nitrogen content of leaves, is attractive to numerous soil organisms, both microflora, and fauna (Evans, 1983). Various types of these organisms at the soil-litter system play an indispensable role in litter breakdown and incorporation, and nutrient cycling (Fig. 1), leading to biological soil fertility (Seastedt, 1984; Reddy 1995; Ananthakrishnan, 1996). The fertility and nutrient supply of the soil under any plantation, and forest system on which plant growth depends is influenced by a multitude of factors (Huttl and Schaaf, 1995), and is the result of the maintenance of various soil environmental conditions including the process of pedogenesis particularly the active biosphere effects, which is known as biological soil fertility. Besides, the soil has its inherent mineralogical fertility, which is the result of chemical composition of parent material of the soil and the rate of weathering and leaching, that release plant nutrients from the mineral fabric of the soil. Both the types of soil fertility-promoting processes, consequently release the contained nutrients for recirculation in the plant-litter soil system on which the optimum growth of plants depends (Maldague, 1970). According to Maldague's concept, there is equilibrium between the fertility processes at any time, but the balance between one and the other may vary from place to place, even with the same type of litter system (Lee, 1985). However, during the initial stage i.e., from the transplantation of saplings till the establishment of the tree-plantation stand and its litter system with litter decomposition process and nutreint release, the trees probably depend for nutrients on the inherent mineralogical soil fertility, or on the exogenous application of fertilizers at the rhizosphere. Few studies have been carried out on the intricacies of soil and litter system particularly its biotic communities, their ecosystem-structure and -function, their role

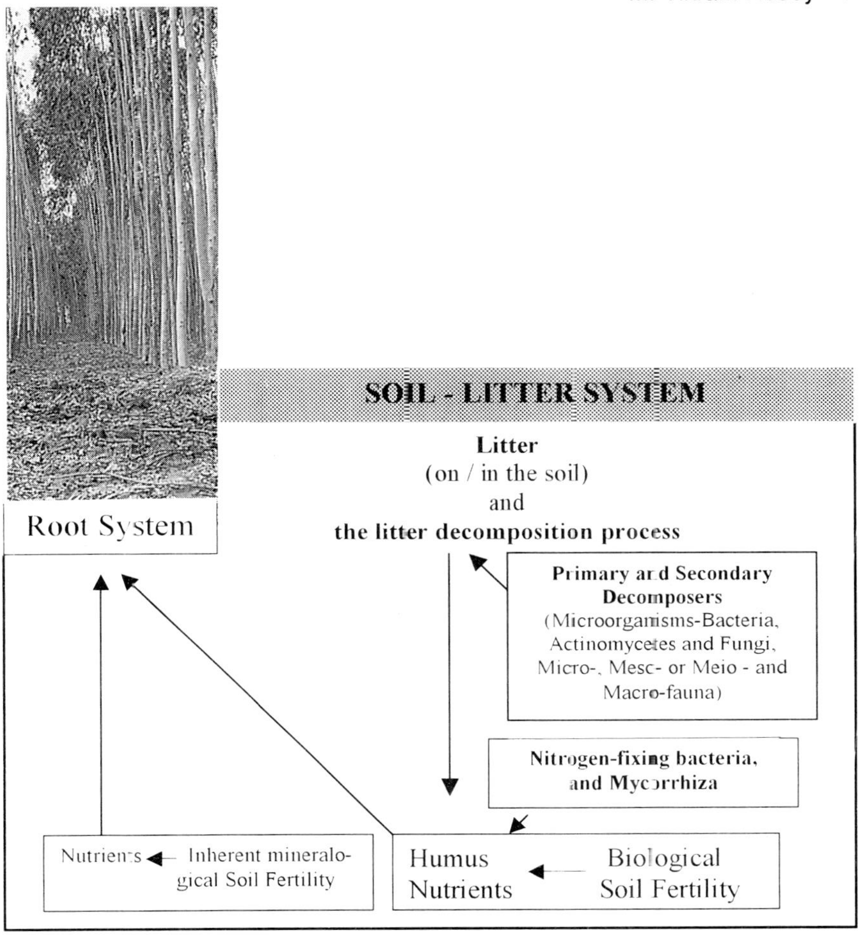

Fig. 1 Soil and litter inhabiting organisms and their crucial role in litter decomposition resulting in Biological Soil Fertility, and Inherent Mineralogical Soil Fertility supplying nutrients to root system of plantation-forest ecosystem.

in pedogenesis and nutrient dynamics under tropical plantation forest ecosystems, though it is known that these organisms affect nutrient turnover playing an indispensable role in tree nutrition through litter breakdown and nutrient release in the rhizosphere, and pedogenesis, and any significant alteration could be expected to alter the tree health, and the effect may be significant in the tree plantations (Evans, 1983; Nambiar and Brown, 1997). Thus, attempts have been made in this volume to bring together the available scattered information on different aspects of structure and function of soil-litter system of various tropical-plantation forests in relation to management in order to put forth a holistic understanding of the system for sustainable management of tropical tree-plantations in the future.

REFERENCES

Ananthakrishnan, T.N., 1996. *Forest Litter Insect Communities Biology and Chemical Ecology.* Oxford & IBH Publ. Co. Pvt. Ltd., New Delhi, Calcutta, 174 pp.

Dixon, R.K., Schroeder, M.P.E., and Winjum J.K., (ed.) 1991. *Assessment of Promising Forest Management Practices and Technologies for Enhancing the Conservation and Sequestration of Atmospheric Carbon and Their Costs at the Site Level.* U.S.E.P.A., Environmental Research Lab., Corvallis, USA, 135 pp.

Evans, J., 1983. *Plantation Forestry in the Tropics.* The English Language Book Society and Clarendon Press, Oxford, 472 pp.

FAO, 1993. Forest Resources Assessment. 1990, FAO, Rome.

FAO, 1995. *Forest Resources Assessment. 1990.* Tropical forest plantation resources, FAO Forestry Technical Paper 128, 81 pp.

FAO, 1997. *State of the World's Forests.* FAO, Rome.

Huttl, R.F., and Scaaf, W., 1995. Nutrient supply of forest soils in relation to Management and site history. *Plant and Soil.* **168–169**: 31–41.

Kohlin, G., and Oswald, M., 2001. Impact of Plantations on Forest use and Forest status in Orissa, India, *Ambio.* **30:** 37–42.

Kurt, F., 1999. Wild animals in a strife-torn world-observations and reflections on 30 years of wildlife conservation in India. *Animal Research and Development.* **30:** 7–43.

Lee, K.E., 1985. *Earthworms: their ecology and relationship with soils and Land use.* Academic Press, Sydney, Orlando, San Diego, New York, London, Toronto, Montreal, Tokyo, 411 pp.

Maldague, M.E., 1979. *"Role des Animaux Edaphiques dans la fertilite des soils forestiers."* Publ. Inst. Nat. Etude Argon. Cong. Ser. Sci. No. 112.

Menon, S., and Bowa, K.S., 1998. Deforestation in the tropics: reconciling disparities in estimates for India. *Ambio.* **27**(7): 576–577.

Moffat, A.S., 1997. Resurgent forests can be greenhouse gas sponges, *Science.* **277:** 315–316.

Nambiar, E.K.S., and Brown, A.G., 1997. *Management of Soil, Nutrient and Water in Tropical Plantation Forests.* ACIAR and CSIRO, Australia and CIFOR, Indonesia, 571 pp.

Noble, I.R., and Dirzo, R., 1997. Forests as human-dominated ecosystems. *Science.* **277:** 522–525.

Nordin, V.J., 1991. Tropical forest—retrospect and prospect. In: A IUFRO Report, XIX World Congress 1990, Montreal, Canada, 543–544.

Reddy, M.V., 1995. *Soil Organisms and Litter Decomposition in the Tropics.* Oxford & IBH Publ. Co. Pvt. Ltd., New Delhi, India, and Westview Press, Colorado, USA, 277 pp.

Seastedt, T.R., 1984. The role of microarthropods in decomposition and mineralization process. *Ann. Rev. Entomol.* **29:** 25–46.

PART—I: LITTER DYNAMICS

CHAPTER 2

THE LITTER SYSTEM IN AFRICAN FOREST-TREE PLANTATIONS

F. Bernhard-Reversat[1] and J.J. Loumeto[2]

[1] *IRD (ORSTOM), Centre d'Ile de France, 32 avenue Henri Varagnat, 93143 Bondy, France*

[2] *Université de Brazzaville, BP 69, Brazzaville, Congo*

ABSTRACT

Africa covers a wide range of climatic regions which give the opportunity to study a wide range of ecologically adapted tree species. Fast growing plantations are made with a reduced number of exotic species, whereas a greater variety of species are planted for timber, often chosen from local species. In the hierarchical model for the control of litter decomposition, climate is the higher factor in the hierarchy. In lowland plantations, temperature is not a limiting factor and the main climatic factor is rainfall, and litter quality will be the second factor in the hierarchy. The aim of the present study is to collect most of the available data on African plantations and to assess the main factors which influence litter dynamics, and the range of variability of the involved processes.

Tree species and rainfall are the main factors controlling litterfall, and litterfall in tree plantations is not basically different from litterfall in natural forests. Most exotic species have a low leaf litter nitrogen content and a high nitrogen use efficiency. Indigenous species have comparatively high nitrogen content. Lignin and phenolics vary with the tree taxa. Standing litter amount is related to climate and is low under dry climate due to low litter production, and in wet climate due to fast decomposition, whereas high amounts are found in medium climate with high production level and medium decomposition rate. Litter fauna generally changes with the establishment of plantations and is related to litter quality.

The decomposition rate is correlated with rainfall, and is not much different in exotic and indigenous plantations when annual rainfall is taken into account. Few data allow to exhibit relationships with litter quality.

Top soil organic matter which originates mainly from litter is highly related to climate and clay content. It appears to be related to the amount of standing litter which results from climate and litterfall. Standing litter could be a relevant measure to evaluate the functioning of the litter system.

Key Words: Africa, tree plantations, litter, soil organic matter.

The interest in studying litters in tropical forests is well established (Proctor, 1983, 1984, Vitousek, 1984) and litter dynamics in African tropical forests have been studied since many years ago (Bartholomew *et al.*, 1953, Nye, 1961, John, 1973, Madge, 1965, Proctor, 1984, Songwe *et al.*, 1988). Although special attention was paid at the beginning to nutrient cycling, the interest in carbon dynamics increased and is now important in litter studies (Schwartz, 1993). Studies on forests and tree plantation were undertaken together, however tree plantations received less attention. (Maheut and Dommergues, 1960). Concern for tropical forest protection and the increased need for land rehabilitation, wood production and pulp production, now emphasize the need for research on tree plantation ecology.

Africa, covering a wide range of climatic regions, from the dry Mediterranean and Sahelian to the wet equatorial, provides the opportunity of studying a wide range of ecologically adapted species. According to the aim of each plantation, those species are divided into fast-growing trees, which provide fuel wood, pulp wood, and are also used for land rehabilitation, and timber trees. Fast growing plantations are made with a reduced number of exotic genus, mainly eucalypts, pines and Australian acacias, which are selected for their growth rate, whereas a greater variety of species are planted for timber, often chosen among local species.

Lavelle *et al.*, (1993) proposed a hierarchical model for the factors controlling litter decomposition, where climate is the highest factor in the hierarchy. According to this model litter systems of highland plantations are highly variable, because temperature variations are dependent on altitude and control litter production and decomposition; consequently the amount of standing litter may show great variations in a reduced space (Morris, 1995, Dames *et al.*, 1998). In the present paper lowland plantations will be mainly considered. In lowland plantations, temperature is not a limiting factor and the main climatic factor is rainfall, and litter quality will be the second factor in the hierarchy (Lavelle *et al.*, 1993, Bernhard-Reversat *et al.*, 1998).

The aim of the present study is to collect most of the available data on African plantations and to assess the main factors which influence litter dynamics, and the range of variability of the involved processes.

Regarding numerous results, it was found relevant to sort the data into genus rather than into species. Eucalypt and pine plantations provide much more data than the other genuses, and their weight in the relationships described hereafter might result in distorted conclusions. However, in order to avoid this drawback an attempt has been made to check the relationships inside the genus.

THE LITTER SYSTEM IN TREE PLANTATIONS

Functions of Litter

Litterfall serves three main functions in the ecosystem (Fig. 1): energy input for soil microflora and fauna, nutrient input for plant nutrition, and material input for soil organic matter building up. The first two functions are completed through decomposition and mineralization, and the third one through decomposition and humification. Those functions are related to the main soil processes, such as biological activity, nutrient cycling and soil structure.

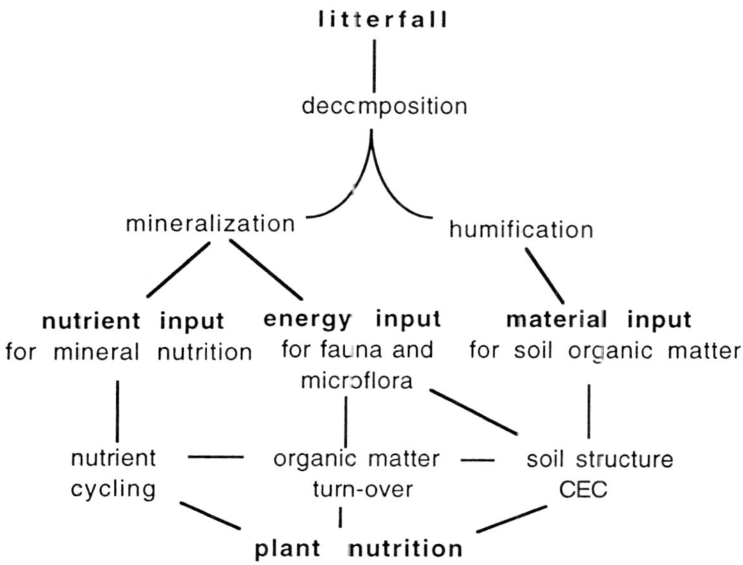

Fig. 1 Simple scheme of the main functions of tree litter.

Plantations Compared to Forests

The characteristics of tree plantations may result in different litter forming processes as compared to natural forests. Tree plantations are generally monospecific and litterfall is dominated by one species, whereas tropical forests are very rich in plant species, at least in humid regions. Consequently the litter quality in tree plantations may lead to nutrient deficiency, extreme decomposition rates, or the accumulation of organic constituents resulting in toxicity for soil living organisms or to allellopathic problems (Chaturvedi and Jha, 1992, Lisanework and Michelsen, 1993, Bernhard-Reversat, 1999).

Sylvicultural practices modify the litter system. Plantations are generally exploited, resulting in nutrients losses by wood exportation and supplying a great amount of residues to the litter system, (Nwoboshi, 1980). Although fertilization is poorly used in African plantations it may bring nutrients. Those practices require nutrient cycling study in order to obtain a nutrient balance (Miller, 1984, Bouillet *et al.*, 1997a). Weeding is generally practised in young plantations and may change the litter system by burying the standing litter (Bouillet *et al.*, 1997b) or by spreading herbicides which may alter the microbial populations (Andariese and Vitousek, 1988).

LITTER INPUT TO THE SOIL (LITTERFALL)

Quantity of Litter

Data were collected from available publications on African tree plantations (Table 1). The amount of litterfall is different according to tree genera (Fig. 2), with the higher litter production for the nitrogen fixing Australian acacias, and the lowest production for conifers (pine

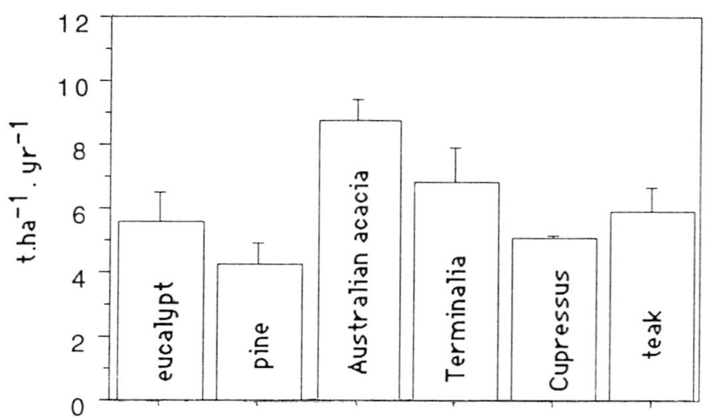

Fig. 2 Average annual litterfal in various genus of planted trees in Africa (data from Table 1).

and cypress) which are often planted in highlands. However a recent study in two plantations of *Pinus kesya* receiving high rainfall in Cameron showed higher litterfall (8.3 and 10.0 kg/ha/year) than the average given at the figure 2 (Njoukam *et al.* 1999). Neither the age of the planted plots nor the age of the trees are related to litter production when the whole data or one genus data are taken into account. However several studies showed that the litterfall of a given species in one site increases with age during and after the juvenile age (Kadeba 1998, Bernhard-Reversat *et al.*, 1999), but the age effect is hidden in the overall data, where variability is introduced by soil, climate and species.

Global relationship with the length of dry season was not observed, but the correlation with rainfall (Fig. 3) was significant (p=0.007). When pine or (eucalyptus) were considered alone, the relationships were still significant (p=0.04). The relationship of litterfall with rainfall was reported for tropical forest Jordan (1988) and according to Lonsdale (1988), precipitations give the best prediction model for leaf litterfall in tropical forests, whereas altitude give the best model for total litterfall, and when worldwide forests are considered latitude give the best model. Meentemeyer *et al.* (1982) find the better relationship with actual evapotranspiration (AET) but this parameter is rarely measured in studies on plantations.

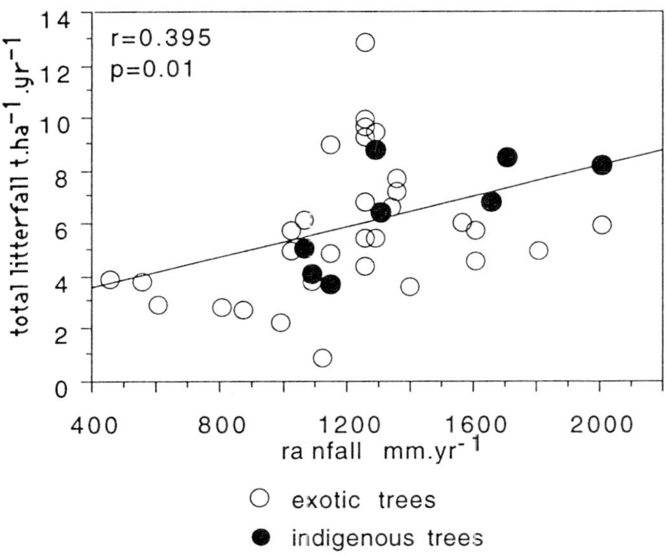

Fig. 3 Relationship between annual rainfall and annual litterfall in African tree plantations (data from Table 1).

Table 1 Data on African forest-tree plantations. K is the authors' value if available, if not K is the ratio litterfall/standing litter. Litter N content is given for leaf litterfall.

Authors	*Tree genus and species*	*Rain* mm/yr	*Plot age* yrs	*Litterfall* /year t/ha	*Litter N* %	*Standing litter* t/ha	*K*	*Top soil* C %
Bernhard-Reversat 1976	*Terminalia ivorensis*	2000	22	8.3	1.88	2.6	3.20	3.02
Bernhard-Reversat 1976	*Terminalia ivorensis*	1700	38	8.6	1.30	3.5	2.50	2.70
Bernhard-Reversat 1993	*Acacia mangium*	1250	7	9.7	1.66	12.0	0.69	0.92
Bernhard-Reversat 1993	*Acacia auriculiformis*	1250	8	10.0	1.65		0.69	0.92
Bernhard-Reversat 1993	*Eucalyptus HS2*	1250	7	5.5	0.65	10.3	0.35	0.71
Bernhard-Reversat 1993	*Eucalyptus PF1*	1250	10			15.5		
Bernhard-Reversat 1993	*Eucalyptus HS2*	1250	10			17.2		
Bernhard-Reversat 1993	*Eucalyptus PF1*	1250	7	6.9	0.66	14.0	0.49	0.71
Bernhard-Reversat *et al.*, 1999	*Eucalyptus PF1*	1250	19	9.4	0.75			0.73
Bernhard-Reversat *et al.*, 1999	*Eucalyptus PF1*	1250	19	12.9				0.78
Bernhard-Reversat unpublished	*Terminalia superba*	1300	32	6.5	1.30	4.6	1.40	1.26
Bernhard-Reversat unpublished	*Eucalyptus camaldulensis*	800	5	2.9	1.00	4.7	0.44	0.82
Bernhard-Reversat unpublished	*Pinus caribaea*	1250	10		0.25	15.9		0.59
Bernhard-Reversat unpublished	*Eucalyptus camaldulensis*	600	8	3.0	0.92	1.5	1.20	0.77
Egunjobi 1974	*Tectona grandis*	1140		9.0	0.89	4.5	2.02	
Egunjobi *et al.*, 1979 a,b	*Pinus caribaea*	1330	9	6.7		19.7	0.34	
Goubiere *et al.*, 1995	*Casuarina equisetifolia*	411	37			60.2		
Harmand 1997	*Acacia polyacantha*	1080	6	4.1	2.12	4.5	0.92	0.47
Harmand 1997	*Eucalyptus camaldulensis*	1080	6	3.9	0.71	12.6	0.39	0.35
Kadeba *et al.*, 1998	*Pinus caribaea*	1250	14	4.5	0.97	15.7	0.29	0.87
Kadeba *et al.*, 1998	*Pinus caribaea*	1800	14	5.0	0.88	17.2	0.29	1.67
Kadeba *et al.*, 1998	*Pinus caribaea*	1560	14	6.1	0.85	18.5	0.33	1.30
Kang *et al.*, 1994	*Treculia africana*	1280	10	8.8				0.96
Kang *et al.*, 1994	*Cordia alliodora*	1280	10	9.6				1.32

Kang *et al.*, 1994	*Gmelina arborea*	1280	10	5.5				1.20
King *et al.*, 1994	*Pinus patula*	885	40		0.83	25.7		0.91
King *et al.*, 1994	*Tectona grandis*	885	20		0.92	25.9		1.04
Knockaert 19 81	*Eucalyptus camaldulensis*	450	9	4.0		9.4	0.42	
Knockaert 1981	*Eucalyptus camaldulensis*	550	49	3.9		21.6	0.18	
Lisanework *et al.*, 1994	*Eucalyptus globulus*	1019	40	5.8	0.75		1.50	3.31
Lisanework *et al.*, 1994	*Cupressus lusitanica*	1019	28	5.0	0.60			3.56
Loumeto unbublished	*Okoumea klaineana*	1650	14	6.9	3.13	2.2	3.15	1.40
Lundgren in Vitousek 1984	*Cupressus*	1060	18	5.2				
Lundgren in Vitousek 1984	*Pinus*	1060	18	6.2				
Maheut *et al.*, 1960	*Tectona grandis*	1600	4	5.8	0.66			0.96
Maheut *et al.*, 1960	*Tectona grandis*	1600	8	4.7	0.94			1.10
Mailly *et al.*, 1992	*Casuarina equisetifolia*	525	34			120.3		0.23
Mailly *et al.*, 1992	*Casuarina equisetifolia*	525	6			23.1		0.08
Nwoboshi 1980	*Tectona grandis*		25	5.0		9.0	0.56	3.02
Ola-Adams *et al.*, 1992	*Tectona grandis*	1140*	18	5.0				
Ola-Adams *et al.*, 1992	*Terminalia*	1140*	13	3.8				
Oliver *et al.*, 1994 unpublished	*Acacia auriculiformis*	1350	5	7.3				1.99
Oliver *et al.*, 1994 unpublished	*Acacia mangium*	1350	5	7.8				1.99
Ronde 1993	*Pinus radiata*	620**	20			14.2		
Ronde 1993	*Pinus radiata*	620**	32			19.4		
Tie Bi and Omont, 1987	*Hevea brasilliensis*	2000	20	6.0		1.2	5.00	1.85
Versfeld and Donald, 1991	*Pinus radiata*	1390	50	3.7	0.64	6.6	0.55	
Wienand and Stock, 1995	*Pinus elliottii*	1113	28	1.0		1.0	1.00	
Wienand and Stock, 1995	*Pinus elliottii*	868	20	2.8		14.0	0.21	
Wienand and Stock, 1995	*Pinus elliottii*	980	25	2.3		7.0	0.35	

*data for Ibadan from Egunjobi 1979. ** data from the Weather Bureau of South Africa for Port Elizabeth.

Tree species and rainfall are the main factors controlling litterfall, and other factors related to sylvicultural practices seem to be less relevant; for instance, a study of the spacing effect in timber plantations in Nigeria did not show any significant difference (Ola-Adams and Egunjobi, 1992)

LITTER QUALITY

Litter quality refers to mineral nutrients as well as to organic compounds and physical features, and is more often studied in leaf litter than in total litter. In plantations, leaves account for the most part of total litter, although bark and twigs may be important in some tree species. Nutrient content is often studied for mineral cycling assessments and organic compounds are being included into recent decomposition studies; physical characteristics are exceptionally taken into account.

PHYSICAL CHARACTERISTICS

The easiest physical feature of litter to be measured is leaf thickness. Two examples of measurements made between the nervures with a thickness gauge by 0.01 mm, one with Australian acacias planted in Senegal, and the other with clones of two eucalypt hybrids planted in Congo, showed that leaf thickness was very significantly different between taxa, here species, hybrids or clones (Fig. 4); in eucalypt, leaf thickness was correlated with in vitro respiration and nitrogen content.

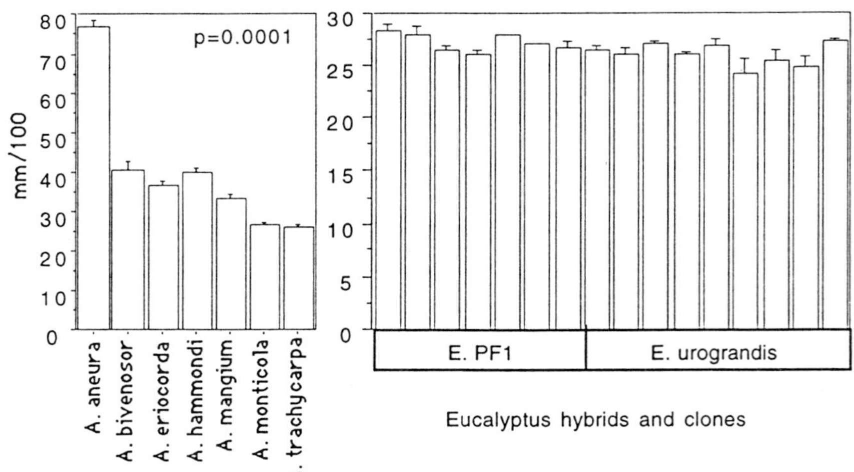

Fig. 4 Leaf blade thickness of litter in various Australian acacias planted in Senegal (left) and in various clones of Congolese eucalypt (right). Means of 10 leaves, measured on moisturized leaves.

Nitrogen Content

Nitrogen content of leaf litter, although expressed as mineral, is the expression of protein content, an important component for litter quality, and is discussed here. The main factor of leaf litter nitrogen variations is tree genus (Fig. 5A) and the ANOVA is highly significant (p<0.0001). Most exotic species are fast growing, and, except nitrogen fixing species, they have a low nitrogen content and a high nitrogen use efficiency (Chapin, 1980, 1983, Smith *et al.*, 1998). Indigenous species have comparatively high nitrogen content. Because of this great variability among genera, other controlling factors are not observed when the whole data are taken into account. However the nitrogen content of eucalypt leaf litter is highly correlated to rainfall: the highest the rainfall the lowest the nitrogen content. Faster tree growth and lower soil nitrogen content could explain this relationship, which is not observed in pine (Fig. 5B).

Organic Compounds

Litter organic compounds are considered to explain decomposition process in temperate as well as in tropical litter (Couteaux *et al.*, 1995, Spain and Le Feuvre, 1987, Cadish and Giller, 1997). Tree plantations received little attention with scarce data (Palm and Sanchez, 1991). The most common compounds which are considered are: soluble organic compounds susceptible to undergo leaching by rain, secondary metabolites including phenolic compounds and tannins, and the leaf fibers, cellulose and lignin. Some interest is sometimes given to terpenes which are assumed to act as allelopathic agents (Del Moral and Muller, 1969).

Data available on litter quality in African tree plantation is scarce. Organic compounds in litters of acacia and eucalypt were measured according to the methods described by Bernhard-Reversat (1999). It was observed that lignin content varied according to genera and was lowest in eucalypts (Fig. 6). Surprisingly, lignin content appears to be closely related to annual rainfall if nitrogen fixing trees are sorted out, and this relationship is still highly significant if eucalypts alone are considered (Fig. 7); this suggests that when water does not limit photosynthesis, an increased synthesis of secondary metabolites occurs (Matsuki, 1997), including those which are at the origin of lignin synthesis; this possible relationship would deserve more investigations, and could be supported by the positive relationship between water-soluble phenolics and rainfall in eucalypt leaf litter (p=0.05), and between methanol soluble phenolics (measured according to Anderson and Ingram, 1993) and rainfall in Australian acacia leaf litter (p=0.003).

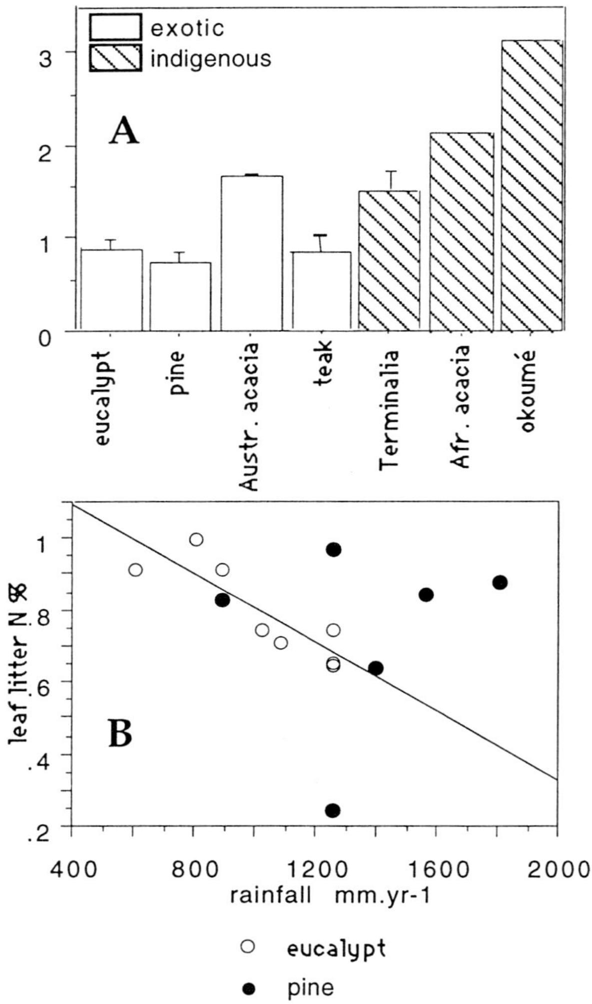

Fig. 5 Leaf litter nitrogen content. A: in various tree genus; B: relationships with annual rainfall in African eucalypt and pine plantations (data from Table 1).

Nutrient Inputs Through Litterfall

Nutrient cycling occurs mainly through litterfall and throughfall and except for potassium in wet forests, litterfall is the main pathway. Nutrient input to the soil by litter in African plantations received much attention (Maheut and Dommergues, 1960, Bernhard-Reversat, 1976,

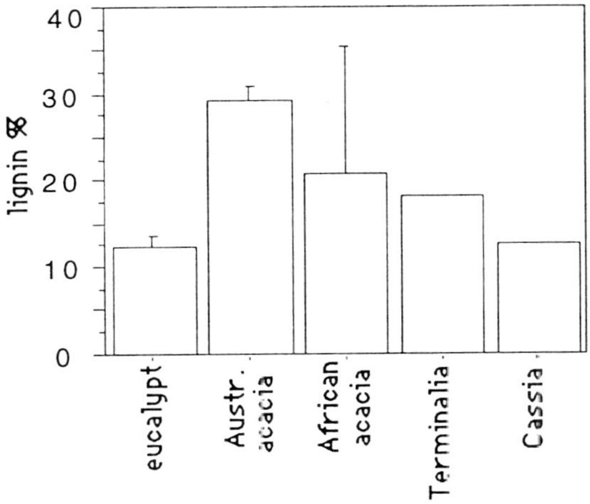

Fig. 6 Lignin content in the leaf litter of some tree genus in African plantations.

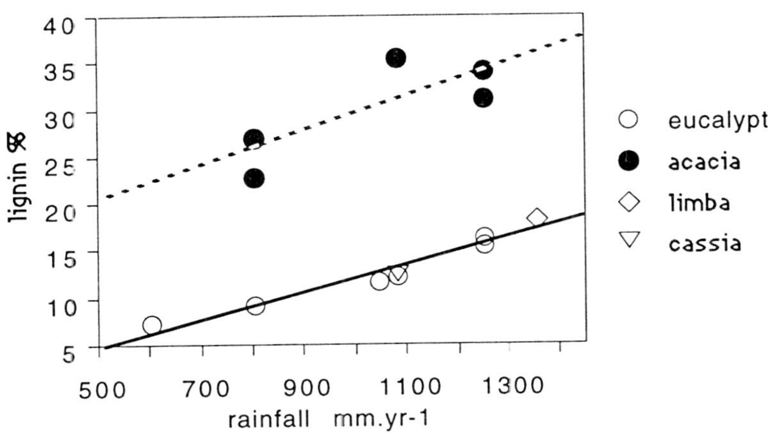

Fig. 7 Lignin content in leaf litter versus annual rainfall. Regression curves for acacias (broken line, not significant) and for other trees (solid line, r=0.974, p<0.0001) in African plantations (data from Bernhard-Reversat, 1997 and unpublished, Harmand, 1998).

1987, Nwoboshi, 1980, Kadeba and Aduayi, 1985, Loubelo, 1990, Kadeba, 1998, Laclau *et al.*, 1999; Njoukam *et al.* 1999); this interest is related to the occurrence of nutrient outputs by log exploitation. Nutrient input is more related to nutrient cycling than to the functioning of litter systems, and moreover a worldwide review was made by Vitousek (1984). According to this author litterfall is related to litter nutrient content for the residuals of climate regression, and plantations did not show different relationships from native forests.

STANDING LITTER ON THE SOIL

Quantity of Standing Litter

The amount of litter on the ground in tree plantations is studied as an indicator of the litter system and the plantation functionning (Table 1). A particular case is that of the filao (*Casuarina equisetifolia*): this exotic tree is planted in dry littoral dune soils, which are free from organic matter and organisms, and litter accumulation is very high, up to 120 t/ha, because of the lack of decomposition (Mailly and Margolis, 1992, Gourbiere and Debouzie, 1995). Highland pine plantations in the Mediterranean mountain climate of South Africa also reach 100 t/ha (Morris, 1995). In tropical climate, excluding the results on filao, standing litter accumulation in plantations is significantly dependent on the length of the dry season (Fig. 8 A), with low accumulation under dry climate due to low litter production, low accumulation in wet climate due to fast decomposition, and high accumulation in medium climate with high production level and medium decomposition rate. Consequently

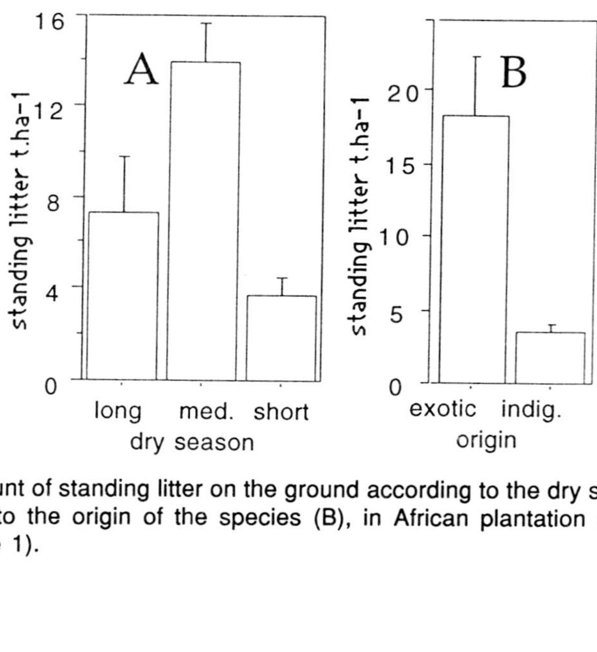

Fig. 8 Amount of standing litter on the ground according to the dry season (A), and to the origin of the species (B), in African plantation (data from Table 1).

no simple correlation can be observed between standing litter accumulation and rainfall. Where exotic species exhibit a wide range of standing litter amounts, the indigenous ones show low litter accumulation because their decomposing organisms are better adapted to the climate they live in, and the average accumulation is higher under exotic trees than under indigenous ones (Fig. 8 B).

Quality of Standing Litter

Like for litterfall, data on the nutrient content of standing litter is given by authors studying biogeochemical cycling (Bernhard-Reversat, 1977, Nwoboshi, 1980, Loubelo, 1990, Harmand, 1997). Available data on organic quality is scarce although there is some natural tropical forests (McKey *et al.*, 1978). Nitrogen in particle size fractions of standing litter was shown to be related to initial litter content, and nitrogen content increases with decreasing particle size in nitrogen poor eucalypt and pine litters whereas it decreases in nitrogen rich acacia litters (Bernhard-Reversat, 1993).

Organic composition of standing litter is related to that of litterfall. However decomposition processes modify its composition, as shown by the comparison of lignin content in the standing litter of exotic plantations with that of natural forests in the Congo (Bernhard-Reversat and Schwartz, 1997); unlike in natural forests, lignin content increased in eucalypt standing litter compared to fresh litter, because of the lack of lignin-decomposing white rot fungi (Fig. 9).

LITTER ASSOCIATED FAUNA

Soil and litter fauna is the main factor of soil fertility (Lavelle *et al.*, 1993). Although both litter decomposition and litter incorporation to soil are dependent on meso and microfauna (Bocock, 1964; Reddy, 1995; Förster *et al.*, 1996, Pereira *et al.*, 1998), data available on African plantations is scant.

Mesofauna

Mesofauna is mainly made up of micro-arthropods. Tian *et al.* (1997) carried out decomposition studies in various vegetations at Ibadan, Nigeria, and assessed the microarthropod mediated decomposition to exhibit a higher rate in crop fields than in wooded stands because micro-arthropods mainly enhance microbial activity and are less important in stands which already have a high microbial activity. Bernhard-Reversat (1993) reported a highly significant increase in the number of micro-arthropods in plantations compared to the native

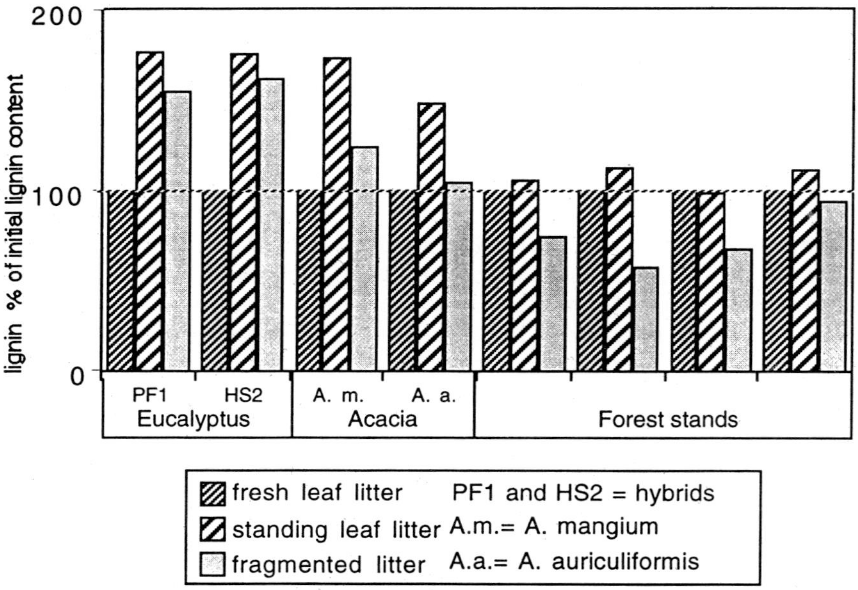

Fig. 9 Lignin content of litterfatl and standing litter fractions, expressed as per cent of the initial lignin content of freshly fallen leaf litter in Congolese plantations and forest stands. (after Bernhard-Reversat and Schwartz, 1997).

savanna in the Congo; no difference in micro-arthropod density occurred between eucalypt plantations and acacia plantations although the litter quality was better under acacia and seemed to increase the diversity of the morphotypes. Maldague (1961) showed that in Zaire the density of micro-arthropods was higher in artificial tree stands than in natural forests. Generally, forested environment appear to be more favourable to microfauna than open stands.

MACROFAUNA

A few studies were made on macrofauna and such data are available for plantations of rubber (Gillot *et al.*, 1995), eucalypt and pine (Dangerfield and Miller, 1996; Mboukou-Kimbatsa *et al.*, 1998), and Australian acacia (Mboukou-Kimbatsa *et al.*, 1998.)

Most results concern both litter and top soil, and show the change in macrofauna density in plantations compared to the native environment, either savanna or natural forest; those changes may concern total macrofauna or special groups as termites, earthworms or millipedes, which may be increased or decreased in density. In the Congo Mboukou-Kimbatsa *et al.* (1998) and Bernhard-Reversat *et al.* (1999) reported an increase in termite, earthworm, and litter group densities in eucalypt plantations of increasing age from 6 to 19 years whereas in the native savanna there were more termites than in plantations and no earthworms (Fig. 10). This increase in density with plantation age suggests an improved soil functioning with age. However in rubber (*Hevea brasiliensis*) plantations, Gillot *et al.* (1996) found that the litter group density increased from 5-year-old to 20-year-old stands, and decreased in a 30-year-old plantation together with decreasing tree productivity, whereas termites decreased since plantations were 10-years-old. The macrofauna population may find a favourable environment in plantation but the foregoing change in soil or litter quality may alter the environment conditions and the functioning of the system; macrofauna density could be a good indicator of the plantation's condition, although

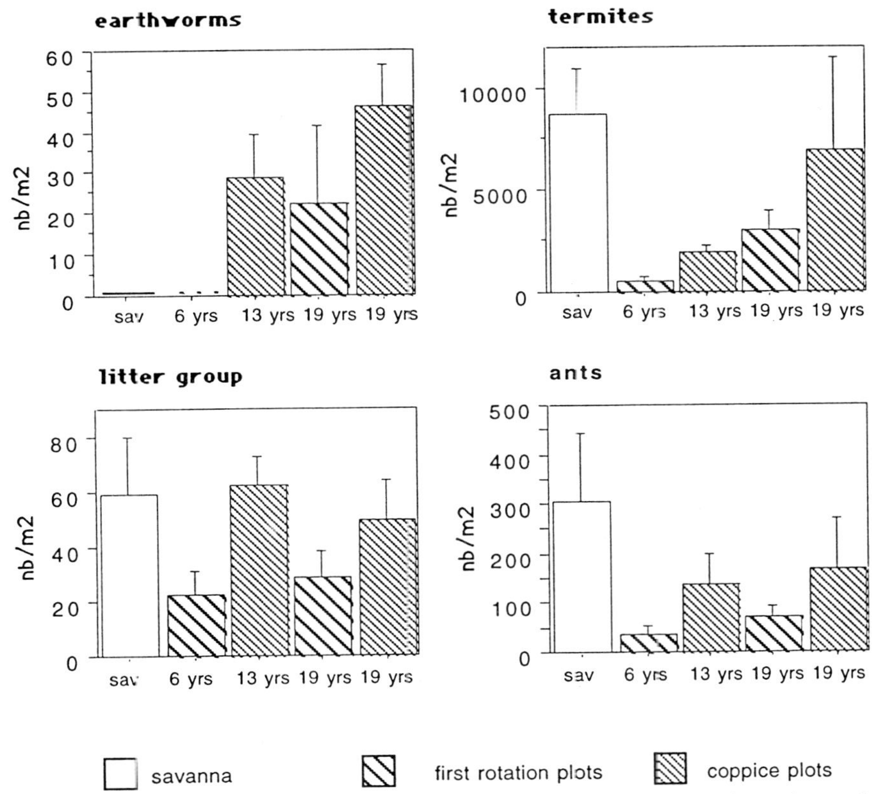

Fig. 10 Macrofauna densities in native savanna and in eucalypt plantations of various ages (Congo, data from Bernhard-Reversat *et al.* 1999).

more data is necessary. The influence of plant species on macrofauna was shown by Dangerfield *et al.* (1995) and Mboukou-Kimbatsa *et al.* (1998).

Relationships of Fauna with Litter Quality and Soil Quality

Litter quality is of critical importance in regulating soil and litter fauna and soil biological functions (Wardle and Lavelle, 1997, Tian *et al.*, 1997). Attempts to relate litter or soil macrofauna decomposing activity to litter quality in some African crops, mainly crop residues and agroforestry prunings, was made by Tian *et al.* (1993, 1995) who showed the important role of fauna in the decomposition of low quality litter. In the eucalypt plantations of the Congo, termites and ant densities were inversely related to the lignin content of the litter, and termite density was negatively correlated to the phenolic compound content (Bernhard-Reversat *et al.*, 1999). Nothing is known about plantations of indigenous trees. Although there is evidence that the monospecificity of the litter affects soil fauna in plantations, the detailed relationships and processes are unknown.

DECOMPOSITION

Litter decomposition received general attention in forest ecosystems including tropical ones (Anderson and Swift, 1983). Two field methods are mainly used, either comparison of litter input and litter standing crop, or weight loss measurements in litterbags.

Decomposition Coefficient

Since the theoretical basis of the calculation of the decomposition coefficient were assessed by Olson (1963) numerous authors used it in decomposition studies. The basic principle for mature tropical forests is to calculate the ratio "annual litter input/standing litter". Because of particular conditions some authors used other formulas (Bernhard-Reversat, 1982). In order to exploit the data of Table 1, we used the K value that was given by the author when available, or if not we used the previous ratio. The variability due to measurement or calculation methods does not prevent assessing some relationships.

The range of variations of K is very wide, from 0.2 in eucalypt plantations in Morocco, to 5 in a rubber plantation in Ivory Coast (Table 1). There is no relationship either with plot age nor with tree age. Genus, origin and climate control decomposition rate. Among the available results, pines and eucalypts have the lowest decomposition rate, whereas the indigenous *Terminalia* showed a high decomposition rate (Fig. 11 A); however more data on planted indigenous species is necessary to check the difference between the low decomposition rate of exotic species and the high decomposition rate of indigenous species,

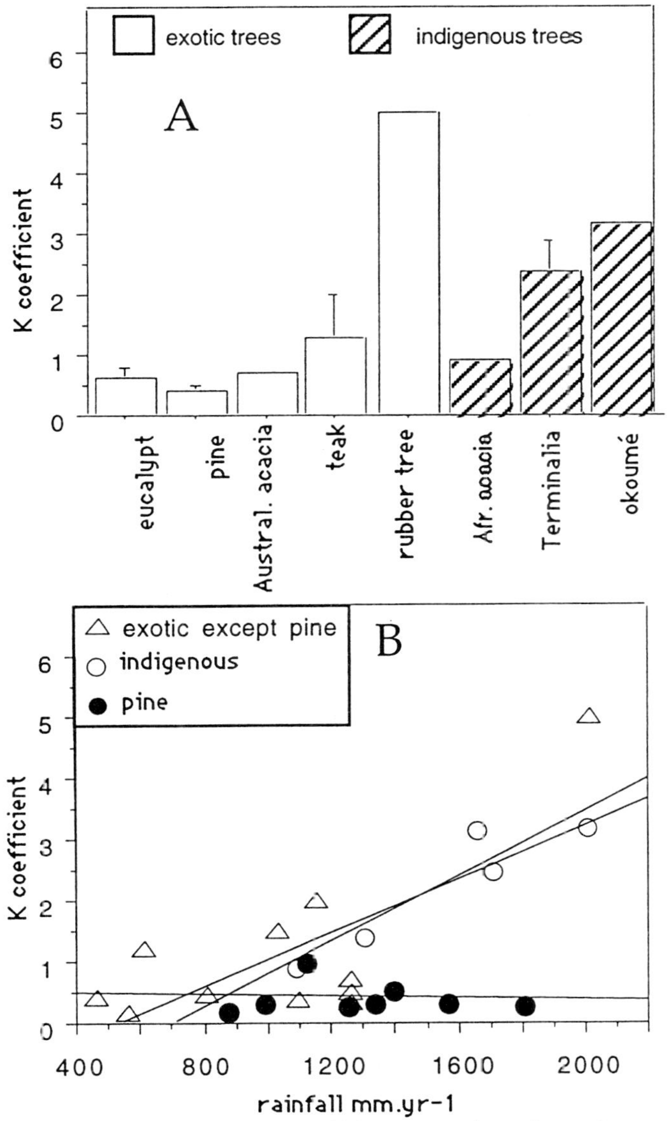

Fig. 11 Decomposition coefficient in African tree plantations: A, according to genus, B, related to rainfall (data from Table 1).

and the comparison is made difficult due to the scarcity of indigenous plantations in the dry regions where exotic species are most often planted.

Differences related to the length of dry season are highly significant (anova: $p<0.0001$) and are due to the high decomposition rate in short dry season stands. The averages for medium and long dry season are not different. The correlation with annual rainfall is significant ($p=0.02$) although pine litter decomposition seems not to be related to rainfall (Fig. 11 B), and the correlation is stronger ($p=0.0008$) when pine is not taken into account; macrofauna density is very low in pine litter and soil (Mboukou-Kimbatsa *et al.*, 1998) and the low decomposition rate of this litter could be due to either physical or chemical litter quality, although data available is little except evidence for low N content. In this case litter quality seems to prevail over climate, unlike in the hierarchical model (Lavelle *et al.*, 1993).

Litterbag Experiments

Litterbags are generally used to compare the decomposition rate of various tree species, to study the influence of macro and micro-fauna and to assess nutrient losses (Lisanework and Michelsen 1994; Loubelo 1990). However the ability of the litterbag technique to provide actual decomposition rate is questioned because weight losses are lower than calculated losses (Swift *et al.*, 1979, Bernhard-Reversat, 1982).

An attempt to put together the results of some litterbag experiments in African plantations is made (Table 2) using either the author's figures or data estimated from their weight loss curves, and percent weight loss at one, three and 12 months were compared. No correlation with the K coefficient occurs after one month decay, but it was observed after three and 12 months. However the litterbag experiments do not show any relationship with the studied factors suggesting that the resulting decomposition rates are highly dependent on sites, seasons and methodological protocols. Experiments conducted in eucalypt plantations in the Congo (Bernhard-Reversat *et al.*, 1999) reported a very significant effect of season; a significant increase of decomposition rate with plantation age was observed and could be related to the decreasing lignin content.

Decomposition and Litter Quality

The decomposition rate as measured by the K coefficient in Table 1 increases significantly with leaf litter nitrogen content ($r=0.623$, $p=0.006$) and a multiple regression shows that nitrogen content is more related to decomposition rate than to rainfall. Nitrogen content was shown to

Table 2 Weight loss per cent of initial weight in in litterbag decomposition compared to the decomposition coefficient K. Weight loss values are those givent by the authors when available, if not, they are estimated according to the authors' curves.

Authors	*Tree genus and species*	*rainfall* mm/ *year*	*weight loss % in 1 month*	*weight loss % in 3 months*	*weight loss% in 12 months*	*K* (*from table* 1)
Bernhard-Reversat 1987 and unpubl.	*Eucalyptus camaldulensis*	800	16	25	25	0.44
Bernhard-Reversat 1993	*Acacia mangium*	1250	21	37	60	0.69
Bernhard-Reversat 1993	*Eucalyptus PF1*	1250	18	25	44	0.49
Bernhard-Reversat 1993	*Eucalyptus HS2*	1250	11	15		0.35
Bernhard-Reversat *et al.*, 1999	*Eucalyptus PF1*	1250	26	32		
Bernhard-Reversat *et al.*, 1999	*Eucalyptus PF1*	1250	22	30		
Bernhard-Reversat unpublished	*Azadirachta indica*	600	29	44	44	
Bernhard-Reversat unpublished	*Eucalyptus camaldulensis*	600	18	36	36	1.20
Bernhard-Reversat unpublished	*Terminalia superba*	1300	13	60	100	1.40
Harmand 1997	*Eucalyptus camaldulensis*	1080			46	0.39
Lisanework *et al.*, 1994	*Eucalyptus globulus*	1019		30	82	1.50
Lisanework *et al.*, 1994	*Cupressus lusitanica*	1019		39	75	
Loumeto unpublished	*Okoumea klaineana*	1650				3.15
Versfeld and Donald, 1991	*Pinus radiata*	1390	8	14	29	0.55

be related to decomposition by several authors in temperate (Aber and Melillo, 1980) and in tropical environment (Laishram and Yadava, 1988), whereas climate is also reported as the main factor of decomposition (Meentemeyer, 1978, Aerts, 1997), and various factors are assumed to act at different decomposition stages (Couteaux *et al.*, 1995). The relationship of litter decomposition to litter organics, particularly lignin and phenolic compounds, were widely studied and some data is reported for other tropical vegetations (Constantinides and Fownes, 1994, Laishram and Yadava, 1988, Palm and Sanchez, 1991) but little is known on African tree plantations whereas data on agroforestry litters is available (Drechsel *et al.*, 1991, Vanlauwe *et al.*, 1996, 1997). In the Congolese eucalypt, Bernhard-Reversat *et al.* (1999) observed in a small number of plots that the decomposition rate was negatively correlated with lignin content during the dry season (p=0.03) suggesting that the lignin content of the litter could control faunal consumption when microbial activity is reduced by water limitation; results on other relationships with litter organic compounds were inconclusive because of the narrow range of variations inside eucalypts, and because decomposition rate increased with the age of plantations.

TOP SOIL ORGANIC MATTER

Through decomposition processes, the consequences of litter quality and quantity influence organic matter building up in the soil (Coleman *et al.*, 1989). According to Anderson and Flanagan (1989) the first hierarchical level of the processes is environmental control, followed by resource type and quality, then organisms, resulting in soil organic matter pool. These authors attribute different scales to this scheme and ecosystem, with total litter, controls total soil organic matter, whereas the composition of litter according to plant species controls patch variations in soil organic matter. We tried to check the influence of climate (rainfall), soil (clay content) and litter on soil organic matter content of some African tree plantations, from the data of Table 1.

The available data does not allow calculating anova for both texture and dry season length. Anova were made separately and are both highly significant (Fig. 12). Increasing annual rainfall increases C content in sandy soils (r=0.738, p=0.0009) but not in clayey soil; more data would be necessary to check these relationships.

The influence of climatic and soil conditions on soil organic matter makes the assessment of control processes by litter difficult; the relationships between soil organic matter and litter amounts are also influenced by litter quality. Neither the genus nor the origin of the trees

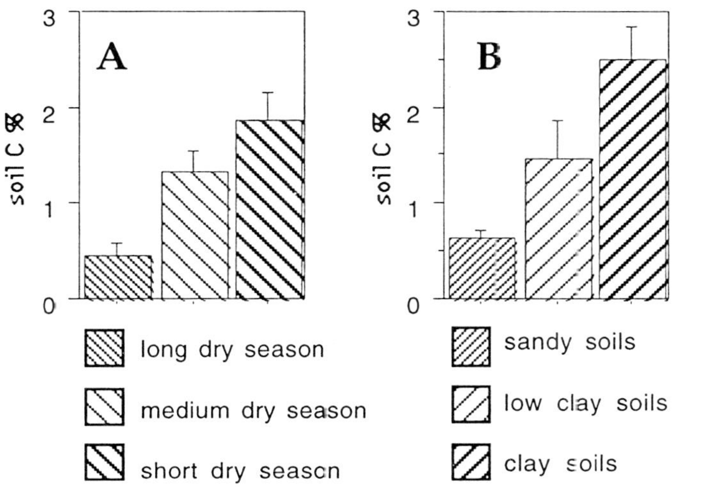

Fig. 12 Soil organic matter content according to the dry season length (A), and clay (B). Sandy soils: clay content < 10%, clay soils: clay content > 20%. Long dry season > 7 months, short dry season < 4 months. (data from Table 1).

is related to soil C with the above data. The age of the plantation (age since the plot was first planted) is well correlated to soil C (p=0.002) which accumulates throughout the years. When plantation age is taken into account in a multiple regression, the amount of standing litter, which result from litterfall and decomposition, is strongly negatively correlated to soil C (p=0.004) showing that low litter decomposition prevents the building up of soil organic matter. However relationships with accumulation can be more strong than relationships with decomposition in young stands under the same climate, and Kotto-Same *et al.* (1997) reported a positive correlation between standing litter stock and total soil C in a succession of fallows after cropping where soil C was mainly related to fallow age.

The light organic particle size fractions were related to the amount of standing litter in Congolese eucalypt plantation. Bernhard-Reversat (1993) showed considerable differences between particle size distribution of top layer soil organic matter under Australian acacia and eucalypt plantations in Congo, with a low level of the medium sized light fraction under eucalypt. The same difference was observed in Senegalese eucalypt plantations compared to neem (*Azadirachta indica*) and *Prosopis* plantations (Bernhard-Reversat, 1987) and refers to different origin and quality of the fractions (Boone, 1994, Gregorich *et al.*, 1996). This suggests that the humification processes are different according to litter quality, which controls the ratio comminution to mineralization.

CONCLUSION: TREE PLANTATIONS AND NATURAL FORESTS

Data on litterfall allows comparison between natural forests and plantations (Vitousek, 1984). Basically no differences appear between natural forest and plantations litterfall in Africa when annual rainfall is taken into account (Fig. 14) although local differences may be observed when the comparison is made on the same site (Table 3). In humid regions, litterfall of nitrogen-fixing fast-growing species and litterfall of timber plantations reach the same amounts than those of natural rain forests (Bernhard-Reversat, 1976, 1993). In dry regions litterfall of fast-growing species is greater than that of natural forest which are adapted to hard environment (Bernhard-Reversat, 1987).

Fast decomposition of organic matter and nutrient poor clays are common in tropical regions and soil organic matter is a very important factor of fertility. The litter system is the main pathway for organic matter from production by vegetation to soil humus. So the setting up of an indicator of the functioning of this pathway would allow

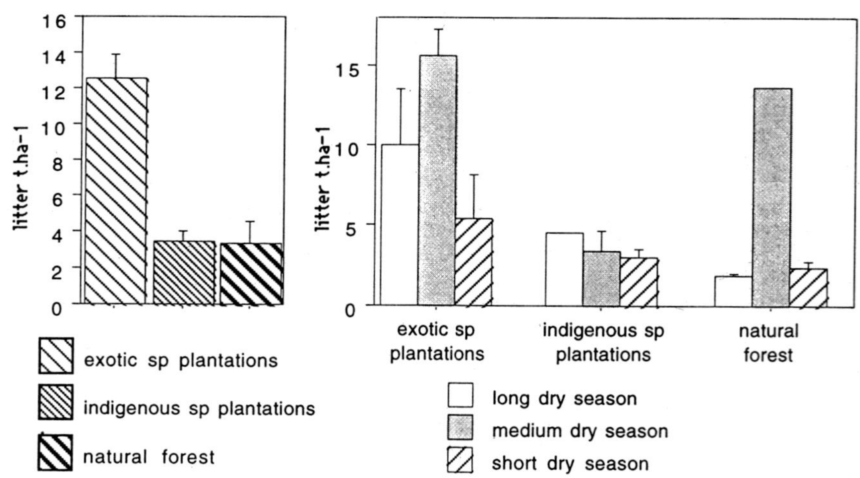

Fig. 13 Comparison of the amounts of standing litter on the ground in some plantations and natural forests (excluding *Casuarina*); A: all sites, B: according to the dry season length (excluding the South African mediterranean climate). (plantation data from Table 1, forest data from Greenland *et al.*, 1960, John 1973, Nye 1961, Madge 1965, Bernhard-Reversat, 1976 and unpublished, King *et al.*, 1994).

Table 3 Comparison of litterfall in tree plantations with litterfall in natural forest on the same or nearby sites

Authors	*dd*	*Rainfall* mm/yr	*Natural forest*	*Tree plantation*
Bernhard-Reversat 1987	Senegal	800	1.9	2.9
King *et al.*, 1994 (leaf litter)	Zimbabwe	885	2.72	3.27–5.08
Bernhard-Reversat 1976	Ivory Coast	2000	9.1	8.2
Bernhard-Reversat 1976	Ivory Coast	1700	8.6	9.6
Schwartz 1993, FBR unpubl	Congo	1300	5 0–5.7	6.4
Lisanework *et al.*, 1994	Ethiopia	1020	10.9	5.0–6.5

evaluations and comparisons of forested ecosystems. The most relevant parameter for soil organic matter building up could be the amount of standing litter on the soil, which integrates litterfall and decomposition, and consequently the climate data, although the relationships are complex. The amount of standing litter is significantly higher in forest plantations than in natural forests. If the origin of the planted species is taken into account, the significance is higher (anova: p=0.001) and it is obvious that only plantations of exotic species are different from natural forest (Fig. 13 A); this is noticeable also when the length of dry season is taken into account although the little data does not permit

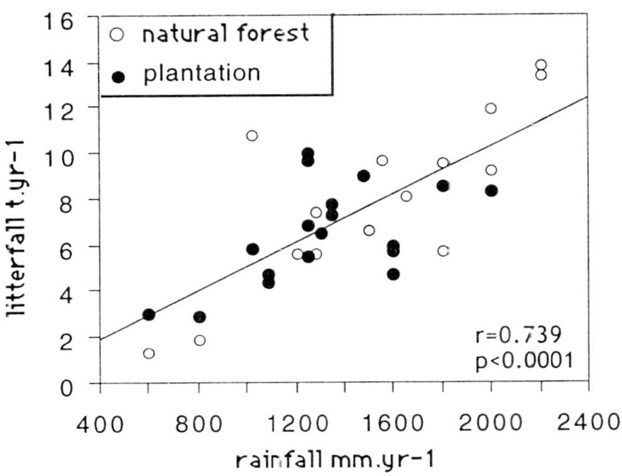

Fig. 14 Litterfall in some African forests and plantations versus mean annual rainfall (data from Table 1 and from Greenland *et al.*, 1960, Madge 1965, John 1973, Devineau 1976, Bernhard-Reversat 1977, 1987 and unpublished, Songwe *et al.*, 1988, Schwartz 1993 and personal comm.).

statistical comparison for each case (Fig. 13 B). This suggests that the adapted decomposers, microflora, mesofauna and macrofauna, are present in indigenous species stands, either planted or in natural stands, whereas some are lacking in exotic species stands. This is also suggested when African eucalypt plantations are compared to Australian natural eucalypt forests (Bernhard Reversat and Schwartz, 1997). The extreme is found in filao (*Casuarina*) plantations on sand dunes in Senegal, where in the absence of an organic soil few organisms decompose the litter, and decomposition was very low resulting in litter accumulation comparable to that of boreal forests (Gourbiere and Debouzie, 1995).

Although many studies have been conducted in tropical Asia and America, few have been reported on the litter system in African forests and plantations. Special attention must be paid to the chemical quality of the litter which plays a key role in the functioning of the litter system. Data collection on the quality of tropical organic resource is on with the aim of improving management of these resources (Gachengo *et al.*, 1998) and it may be assumed that more data on litter quality would contribute to more evidence of the dynamics of organic matter in African plantations and forests, and enhance our understanding of the processes controlling sustainability of tree plantations.

ACKNOWLEDGEMENTS

Data on the Congolese plantations originates from studies granted by CIFOR. R. Oliver, F. Ganry (CIRAD) and D. Schwartz IRD are acknowledged for giving unpublished data.

REFERENCES

Aber, J.D., and Melillo J.M., 1980. Litter decomposition: measuring relative contributions of organic matter and nitrogen to forest soils, *Canadian. J. Bot.*, **58**: 416–421.

Aerts, R., 1997. Climate, leaf litter chemistry and leaf litter decomposition in terrestrial ecosystems: a triangular relationship, *Oikos,* **79:** 439–449.

Andariese, S.W., and Vitousek, P.M., 1988. Soil nitrogen turnover is altered by herbicide treatment in a north Carolina piedmont forest soil, *For. Ecol. Manage.,* **23:** 119–25.

Anderson, J.M., and Flanagan, P.W., 1989. Biological processes regulating organic matter dynamics in tropical soils, In: *Dynamics of soil organic matter in tropical ecosystems,* (D.C. Coleman, J.M. Oades, and G. Uehara, eds.), Niftal Project, Univ. Hawaii, Honolulu, 97–124 pp.

Anderson, J.M., and Ingram, J.S.I., 1993. *Tropical Soil Biology and Fertility: A handbook of methods,* CAB International, Oxon, 221 pp.

Anderson, J.M., and Swift, M.J., 1983. *Decomposition in tropical forests. Tropical Rain Forest: Ecology and resource management symposium,* (S.L. Sutton, T.C. Whitmore, A.C. Chadwick, eds.), Special publ. 2, British Ecol. Soc. Blackwell Sci, Oxford.

Bartholomew, W.V., Meyer, J., and Laudelout, H., 1953. Mineral nutrient immobilization under forest and grass fallow in the Yangambi (Belgian Congo) region, with some preliminary results on the decomposition of plant material on the forest floor, *Publ. INEAC. Ler. Sc.*, **57:** 2–27.

Bernhard-Reversat, F., 1976. Essai de comparison des cycles d'éléments minéraux dans les plantations de framiré (*Terminalia ivorensis*) et en forêt naturelle de Côte d'Ivoire, *Bois Forêts Trop.*, **167:** 125–38.

Bernhard-Reversat, F., 1977. Recherches sur les variations stationnelles ses cycles biogéochimiques en forêt ombrophile de Côte d'lvoire. Cah. ORSTOM, ser. Pédol, **15:** 175–189.

Bernhard-Reversat, F., 1982. Measuring litter decomposition in a tropical forest ecosystem: comparison of some methods, *Int. J. Ecol. Environ. Sci.*, **8:** 63–71.

Bernhard-Reversat, F., 1987. Litter incorporation to soil organic matter in natural and planted tree stands in Senegal, *Pedobiologia*, **30:** 401–417.

Bernhard-Reversat, F., 1993. Dynamics of litter and organic matter at the soil-litter interface in fast-growing tree plantations on sandy ferralitic soils (Congo), *Acta Oecol.*, **14:** 179–195.

Bernhard-Reversat, F., 1999. Change in CO_2 release relationships with initial litter quality during early laboratory decomposition of tropical leaf litters, *Eur. J. Soil Biol.*, **34:** 117–122.

Bernhard-Reversat, F., Harmand, J.M., and Uguen, K., 1998. Les litières et la dynamique de l'azote dans divers biotopes à Acacia d'Afrique occidentale et centrale, In: *L'Acacia au Sénégal*, (C. Campa, C. Grignon, M. Gueyye, and S. Hamon, eds.), Editions de l'Orstom, Paris, 205–219 pp.

Bernhard-Reversat, F., Laclau, J.P., Loubana, P.M., Loumeto, J.J., Mboukou-Kimbatsa, I.M.C., Reversat, G., 1999. Changes in biological factors of fertility in managed *Eucalyptus* plantations grown on a savanna soil in Congo, The International Workshop on the "*Rehabilitation of Degraded Tropical Forestry Ecosystems*". 2–4 November, 1999, CIFOR in Bogor, Indonesia.

Bernhard-Reversat, F., and Schwartz, D., 1997. Change in lignin content during litter decomposition in tropica forest soils (Congo): comparison of exotic plantations and native stands C.R. Acad. Sci. Paris, *Sci. Terre Planetes.*, **325:** 427–432.

Bocock, K.L., 1964. Changes in the amounts of dry matter, nitrogen, carbon and energy in decomposing woodland leaf litter in relation to the activities of the soil fauna, *J. Ecol.*, **52:** 273–284.

Boone, R.D., 1994. Light-fraction soil organic matter: origin and contribution to net nitrogen mineralization, *Soil Biol. Biochem.*, **26**: 1459–1468.

Bouillet, J.P, Nizinski, G, Nzila, J.D., and Ranger, J., 1997a. The sustainability of *eucalyptus* commercial plantations: the Congolese approach, In: *Actes du meeting IUFRO, Silviculture and genetic improvement of Eucalyptus*, Salvador, Bahia, Brazil, Aôut 1997; volume 4: Environmental and social impacts of eucalypt plantations, 232–237.

Bouillet, J.P, Ognouabi, N., and Bar-Hen, A., 1997. Influence of soil preparation and weeding on the root development of an hybrid *Eucalyptus* in the Congo, In: *Actes du meeting IUFRO, Silviculture and genetic improvement of Eucalyptus*, Salvador, Bahia, Brazil Aôut 1997; volume 3: Silviculture, productivity and utilization of eucalypt, 252–257.

Cadish, G., and Giller, K.E., (eds.), 1997. *Driven by nature, plant litter quality and decomposition*, CAB International, Oxon, 409 pp.

Chapin, F.S., 1980. The mineral nutrition of wild plants, *Ann. Rev. Ecol. Syst.*, **11:** 233–260.

Chapin, F.S., 1983. Patterns of nutrient absorption and use by plants from natural and manmodified environments, *Plant*, 176–187.

Chaturvedi, O.P., and Jha, A., 1992. Studies on allelopathic potential of an important agroforestry species, *For. Ecol. Manage.*, **53:** 91–98.

Coleman, D.C., Oades, J.M., and Uehara, G. (eds.), 1989. *Dynamics of soil organic matter in tropical ecosystems*, Niftal Project, Univ Hawaii, Honolulu, 249 pp.

Constantinides, M., and Fownes, J.H., 1994. Nitrogen mineralization from leaves and litter from tropical plants: relationship to nitrogen, lignin and soluble polyphenol concentrations, *Soil Biol. Biochem.*, **26**: 49–55.

Couteaux, M.M., Bottner, P., and Berg, B., 1995. Litter decomposition, climate and litter quality, *Tree*, **10**: 63–66.

Dames, J.F., Scholes, M.C., and Straker, C.J., 1998. Litter production and accumulation in *Pinus patula* plantations of the Mpumalanga province, South Africa, *Plant and Soil*, **203**: 183–190.

Dangerfield, J.M., and Miller, A.E., 1996. Millipede fecal pellet production in selected natural and managed habitats of Southern Africa: Implications for litter dynamics. *Biotropica*, **28**: 113–120.

Del Moral, R., and Muller, C.H., 1969. The Allelopathic effect of *Eucalyptus Camaldulensis*, *American Midland Naturalist*, **83**: 254–282.

Devineau, J.L., 1976. Données préliminaires sur la litière et la chute des feuilles dans quelques formations forestières semi-décidues de moyenne Côte-d'Ivoire, *Oecol. Plant.*, **11**: 375–395.

Drechsel, P., Glaser, B., and Zech, W., 1991. Effect of four multipurpose tree species on soil amelioration during tree fallow in Central Togo, *Agroforestry systems*, **16**: 193–202.

Egunjobi, J.K., 1974. Litterfall and mineralization in a teak *Tectona grandis* stand *Oikos*, **25**: 222–226

Egunjobi, J.K., and Onweluso, B.S., 1979. Litter fall, mineral turnover and litter accumulation in *Pinus caribaea* at ibadan Nigeria, *Biotropica*, **11**: 251–255.

Förster, B., Eder, M., Morgan, E., and Knacker, T., 1996. A microsome study of the effects of chemical stress, earthworms and microorganisms and their interactions upon litter decomposition, *Eur. J. Soil Biol.*, **32**: 25–33.

Gachengo, C., Palm, C.A., Adams, E., Giller, K.E., Delve, R.J., and Cadisch, G., 1998. *Organic resource database*, TSBF 1998 Annual Report, TSBF, Nairobi.

Gillot, C., Lavelle, P., Blanchart, E., Keli, J., Kouassi, P., and Guillaume, G., 1995. Biological activity of soil under rubber plantations in Côte d'Ivoire, *Acta Zool. Fennica*, **196**: 186–189.

Gourbiere, F., and Debouzie., D., 1995. Spatial distribution and estimation of forest floor components in a 37-year-old *Casuarina equisetififolia*, (Forst.) plantation in coastal Senegal, *Soil biol. Biochem.*, **27**: 297–304.

Gregorich, E.G., Monreal, C.M., Schnitzer, M., and Schulten, H.-R., 1996. Transformation of plant residues into soil organic matter: chemical characterization of plant tissue, isolated soil fractions, and whole soils, *Soil Science*, **161**: 680–693.

Greenland, D.J., and Kowal, J.M.L., 1960. Nutrient content of the most tropical forest of Ghana, *Plant and Soil*, **112**: 154–174.

Harmand, J.M., 1998. *Rôle des espéces ligneuses à croissance rapide dans le fonctionnement biogéochimique de la jachère. Effet sur la, restauration de la fertilité des sols ferrugineux tropicaux, cas du Bassin de la Bénoué au Nord-Cameroun*, Thèse université Paris VI, CIRAD, Département Forêt, France, 212 pp + annexes.

John, D.M., 1973. Accumulation and decay of litter and net production of forest in tropical west Africa, *Oikos*, **24**: 430–435.

Jordan, C.F., 1989. Are process rates higher in tropical forests?, In: *Mineral nutrient in tropical forest and savanna ecosystems*, (J. Proctor, ed.), Blackwell Sci. Publ., Oxford, 205–215 pp.

Kadeba, O., and Aduayi, E.A., 1985. Impact on soils of plantations of *Pinus caribaea* stands in natural tropical savannas, *For. Ecol. Manage.*, **13**: 27–39.

Kadeba, O., 1998. Above ground nutrient dynamics of Caribbean., Pine (*Pinus caribaea*) plantation ecosystems, In: *Soil tropical forest ecosystems: characteristics, ecology and management*, O., Kadeba, A., Schulte, *et al.*, (eds.), 125–132 pp.

Kang, B.T., and Akinnifesi, F.K., 1994. Performance of selected woody agroforestry species grown on an alfisoil and an ultisoil in the humid lowland of west Africa, and their effect on soil properties, *J. Trop. Forest Sci.*, **7:** 303–312.

King, J.A., Campbell, B.M., and Ladipo, D.O., 1994. Soil organic matter relations in five land cover types in the Miombo region (Zimbabwe), *For. Ecol. Manage.*, **67:** 225–239.

Knockaert, C., 1981. Production de litiere dans quatre plantations *d' Eucalyptus camaldulensis* et dans un peuplement naturel de quercus suber, *Annales Recheche For. Maroc.*, **21:** 351–373.

Kotto-Same, J., Woomer, P.L., Moukam, A , and Zapfack, L., 1997. Carbon dynamics in slash-and-burn agriculture and land use alternatives of the humid forest zone in Cameron, *Agr. Ecosyst. Environment*, **65:** 245–256.

Laclau, J.P., Bouillet, J.P., and Ranger, J., 1999. The dynamics of biomass and nutrient accumulation in a clonal *Eucalyptus* plantation in the Congo, *For. Ecol. Manage.*, (in press).

Laishram, I.D., and Yadava, P.S., 1988. Lignin and nitrogen in the decomposition of leaf litter in a sub-tropical forest ecosystem at Shiroy hills in north-eastern India, *Plant and Soil*, **106:** 64.

Lavelle, P., Blanchart, E., Martin, A., and Martin, S., 1993. A hierarchical model for decomposition in terrestrial ecosystems: application to soils of the humid tropics, *Biotropica*, **25:** 130–150.

Lisanework, N., and Michelsen, A., 1993. Allelopathy in agroforestry systems : The effects of leaf extracts of *Cupressus lusitanica* and three eucalyptus spp. on four Ethiopian crops, *Agroforestry*, **21:** 63–74.

Lisanework, N., and Michelsen, A., 1994. Litterfall and nutrient release by decomposition in tree plantations compared with a natural forest in the Ethiopian highland, *For. Ecol. Manage.*, **65:** 149–164.

Lonsdale, W.M., 1988. Predicting the amount of litterfall in forests of the world, *Ann. Bot.*, **61:** 319–324.

Loubelo, E., 1990. *Etude comparative de quelques éléments du fonctionnement de deux peuplementd d'Eucalyptus au Congo*, Thesis, Université de Rennes, 141 pp.

Madge, D.S., 1965. Leaf fall and litter disappearance in a tropical forest, *Pedobiologia*, **5:** 273–288.

Maheut, J., and Dommergues, Y., 1960. Les teckeraies de casamance capacité de production des peuplements caractéristiques biologiques et maintien du potential productif des sols, *Bois Forêts Trop.*, **70:** 25–42.

Mailly, D., and Margolis, H.A., 1992. Forest floor and mineral soil development in *Casuarina equisetifolia* plantations on the coastal sand dunes of Senegal, *For. Ecol. Manage.*, **55:** 259–278 pp.

Maldague, M.E., 1961. *Relation entre le couvert végétal et la microfaune. Leur importance dans la conservation biologique des sols tropicaux*, Publ. INEAC ser. sci n° 190, 122 pp.

Matsuki, M., 1997. Regulation of plant phenolic synthesis: from biochemistry to ecology and evolution, *Aust. J. Bot.*, **44:** 613–634.

Mboukou-Kimbatsa, L.M.C., Bernhard-Reversat, F., and Loumeto, J.J., 1998. Change in soil macrofauna and vegetation when fast growing trees are planted on savanna soils, *For. Ecol. Manage.*, **110:** 1–12.

McKey, D., Waterman, P.G., Mbi, C.N., Gartlan, J.S., and Struhsaker, T.T., 1978. Phenolic content of vegetation in two African rain forests: Ecological implications, *Science*, **202:** 61–64.

Meentemeyer, V., 1978. Macroclimate and lignin control of litter decomposition rates, *Ecology*, **65:** 465–472.

Meentemeyer, V., Box, E.O., and Thompson, N.R., 1982. World patterns and amounts of terrestrial plant litter products, *Bioscience*, **32:** 125–128.

Miller, H.G., 1984. *Dynamics of nutrient cycling in plantation ecosystems. Nutrition of Plantation Forests*, Academic Press, London, 54–78.

Morris, A.R., 1995. Forest floor accumulation, nutrition and productivity of *Pinus patula* in the Usutu forest, Swaziland, *Plant and Soil*, **168–169:** 271–278.

Njoukam, R., Oliver, R., and Peltier R., 1999. Restitution minérale au sol par la litiére dans les plantations de *Pinus kesya* Royle ex-Gordon dans l'ouest du Cameroun, *Ann For. Sci.*, **56:** 431–439.

Nwoboshi, L.C., 1980. Nitrogen cycling in a teak plantation ecosystem in Nigeria, In: *Nitrogen cycling in West Africa*, (T. Rosswall, ed.), SCOPE/UNEP, Stockholm, 353–361 pp.

Nye, P.H., 1961. Organic matter and nutrient cycles under moist tropical forest, *Plant and Soil*, **13:** 333–346.

Ola-Adams, B.A., and Egunjobi, J.K., 1992. Effect of spacing on litterfall and nutrient content in stands of *Tectona grandis* Lin. f. and *Terminalia superba* Engl. & Diels, *Afr. J. Ecol.*, **30:** 18–32.

Oliver, R., and Ganry F., 1994. *Etude des modifications de fertilité induites par une jachére*, unpublished report MRT, 30 pp. Personal comm.

Olson, J.S., 1963. Energy storage and the balance of producers and decomposers in ecological systems, *Ecology*, **44:** 322–331.

Palm, C.A., and Sanchez, P.A., 1991. Nitrogen release from the leaves of some Tropical legumes as affected by their lignin and polyphenolic contents, *Soil Biol. Biochem.*, **23**: 83–88.

Pereira, A.P., Graqa, M.A.S., and Molles, M., 1998. Leaf litter decomposition in relation to litter physico-chemical properties, fungal biomass, arthropod colonization, and geographical origin of plant species, *Pedobiologia*, **42:** 316–327.

Proctor, J., 1983. Tropical forest litterfall. I. Problems of data comparison, In: *Tropical rain forests: ecology and management*, (S.L., Sutton, T.C. Whitmore, and A.C. Chadwick, eds.), Blackwell Scient. Publ., Oxford, 267–273 pp.

Proctor, J., 1984. Tropical-forest litterfall. II. The data set, In: *Tropical rain forest: the Leeds symposium*, (A.C. Chadwick, and S.L. Sutton, eds.), Leeds Philosophical and Literacy Soc., Leedspp.

Reddy, M.V., 1995. Soil Organisms and Litter Decomposition in the Tropics, Oxford & IBH Publ. Co. Pvt. Ltd., New Delhi, Calcutta, 274 pp.

Ronde, C. de, 1993. Spatial variations of forest floor loading in pine stands of the Tsitsikamma, South Afr. For. J. n°, **166:** 1–7.

Schwartz, D., 1993. Les retombées de litière. en tant que source du carbone et de l'azote du sol. Quantification et periodicité des apports en relation avec les caracteristiques climaticiues et edaphiques dans deux parcelles de forêt dense à Dimonika (Mayombe, Congo), In: *Echanges Forêt-Atmosphère en milieu Tropical Humid*, (Cros, B.; J. Diamouangana, M. Kabala, eds.), 141–158 pp.

Smith, K., Gholz, H.L., and de Assis, O.F., 1998. Litterfall and nitrogen-use efficiency of plantations and primary forest in the eastern Brazilian Amazon, *For. Ecol. Manage.*, **109:** 209–220.

Songwe, N.C., Fasehun, F.E., and Okali, D.U.U., 1988. Litterfall and productivity in tropical rain forest, Southern Bakundu Forest Reserve, Cameron, *J. Trop. Ecol.*, **4**: 25–37.

Spain, A.V., and Le Feuvre, R.P., 1987. Breakdown of four litters of contrasting quality in a tropical Australian rain forest, *J. Appl. Ecol.*, **24:** 279–288.

Swift, M.J., Heal, O.W., and Anderson, J.M., 1979. *Decomposition in Terrestrial Ecosystems*, Studies in Ecology 5. Blackwell Scient. Publ., 372 pp.

Tian, G., Brussard, L., Kang, B.T., and Swift, M.J., 1997. Soil fauna mediated decomposition of plant residues under constrained environmental and residue quality conditions, In: *Driven by nature, plant litter quality and decomposition*, (G. Cadisch, and K.E. Giller eds.), CAB International, Oxon, 125–134 pp.

Tian, G., Brussard, L., and Kang, B.T., 1993. Biological effect of plant residues with contrasting chemical composition under humid tropical conditions: effect on soil fauna, *Soil Biol. Biochem.*, **25:** 731–737.

Tian, G., Brussard L., and Kang B.T., 1995. Breakdown of plant residues with contrasting chemical composition: effect of earthworms and millipedes, *Soil Biol. Biochem.*, **27:** 277–280.

Tie Bi, T., and Omont, H., 1987. Etude des sols de parcelles d'Hévéa en basse Côte d'Ivoire, *Caoutchoucs et Plastiques* n°, **674:** 114–116.

Vanlauwe, B., Nwoke,. O.C., Sanginga, N., and Merckx, R., 1996. Impact of residue quality on the C and nitrogen mineralization of leaf and root residues of three agroforestry species, *Plant and Soil*, **183:** 221–231.

Vanlauwe, B., Sanginga, N., and Merckx, R., 1997. Decomposition of four *Leucaena* and *Senna* prunings in alley crowing systems under sub-humid tropical conditions, *Soil Biol. Biochem.*, **29**: 131–137.

Versfeld, D.B., and Donald, D.G.M., 1991. Litterfall and nutrient release in mature *Pinus radiata* in the south-western Cape, *South Afr. forestry J.*, **156:** 61–69.

Vitousek, P.M., 1984. Litterfall, nutrient cycling, and nutrient limitation in Tropical Forest, *Ecology*, **65:** 285–298.

Wardle, D., and Lavelle, P., 1997. Linkage between soil biota, plant litter quality and decomposition, In: *Driven by nature, plant litter quality and decomposition*, (G. Cadisch and K.E. Giller, eds.), CAB International, Oxon, 107–124 pp.

Wienand, K.T., and Stock, W.D., 1995. Long term phosphorus effects on the litter dynamics of an age sequence of *Pinus elliottii* plantations in the southern Cape of South Africa, *For. Ecol. Manage.*, **75:** 135–146.

CHAPTER 3

LITTER PRODUCTION AND NUTRIENT DYNAMICS IN TROPICAL FOREST PLANTATIONS OF INDIA

J.S. Samra* and A. Raizada

Central Soil and Water Conservation Research and Training Institute, 218 Kaulagharh Road, Dehradun–248195, U.P., India

**Present address: Natural Resource Management, Indian Council of Agricultural Research, Krishi Bhawan, New Delhi–110001, India*

ABSTRACT

Rehabilitation of degraded forest land by monocultures of native and exotic species has been a topic of debate in several forums. It is presumed that these plantations through their 'niche' effect can help in the recruitment of local species. This is facilitated by the ameliorative action of litter decomposition, its mineralization and subsequent availability. Plantations of different species return varying amounts of macro elements due to their in-built mechanism of nutrient withdrawal prior to leaf senescence. In this chapter we describe the quantities of leaf litter and nutrients returned annually by plantations of different species and their effect on soil organic matter levels and nutrient status.

In the Nilgri's (Southern India) *Eucalyptus globulus* plantations returned 1.93 t ha^{-1}, yr^{-1} of leaf litter which was less than that produced from native *shola* forests (2.12 t ha^{-1} yr^{-1}). Investigations on five different *Eucalyptus* plantations revealed that plantation soil nutrient status began to represent temperate forest soils. In the arid regions, organic matter increased by 54 per cent under *Prosopis cineraria* trees, while in the semi-arid tract, soil pH was modified and organic matter increased under *Eucalyptus* monoculture. In the sub-humid eastern part of India organic matter content in soil was highest under *Tectona grandis* (teak) plantations, while *Acacia auriculiformis* plantations resulted in an increase in soil pH values.

It was observed that *Eucalyptus tereticornis* plantation returned more calcium and phosphorus than temperate hardwoods in the Himalayan foothills, while *Shorea robusta* (Sal) monocultures resulted in a decrease in soil pH. *Pinus roxburghii* plantations retranslocated the largest amount of nutrients prior to needle drop, in comparison to teak, sal and *Eucalyptus*. Litter production in conifers raised in the Himalayas ranged from 2.4 to 5 t ha^{-1} yr^{-1} with quantities increasing with age in *Pinus roxburghii* plantations, while in *Pinus patula* it ranged from 1.47 to 3.98 t ha^{-1} yr^{-1}. In the 'terai' *Populus* (Poplar) plantations returned substantial quantities of nutrients with litter production varying from 1.35 to 4.15 t ha^{-1} yr^{-1} in plantations of different ages.

Significant changes in salt affected soils were observed under different tree plantations. Annual litter production and nutrient cycling was highest in *Acacia nilotica*. *Prosopis juliflora* was more effective in reducing soil pH from 9.5 to 8.9, in comparison to *Dalbergia sissoo, A. nilotica* and *Eucalyptus*.

The use of perennial deciduous or evergreen species has significant effects on soil properties and nutrient availability. Suitable species combinations consisting of exotic and indigenous species can greatly enhance soil amelioration and facilitate the migration of species.

Key Words: Forests, plantations, regions, litter, soils, nutrients.

Forests in the tropical regions of India are in a critical state due to continuous exploitation and unscientific management over the last century and in many areas are on the verge of extinction, where their last remnants barely survive. The rehabilitation of these degraded forests is a major challenge facing ecologists, foresters and all those who are concerned with environmental and livelihood securities. The difficulties associated with their rehabilitation are site-specific and are a function of initial forest structure and species composition, soil and climate. The resilience of ecosystem processes such as energy flows and biogeochemical cycling, history and intensity of disturbances and stresses that might have inhibited or slowed down natural forest community development are important fluxes that need to be investigated in order to formulate successful rehabilitation programmes.

Degraded forests in a large part of sub-tropical and tropical regions are characterized by eroded and hence nutrient deficient soils, hydrologic instability, reduced primary productivity and increasingly lowered biological diversity at both above-ground and below-ground levels. Persistent physical, chemical and biological barriers have prevented natural succession in various forest types and have even resulted in the retrogression of vegetation to unstable climaxes. Vegetation related barriers to natural forest succession include-low propagule availability (seeds or root stocks), seed collection (for nurseries and human consumption) and predation, absence of microhabitats for plant establishment, nutrient deficient soils, absence of root symbionts (VAM, *Rhizobium*), seedling die-back and predation, drought, fire and competition from old vegetation.

Forest ecosystem rehabilitation in India has traditionally followed a single objective-of afforestation with suitable exotic and indigenous species. More often than not, these plantations are either consumed for fuelwood supplies or are harvested frequently for commercial needs (paperpulp, plywood and small timber). Forest plantations play a positive role towards ecosystem rehabilitation by facilitating direct natural successional processes. Several studies have indicated that plantations of native and exotic species adapted to harsh conditions of degraded sites, reverse the process of degradation processes by stabilizing soils, increasing organic matter, moderation of pH and improvement of soil fertility (Singh *et al.*, 1990; Parrotta, 1992) and amelioration of problem soils (Shukla and Misra, 1993). Through their localized effects on microclimate and biotic conditions, these plantations facilitate recruitment, survival and growth of native tree species in the understorey, thus acting as 'foster ecosystems' (Lugo, 1988).

The actual fôrest area in India is 43.5 million ha (23.3 per cent of the total geographical area) out of which 19.5 per cent has forest cover and the remaining is barren. Only 39 per cent of the forest cover has a crown density of 10–40 per cent (FSI, 1995). The degraded forests on 57 per cent of the area need urgent attention for enhancing productivity and conservation of related natural resources such as soil, water and biodiversity etc. A major part of the tôtal wood removed from the forests is used as fuelwood. The current sustainable production of fuelwood is 17 million tonnes from forests and 98 million tonnes from non-forest areas. There is a net deficit of 86 million tonnes of fuelwood, (Rai and Chakrabarti, 1996) which however continues to be removed from the forest areas resulting in their over-exploitation and consequent degradation.

In view of the increasing livelihood requirements and threats to environmental security, large outlays are being made for raising forest plantations all over the country. Inventories on nutrient dynamics and soil quality are important for predicting site degradation and site amelioration, consequent to raising of indigenous and exotic plantations and logging of trees. There is a paucity of information on the impact of plantations on soil nutrient and organic matter status, inspite of massive investments in afforestation. In this article we have attempted to review the available information on nutrient dynamics vis-a-vis tree plantations in the subtropical and tropical regions of India. The general patterns and trends of nutrient inputs, turnover and humus formation, and the implications of raising plantations are discussed, with inputs from various investigations over the last two decades.

LITTER PRODUCTION FROM STANDING CROP

Man's interest in dead plant material is shown by the fact that there is often a special word for it like litter (English), streu (German), forna or stro (Swedish). Fallen litter consists of leaf fall, shoot fall (crown trimming and branch pruning), bark fall and bud scale fall. Litter represents an important link in organic production—decomposition cycle and is therefore an important fundamental process (Meentemeyer *et al.*, 1982). Apart from being an important source of nutrients, litter fall results in the formation and renewal of the forest floor. The litter mass on the forest/plantation floor is an important structural component of the ecosystem. It protects the soil against erosion, influences forest hydrology (Singh and Gupta, 1977), forms humus which is a critical component of soil development, influences establishment and growth of seedlings, and hence facilitates the development of forest communities.

Leaf fall is a complex eco-physiological process, influenced by many external and internal factors and it would be difficult to draw a sharp line of distinction (Kozlowski, 1973). External factors like temperature, water stress and day length are closely associated with leaf/litter fall in the tropics. The effect of water stress is clearly seen in climates subjected to seasonal drought or sometimes in species growing in unsuitable sites (Raizada and Srivastava, 1986). Shedding of mature and old leaves in evergreen plants seems to be the function of leaf age, as they lack in seasonal periodicity, and competition for nutrients and/or plant hormones between the young and old leaves. The broad leaved evergreen species, shed their leaves of the previous year's foliage at a time coinciding with flowering and/or appearance of new leaves. This type of leaf fall usually extends over a period of few weeks beginning shortly after the expansion of new leaves. In sal (*Shorea robusta* Gaertn.) a semi-evergreen species, leaf fall precedes the emergence of new foliage and it resemble a deciduous tree for a brief period.

The prime controlling factor for development of organic matter is climate, through its influence on litter production and decomposition of litter. The key for modelling certain ecosystem dynamics is to quantitatively relate the process of litter production and decomposition to climate. It is therefore possible to model a process using information rich variables to predict information 'poor' variables. This approach has been used by Box (1978) to produce maps of primary production rates and by Meentemeyer (1977) to produce maps of litter decomposition rates for USA. Using computer procedures the litter accumulation for global land area was estimated to equal an equivalent of 75×10^9 mt C. The estimates of global soil carbon are now converging near 1500×10^9 mt C, which means that only 5 per cent of the globe's total soil carbon is found in surface litter (Box, 1978).

DECOMPOSITION OF DETRITUS AND FORMATION OF SOIL ORGANIC MATTER

Decomposition of forest debris or detritus is the sum total of a large number of inter-related processes, by which, organic matter is broken down to smaller particles and soluble forms of nutrients for uptake by plants. The rate at which decomposition proceeds is largely regulated by the prevailing climate and the microclimate, that is affected both by the resource (debris) quality and factors of the edaphic complex (Anderson and Swift, 1983). Soil physico-chemical factors which affect decomposition rates include moisture, aeration, O_2, CO_2, pH and soil temperature.

During decomposition, resistant organic substances are produced by microbes which accumulate in the soil as humus. Many ecologists prefer to use the term mineralization for those decomposition events that actually release inorganic compounds (e.g. CO_2, H_2O, NH_4^+, Ca^{-+}) from organic matter. The physical reduction and mixing up of litter is largely carried out by an abundance of soil animals, ranging from microscopic nematodes to large earthworms (Hole, 1981). The result is an increased surface area for microbial attack and the movement of litter to more constant conditions of temperature and moisture in the lower profile.

ROLE OF SOIL MICROFAUNA IN DECOMPOSITION OF PLANT DETRITUS

Soil microfauna are very selective in their choice of detritus. Springtails and mites are prevalent in coniferous forests, while earthworms that are intolerant of coniferous litter, are more abundant in temperate deciduous and tropical forests (Swift *et al.*, 1979). Earthworms are effective in mixing the litter material through the soil profile and their movements leave open channels for air and water (Aina, 1984). Termites are most abundant in warm temperate and tropical forests, where their wood-tunneling activity causes rapid disappearance of fallen logs (Gentry and Whitford, 1982).

The transformation of litter to humus occurs sequentially as organic matter is fragmented and mixed through the forest floor. The C : N ratio of the residue decreases rapidly and equals bacterial and fungal tissue over a time period. Bacteria and fungi typically have high nutrient requirement; some nutrient ions from the detrital mass are retained for protein synthesis and other components. Retention of nutrients in soil microbial biomass is known as immobilization. While most of the nutrients are lost by leaching or absorption in the tissues of the microbes, calcium, a cell wall constituent, is lost concurrently with the physical disappearance of litter mass. Sometimes calcium may be immobilized as calcium oxalate in the fungal tissues.

The microbial populations die due to non-availability of organic rich litter and the accumulated N is released as NH_4^+ from dead microbial tissues. The mineralization process commences with C : N ratios near 30 : 1 (Berg and Staaf 1981), with variations occurring due to substrate quality. Thus, N is mineralized as a result of the balance between the degradation of organic substances for energy and proteins synthesis by microbes.

THE FORMATION AND NATURE OF FOREST HUMUS

The term forest 'humus' refers to the entire organic portion of the soil profile. It includes undecomposed leaves, needles and twigs or 'litter', partly decomposed remains or 'duff' and finely divided, decay resistant residue or amorphous humus. In many situations the humus gets mixed with the mineral soil partly by the activity of organisms and partly through infiltration when it is called incorporated humus.

Humus accumulates in the lower horizon of forest soils and is often complexed with clay mineral components. Humus substances have a high concentration of N in amino ($-NH_2$) groups, but these are still resistant to microbial attack, suggesting that humus results from the formation of stable lignoprotein complexes. The humus layer is of immense ecological significance and is an important agent in the process of soil development. Humus is a seat of useful organisms, source of nutrients, growth substances and buffering colloids and therefore plays a leading part in the maintenance of soil fertility in nurseries, plantations and forests. Depending on edaphic conditions and detritus quality, the reaction of the humus layer may vary from being acidic (pH 5.0 to 3.0) to alkaline in certain cases (pH nearing 8.0). Humus contains an overwhelming proportion of N and P in the soil pool. Since humic substances are stable, this large nutrient pool turns over very slowly and serves as a major source of exchange sites for ions in the soil solution.

Inspite of the relatively small quantity of humus in a plantation and/or forests, the importance of this layer cannot be underestimated. Humus in an undisturbed forest soil profile, exceeds the combined content of organic matter in the forest floor and aboveground biomass (Schlesinger, 1977). Apparently this would be there even in plantations, except that detritus quality from the monoculture would bring in a certain amount of skewness towards the availability of one or more nutrients. These variations affect soil quality and hence seedling recruitment of the native forest vegetation in the long run.

RESEARCH EFFORTS IN DIFFERENT REGIONS OF INDIA

SOUTHERN HILLY REGION (NILGIRIS)

Tropical pines were introduced in the Nilgiris (Udhagamandalam, Tamil Nadu) for large scale afforestation from 1967 onwards (Chowdhary *et al.*, 1982). The weeping pine (*Pinus patula* Schl and Chain), has been found to be successful in the region, specially in 'frost holes'.

Investigations by Malhotra *et al.* (1987) on litter production from an age series of *P. patula* plantation revealed that substantial amounts of needles were shed with increasing age (Table 1). The annual return of nutrient through litterfall was more for N and Ca and lowest for P.

Table 1 Annual return of major nutrients (kg/ha/year) to the floor from *P. patula* plantations at different ages in the Nilgiris.

	Nutrient content (kg/ha/year)				
Age (yrs)	*N*	*p*	*K*	*Ca*	*Mg*
6	16.6	0.2	2.5	17.9	2.3
8	70.0	1.8	3.3	64.5	11.0
10	64.7	2.7	7.7	65.8	4.5
12	89.3	3.8	8.6	72.8	15.0
14	75.4	1.0	12.5	79.8	10.0

The litter production, nutrient return and improvements in soil characteristics in twelve-year old plantations of *Eucalyptus globulus* (blue gum) and *Acacia mearnsii* (black wattle) were compared with natural 'shola' forests of the Nilgiri region (Venkataraman *et al.*, 1983). The total annual leaf litter accumulation was 1935 kg/ha in blue gum and 960 kg/ha in black wattle, and 2121 kg/ha in 'shola' forests, and these quantities were lower in comparison to sal, teak and chir recorded from other regions. Nitrogen return was highest from blue gum (31 kg/ha/yr) while shola forests returned the highest amount of calcium (34 kg/ha/yr) and phosphorus (1.3 kg/ha/yr) (Table 2). Organic matter content in the 0–15 cm soil layer was highest (1956 kg/ha) in shola forests and least in black wattle plantations.

Soil chemical properties in the 0–15 cm layer after 12 years were improved by these plantations (Table 3). Organic carbon increased by 89 per cent over the original soil in the 0–15 cm layer under *E. globulus* and by 100 per cent in the 0–15 cm layer under *P. patula,* indicating clearly that humus formation occurred in substantial quantities.

Table 2 Nutrient returns (kg/ha/yr) and organic matter (kg/ha) in three forest systems in the Nilgiris.

Vegetation	*Nutrients* (kg/ha/yr)					
	N	*P*	*K*	*Ca*	*mg*	*Organic Matter* (kg/ha)
Shola	23.1	1.3	11.0	34.0	5.0	1956
E globulus	31.0	0.8	4.3	18.8	2.1	1548
A. mearnsi	22.1	0.5	3.4	3.3	0.9	662

Table 3 Changes after 12 years in soil chemical properties under two forest plantations raised in the Nilgiris.

Location	*Depth(cm)*	*Organic carbon (%)*	*Nitrogen (%)*	*Exch. (m.eq.%)*	
Original soil	0–15	2.86	0.10	9.7	1.1
E. globulus	0–15	5.41	0.61	16.4	1.2
	15–30	3.23	0.30	11.3	0.8
	30–60	1.80	0.18	8.0	0.6
A. mearnsii	0–15	5.71	0.65	13.8	1.7
	15–30	3.20	0.35	10.2	1.1
	30–60	1.91	0.22	6.8	0.7

Soil enrichment in different soil layers by plantations of five *Eucalyptus* species was evaluated in lateritic soils in the Nilgiris (Negi *et al.*, 1988). Their investigations revealed that forest floor behaviour was similar to that of tropical forests and soil nutrient status was akin to temperate forests.

Organic matter and nutrient status of the forest floor in five plantations are given in Table 4. The organic matter content varies from 1330 to 5620 kg/ha in the litter layer and 2090 to 6824 kg/ha in the decomposing layer. The amount of organic matter and nutrient accumulation in *Eucalyptus* in the Nilgiris was lower than many other forest stands (Waring and Schlesinger, 1985). The nutrient concentrations in green leaves of the evergreen *Eucalyptus* was lower when compared with deciduous species as a nutrient conserving mechanism (Jordan, 1985). The trend of nutrient concentration in the top litter layer was in the order Ca > N > K > Mg> P, in all the five species. The nutrient concentration in the decomposing litter layer was generally higher than the top layer, indicating its breakdown and accumulation in microbial biomass.

Table 4 Organic matter and nutrients returned (kg/ha/yr) in the top litter layer (L) and decomposing layer (F) in five *Eucalyptus* plantations in the Nilgiris.

Stands	*Organic matter*			*Nutrients returned*				
	L	*F*	*Total*	*N*	*P*	*K*	*Ca*	*Mg*
E. acemoides	2068	6808	8876	43	3	12	34	2
E. eugenoides	1692	6160	7852	47	2	10	27	3
E. paniculata	5020	6428	11448	63	4	20	160	5
E. pilularis	1746	6824	8570	30	2	6	18	2
E. propinqua	1330	2090	3420	22	2	4	22	1

The total amount of K in the soil (0–25 cm) was highest followed by N > Ca > Na > P > Mg in *E. acemoides, E. eugenoides, E. paniculata* and *E. propinqua*. However, the trend was N > Ca > K > Na > Mg > P. In case of *E. pilularis*. The C : N ratio and amount of organic matter was highest in *E. propinqua* while the other species showed marginal differences with each other (Table 5).

Table 5 Organic matter (t/ha) and soil nutrient status (0–25 cm soil depth) in different stands of *Eucalyptus* in the Nilgiris.

Stands	*C : N ratio*	*Organic matter (t/ha)*	*Nutrients (kg/ha)*					
			N	*P*	*K*	*Ca*	*Mg*	*Na*
E. acemoides	8	127	8767	646	9871	4160	291	1167
E. eugenoides	7	107	9068	462	11580	4815	420	1136
E. paniculata	8	133	9564	596	12668	5118	168	1142
E. pilularis	9	103	6969	407	2822	4719	508	1299
E. propingua	15	144	5641	457	15514	5407	168	992

The impact of plantations of *E. globulus and P. patula* plantations on soil quality and nutrient cycling in the hills of Udhagamandalam were evaluated and compared with natural broad leaved shola forests (Jeeva and Ramakrishnan, 1997). Mean annual litter production increased with age, the *E. globulus* producing more litter in the 2 and 6 year old stands than *P. patula* (Table 6). Litter production from shola forests was less than that produced from 10 year old stands of *E. globulus and P. patula*. Peak leaf fall in *P. patula* occurred during the summer months (April-May) and a small period in November, December and April, peaking in January. On the other hand leaf fall in the shola forests occurred during the winter months, reaching a peak in January.

Table 6 Annual litter fall (±SE) in three forest types in the Western Ghats.

Forest type	*Age of plantation* (years)	*Litter fall* (kg/ha/yr)
Eucalyptus globulus	2	1925 ± 161
	6	3509 ± 270
	8	5326 ± 472
	10	5969 ± 492
Pinus patula	2	1023 ± 97
	6	2675 ± 190
	8	5271 ± 498
	10	6246 ± 576
Shola forests	–	2385 ± 240

The quantity of nutrients returned annually to the plantation floor increased with stand age *Eucalyptus globulus* litter had higher amounts of different nutrients than *P. patula* of similar age, the differences being more clear in the younger plantations. Nutrient return by the 'shola' forests was generally lower, except for calcium which was higher than in a 10 year old *P. patula* plantation. The trend of nutrient return in order was-N > Ca > K > Mg > P for *E. globulus,* N > Ca > Mg > K > P for *P. patula* and Ca > N > K > Mg > P for the shola forests (Table 7).

The forest floor nutrient content determined in 10 year old plantations was generally higher with *P. patula,* followed by shola forests and least with *E. globulus* forests. The variability of nutrient content on the plantation/forest floor was as follows: Ca > N > K > Mg > P for *E. globulus;* N > Ca > Mg > K > P for *P. patula* and Ca > N > K > Mg > P for shola forests (Table 8).

Table 7 Amount of nutrients returned (kg/ha/yr) to the forest floor in three forest types in the Nilgiris.

Forest type	*Age (yrs)*	*N*	*P*	*K*	*Ca*	*Mg*
E. globulus	2	22.8	1.3	9.7	22.4	3.5
	6	48.3	2.2	15.9	48.1	8.6
	8	63.3	2.7	32.9	78.2	12.8
	10	73.6	3.7	33.2	72.8	17.7
P. patula	2	8.1	0.6	6.3	6.7	7.1
	6	18.8	1.7	14.3	19.1	20.9
	8	3.5	2.9	30.2	39.8	36.6
	10	60.2	3.3	40.3	49.3	46.5
Shola forests	–	47.4	2.4	21.16	68.4	18.5

Table 8 Nutrient content (kg/ha) of the forest floor in three forest types in the Western Ghats.

Forest type	*Age (yrs.)*	*N*	*P*	*K*	*Ca*	*Mg*
E. globulus	10	97.5	10.6	74.5	123.6	45.7
P. patula	10	188.5	22.7	109.8	158.8	128
Shola forests	–	173.7	11.2	86.5	183.4	63.8

The annual rate of nutrient cycling in the soil was maximum in Shola forests, followed by *P. patula* and *E. globulus*. The total quantity of nutrients in the soil decreased with age under *E. globulus* compared with that under *P. patula*. It was concluded that nutrient cycling under shola forests is under steady state, and under short rotation forestry with fast growing exotics and harmful practice of removal of litter, nutrient losses from the soil pool would be substantial.

Semi-arid Regions

The impact of *Eucalyptus* hybrid monocultures raised on eroded slopes in a red soil watershed was investigated (Raizada and Jayaram, 1995). Significant changes were noticed with regard to soil pH, organic carbon, available N and available P_2O_5. Organic carbon content in 0–7.5 cm soil depth increased significantly in the 10 year old plantation compared with 6 year old stand and control. Increases in available N and P_2O_5 were also significant in the surface soil layer (Table 9).

The effect of alley cropping of Leucaena with field crops on soil properties, was evaluated at Kota, Rajasthan (Prasad, 1994). The improvements were more pronounced in the top soil (0–15 cm) in the 15–30 cm layer. Leucaena increased organic carbon from 0.37 per cent to 0.43 per cent, available N from 163.5 to 191.5 kg/ha, P from 8.2 to 15 kg/ha and decreased soil pH from 7.6 to 7.4 in the 0–15 cm soil layer (Table 10).

Arid Zone of Rajasthan

Agarwal *et al.* (1993) observed marked differences in nutrient status of the soil under the canopy of *Prosopis cineraria* (Linn. Macbride) in Rajasthan (Table 11). There was an increase of 54 per cent in organic matter, 90 per cent in total N, 75 per cent in total Sulphur and 43 per cent in total P. These improvements were attributed to mineralization of litter fall over several years.

Table 9 Impact of *Eucalyptus* hybrid plantations on soil characteristics in semi-arid red soils of Karnataka.

Soil parameter	*Age (years)*	*Soil depth (cm)*		
		0–7.5	7.5–15.0	15–30.0
pH (1:2.5)	Control	6.85	6.90	7.03
	6	6.60	6.55	6.55
	10	6.00	7.10	7.10
	F test	**	**	**
Organic carbon(%)	Control	0.96	0.89	0.86
	6	0.91	0.83	0.71
	10	1.30	1.18	0.97
	F test	**	**	**
Available N (kg/ha)	Control	486.65	399.57	356.88
	6	437.13	368.83	344.92
	10	744.66	674.48	479.82
	F test	**	**	**
Available P_2O_5, (kg/ha)	Control	6.52	6.21	6.21
	6	11.17	8.70	8.07
	10	16.76	9.62	9.31
	F test	**	NS	**

*significant at 5% level; **significant at 1% level; NS – not significant.

Table 10 Changes in soil properties with *Leucaena* alley cropping after 10 years in Kota, Rajasthan.

Parameter	*Soil depth (cm)*	*pH*	*Organic matter (%)*	*Available nutrients (kg/ha)*		
				N	*P*	*K*
Initial	0 – 15	7.6	0.37	163.5	8.2	460
	15 – 30	7.8	0.36	156.4	8.15	410
Leucaena alleys	0 – 15	7.4	0.43	191.5	14.98	500
	15 – 30	7.5	0.40	190.3	11.20	450
Bare field (control)	0 – 15	7.6	0.38	165.0	14.0	450
	15 – 30	7.7	0.36	156.6	12.32	375

Sub-humid Eastern Region

Nutrient dynamics in plantations and natural forests of sal (*Shorea robusta*), teak (*Tectona grandis*) and *Eucalyptus* in pure and mixed stand, were determined in the lateritic soils of West Bengal (Singh *et al.*, 1988). Organic matter in soil was maximum (1.54%) under teak and least

Table 11 Characteristics of surface soil (0–15 cm) under *P. cineraria* canopy compared with an open site in Jodhpur, Rajasthan.

Characterstics	*Open site*	*Under tree canopy*
pH	8.2	8.0
Organic matter (%)	0.37	0.57
Total N%	0.020	0.038
Total S%	0.016	0.028
Total P%	0.028	0.038
Available N (kg/ha)	190.0	250.0
Available P (kg/ha)	7.7	22.4
Available K (kg/ha)	370.0	633.0

Table 12 Changes in physico-chemical properties of lateritic soils under different plantations in West Bengal.

Species	*Age* (*yrs*)	*pH*	*O.C.* (%)	*CEC* (*m.eq./ 100 g soil*)	*Exchangeable cations* (*m.eq.%*)			
					Ca	*Mg*	*Na*	*K*
Sal coppice	–	5.4	0.71	9.41	1.94	1.04	0.05	0.52
Teak forest	–	5.6	1.31	10.13	7.01	1.09	0.04	0.60
Sal plantation	15-23	5.2	0.20	6.56	2.32	1.06	0.09	0.28
Teak plantation	18-25	5.7	1.54	13.03	9.53	1.71	0.04	0.72
Mixed plantation	14-24	5.3	0.77	8.43	2.43	0.98	0.09	0.41

(0.20%) under sal plantation (Table 12). Exchangeable Ca was higher than other cations (Mg, Na and K) in the teak plantation, indicating that distribution of Ca is species specific. The differences in clay content were marginal in different plantations, but higher values of soil characters in the teak plantation resulted in more nutrient return than that recorded from sal coppice stand. There was a positive correlation between depth and exchangeable cations (Ca, Mg and K) in teak stands, indicating higher leaching loss of nutrients than in other plantations.

Eucalyptus tereticornis plantations raised in Midnapore district of West Bengal (Nandi *et al.*, 1991) did not result in significant variations in soil properties due to continued biotic disturbances in the area. In Tripura, Chakraborty and Chakraborty (1989) observed that *Acacia auriculiformis* plantations resulted in a decline of soil pH from acidic (5.9) to slightly alkaline (7.6) under lateritic soil conditions. Electrical conductivity increased from 28.0 to 48.4 in mhos/cm, while water holding capacity increased from 23.0 to 32.7% (Table 13).

Table 13 Changes in soil properties under *A. auriculiformis* plantations under lateritic soils in Tripura.

Age (years)	*pH (m mhos/cm)*	*E.C.*	*W.H.C.* (%)	*O.C.* (%)	*Total N* (%)
2	6.7	29.5	24.1	0.96	0.37
3	6.7	43.4	27.6	2.27	0.46
4	7.6	48.4	32.7	2.70	0.50
Control	5.9	27.2	23.0	0.81	0.36

Sub-Humid Regions of Northern India

Anthropogenic factors and increasing pressure on agricultural land has led to large scale destruction of forests in the northern sub-tropical region. Only a few traces of the original forest vegetation of sal (*Shorea robusta*) and its associates are now left in the region and even these are under constant threat. Intensive afforestation efforts have been made during the pre- and-post independence era and led to the introduction of fast growing exotic species of *Eucalyptus* (for paper and pulp) and *L. leucocephala* (for fuelwood and fodder). In recent years poplar (*Populus* sp.) and its numerous clones have been introduced for paper pulp, plywood and veneers. The ecological impact of these exotic plantations have not been adequately studied. In the western Himalayas such exotic plantations became a hindrance for natural forest succession and led to a reduction in the area under natural forests.

Singhal *et al.* (1975) concluded that litter of *Eucalyptus* humifies easier and faster than natural sal under Doon Valley conditions and soil organic matter under *Eucalyptus* species was more hydrolyzable. The translocation of the humified material was more pronounced under *Eucalyptus* although to a limited depth.

The litter production in *Eucalyptus* hybrid plantations was 3377, 3801 and 6207 kg/ha/yr from 5, 7 and 10 year old plantations at Dehradun (George, 1982). Leaf fall contributed to the bulk of the litter (66–76%) followed by twigs (16–24%) and bark (5–10%). The trend of nutrient return was in the order of Ca > N > K > Mg > P. The amount of calcium returned was 40.2, 42.8 and 73.2 kg/ha, while N values were 29.8, 32.1 and 59.2 kg/ha, in 5, 7 and 10 year old plantations, respectively. Singh (1984) determined quantity of nutrients returned from 5–9 year old *E. tereticornis* plantations in U.P. (Table 14). These plantations returned higher amounts of Ca and P than temperate hardwoods, deciduous forests as well as conifers, except that of *Terminalia* - *S. robusta* community which returned more P. Nitrogen release in temperate and coniferous forests was distinctly higher than *Eucalyptus*.

Table 14 Nutrient return (kg/ha/yr) though litter fall in *E. tereticornis* plantations in Uttar Pradesh.

Age (years)	*N*	*P*	*Ca*
5	14.2	8.6	98.2
6	26.0	11.6	140.8
7	22.8	16.2	217.5
8	30.5	20.8	248.5
9	34.6	13.8	334.5

The impact of sal and *E. camaldulensis* monoculture was investigated on soil properties in a natural sal growing area in the Himalayan foothills (Jha and Pande, 1984). The soil moisture and organic matter content (up to 90 cm depth) were highest in a natural sal forest, while organic matter content in *Eucalyptus* stands was closer to natural sal forest, than to sal monocultures. Pure *Eucalyptus* stands showed a substantial increase in soil pH, while sal monoculture resulted in a decrease, when compared with natural sal forest (Table 15). Total N and P were highest in natural sal in the surface layer (0–15 cm depth), followed by *Eucalyptus* and sal. The total K was highest in *Eucalyptus.* The findings suggest that raising *Eucalyptus* monocultures in areas previously occupied by natural sal forests resulted in higher levels of organic matter accumulation than sal plantations.

Table 15 Changes in soil pH and organic matter at different depths in three forest types in Himalayan foothills.

Site	*Depth (cm)*	*pH*	*Organic matter %*
S. robusta forest	0 – 30	6.2	4.52
	30 – 60	6.2	1.86
	60 – 90	6.2	0.83
S. robusta plantation of 1926	0 – 15	5.4	1.31
	15 – 30	5.8	1.17
	30 – 60	5.9	1.44
	60 – 90	6.0	1.58
E camaldulensis plantation of 1926	0 – 15	6.3	1.90
	15 – 30	6.4	1.62
	30 – 60	6.6	1.80
	60 – 90	6.9	1.68

Singh *et al.* (1990) reported (Table 16) that the amount of organic carbon, total N, available P and K were higher in soils under *Eucalyptus* hybrid plantations and lowest in agricultural lands, except in the case of *P. roxburghii* where the values of available P and K were still lower. The highest C : N ratio was recorded under *L. leucocephala* and least under *P. roxburghii.*

Decomposition patterns of *S. robusta* and *E. camaldulensis* litter revealed 67 per cent and 82 per cent weight loss respectively, at Dehradun (Bahuguna *et al.,* 1990). The mobility of nutrients from decomposing litter was in the order: Mg > K > P > Ca > N for sal and K > Mg > P > Ca > N for *E. camaldulensis.* Open, light canopy in *Eucalyptus* allows more solar radiation and consequent rise in temperature, which hastens decomposition of litter. Continuous litterfall in *Eucalyptus* facilitates steady rates of breakdown, while in sal (a deciduous species) a peak period of litterfall may inhibit the activities of the microbial population. Sal is a stable sub-climax in the Doon Valley, the plantation floor is more stable and regulated and its litter decomposition rates are more steady. The rapid litter breakdown and nutrient release makes *Eucalyptus* a potential species for afforestation on poor sites.

Table 16 Influence of various tree species on characteristics of deep alluvial soils at Dehradun (Uttar Pradesh).

Species	*Soil depth (cm)*	*Organic carbon (%)*	*Total N (%)*	*Available N (ppm)*	*Available P (kg/ha)*	*Available K (kg/ha)*
Tectona grandis	0 – 10	3.54	0.341	569.2	42.6	345
	10 – 30	1.51	0.153	399.6	20.8	85
Pinus roxburghii	0 – 10	2.94	0.251	504.8	27.7	293
	10 – 30	1.40	0.140	245.6	15.7	188
Shorea robusta	0 – 10	1.91	0.162	386.6	43.3	303
	10 – 30	1.101	0.109	189.4	20.6	170
Dendrocalamus	0 – 10	2.21	0.208	414.4	28.2	295
strictus	10 – 30	1.981	0.192	240.0	18.9	128
L. leucocephala	0 – 10	2.93	0.300	576.8	43.3	372
	10 – 30	1.10	0.125	427.1	24.8	382
E tereticornis	0 – 10	3.62	0.377	537.8	45.1	407
	10 – 30	1.09	0.155	380.1	38.7	168
Agricultural land	0 – 10	0.763	0.086	1204.6	76.1	155
	10 – 30	0.617	0.082	523.0	72.80	132
C.D. (0.05)	0 – 10	0.024	0.007	1.54	0.5	6.1
	10 – 30	0.011	0.008	2.32	0.3	3.8

Litter production and nutrient conservation patterns in four plantations of sal, teak, pine and *Eucalyptus* revealed that the highest quantity of litterfall was recorded in sal which was only 3 per cent of the total above ground biomass (Pande and Sharma, 1993). In contrast, *Eucalyptus* shed 12.2 per cent of its above ground biomass as leaf litter (Table 17).

Table 17 Litter production in different tree plantations at Dehradun (Uttar Pradesh).

Plantation	*Age (years)*	*Litter production (t/ha)*	*Total above ground biomass (t/ha)*	*Litter % of above ground biomass*
S. robusta	67	11.27	390.0	3.0
T. grandis	67	6.95	129.58	5.36
Eucalyptus hybrid	21	7.07	57.91	12.21
P. roxburghii	67	9.7	141.55	6.84

The annual conservation of N was highest in pine followed by sal, teak and eucalyptus. The trend was in order: teak > sal > eucalyptus > pine. The nutrient return through leaf fall in sal conserved the highest amount of all nutrients except nitrogen, which was highest in pine. *Eucalyptus*, in general, conserved the least amount of nutrients (Table 18). All the species showed unimodal leaf fall pattern, except *Eucalyptus* where it was bimodal. The peak leaf fall period was March–April in sal, and April–May in teak. In Eucalyptus two peaks were observed, first during Oct.–Nov. and again in April–May. Litter production trends revealed a decreasing trend with increasing biomass. Litter nutrient concentrations were higher in deciduous species like sal and teak than in the evergreen species like pine and eucalyptus. The highest quantity of nutrients were returned by sal mainly through Ca and N.

Table 18 Annual nutrient return (kg/ha) through leaf fall from four tree plantations at Dehradun (UP).

Nutrients	*Sal*	*Teak*	*Pine*	*Eucalyptus*
N	72.35	45.68	25.42	52.17
P	4.36	2.30	3.94	4.17
K	23.25	27.18	26.56	38.07
Ca	83.10	62.81	30.27	84.10
Mg	12.63	8.47	6.31	8.98
Total	195.69	146.44	92.50	187.49

The water holding capacity in irrigated *Eucalyptus* plantations in Gujarat, was lower than bare plots, (Table 19) after 4 years (Srivastava, 1993). However, soil pH did not change with increasing plantation age.

Organic carbon during the 4th and 5th year were lower than control, but increased in the 6th and 7th year. There were no significant changes in the quantities of available P but variations occurred in the amount of available K.

Table 19 Changes in soil properties under *Eucalyptus* plantations of various ages in Gujarat.

Age	*pH* (1: 2.5)		*WHC* (%)		*Organic carbon* (%)	
(*years*)	*Control*	*Plantation*	*Control*	*Plantation*	*Control*	*Plantation*
4	7.7	6.8	43.3	52.7	0.45	0.34
5	7.2	6.6	61.0	51.0	0.22	0.35
6	6.6	6.4	54.0	46.0	0.57	0.65
7	6.6	6.6	62.6	56.0	0.73	0.62

The physico-chemical properties of soil were monitored after 6 years of planting of ten tree species (indigenous and exotic) on soil properties in the Tarai region (Mallick and Prakash, 1996). There was no effect on soil texture, but bulk density increased, particle density decreased and total porosity increased by 4–11 per cent. Changes in chemical properties were significant in the surface layer (0–20 cm). Soil pH decreased by 3–5 per cent while organic carbon increased by 33–73 per cent and available plant nutrients by 8–30%. In another study from the 'tarai' region Joshi *et al.* (1997) determined litter fall from nine plantations of six species. Leaf litter contributed a major portion (76–98%) of the total litter production and the rest was contributed by other woody or reproductive parts (Table 20).

Table 20 Annual litter production (kg/ha/yr) in different plantations in the tarai region.

Plantation	*Age (years)*	*Leaf*	*Wood*	*Reproductive*	*Total parts*
Eucalyptus	5	31.03.0	181.0	28.6	3312.6
Grevillea robusta	17	4594.3	50.7	69.9	4714.9
Leucaena	10	4351.5	536.8	835.7	5723.8
P. roxburghii	18	6015.0	57.8	63.6	6136.2
	22	7485.6	316.8	259.1	8061.5
P. deltoides	5	7756.8	262.6	–	8019.4
	6	5235.5	138.0	–	5373.5
	17	5478.3	428.7	–	5907.0
T. grandis	21	6802.1	60.6	242.1	7104.8

Litter production was related to various abiotic factors such as evaporation, rainfall, temperature and wind velocity. The maximum input of N through leaf litter occurred in *Leucaena* (96 kg/ha) and least in *Eucalyptus* (25.8 kg/ha). The total N input in different plantations followed the order: *Leucaena* > 17 year *Populus* > 6 year *Populus* > *Tectona* > 5 year *Populus* > 22 year *Pinus* > 18 year *Pinus* > *Grevillea* > *Eucalyptus*. The P input was maximum in teak and least in *Eucalyptus*, while *Populus* litter returned the highest amount of K.

Annual litter production of 4164 kg/ha was observed in a *Dalbergia sissoo* plantation in U.P., which returned (kg/ha/yr) in the following order: Ca (132) > N(67) > K(16) > Mg(8) > P(3), (Sharma *et al.*, 1988). The major contribution was from leaf litter, which accounted for 89% of the total litter production.

In order to meet the growing demand of raw material for paper and plywood industry, a number of exotic species like *Eucalyptus*, *A. auriculiformis*, *P. patula* and various clones of *Populus deltiodes* have been introduced in India. Poplars cultivation in agroforestry practices is now widespread in the tarai, western U.P., Punjab and Haryana.

Litter production and nutrient dynamics in a 14-year old poplar (*P. deltiodes*) plantation was studied in Dehradun by Raizada and Srivastava (1986). They observed a bimodal litter fall pattern amounting to 3.08 t/ha/yr, of which 2.85 t/ha/yr was contributed by leaf litter. Leaf litter nutrient concentration were negatively correlated with the magnitude of leaf fall, except for Mg, due to the deciduous nature and quantity of leaf fall. Concentrations of Ca and N were higher than K, Mg and P. Nitrogen and K showed significant mobility prior to leaf fall, suggesting retranslocation into woody tissues, while magnesium appears to be immobile. The return of nutrients though leaf fall amounted to 29.4 kg N, 29 kg Mg, 23.8 kg Ca, 21.3 kg K and 2.8 kg P (Raizada and Srivastava, 1991a). Poplar leaf litter decomposed rapidly, with 53 per cent weight loss occurring in the first six months. Nutrient release from decomposing leaves occurred in the order K > Ca > Mg > N > P, indicating immobilization of N and P in microbial tissue (Raizada and Srivastava, 1991b).

In Punjab, Ralhan *et al.* (1996) reported production of 4.15 t/ha of litter from a 4 year old stand of *P. deltoides* (clone G 3) planted for agroforestry practices. There was a decline in the 5th year due to pruning of the lower branches (Table 21). The returns of N, P, K and Ca were highest in the 4 year old stand, these being 39.7, 8.2, 20.1 and 24.8 kg/ha, respectively.

Table 21 Annual litter production from *P. deltoides* plantations in agroforestry systems in Punjab.

Age (years)	*Total litter (kg/ha/yr)*	*Mean*
2	3432.8	684.1
3	3063.2	612.8
4	4158.8	831.8
5	3530.8	706.3

Populus deltiodes (clone G 3) produced 1354, 1942, 2888 and 2472 kg/ha of litter from 3, 5, 7 and 9 years old plantations in the U.P. Tarai (Tandon et al., 1991). The leaf litter contributed 70–85 per cent of the total, and the remaining by twig litter. Calcium was returned in the largest quantities, followed by N, Mg, K and P in that order (Table 22).

Table 22 Return of nutrients (kg/ha) from annual litter fall in poplar plantations of various ages in U.P. Tarai.

Age (years)	*N*	*P*	*K*	*Ca*	*Mg*
3	10.74	0.27	4.14	15.87	1.74
5	21.51	0.54	4.45	31.17	2.37
7	27.95	0.65	11.20	46.57	3.35
9	22.76	0.73	7.83	34.88	2.50

In U.P. (Meerut district), Sharma (1999) determined litter production from an irrigated 3 and 4 year old plantation of P. deltoides. An annual leaf litter production of 914 and 1292 kg/ha was recorded. The returns of nutrients by leaf litter was in the following order: Ca > N > Mg > K > P, the values (kg/ha) being 18.3–24.3, 11.2–12.5, 7.8–10.0, 7–9 and 2.4–2.6, in the same order.

Old river bed land are formed due to erosion in stream banks and course meandering by seasonal streams and rivers in the sub-tropical region. These lands can be rehabilitated and stabilized by afforestation with suitable tree species. Improvements of soil characterstics through organic matter and nutrient enrichment by trees has made it possible to introduce silvi-pastoral and pastoral systems on these sites. Raizada et al. (1996) determined the impact of these plantations on soil characterstics of old river bed lands in the Doon valley.

Nutrient concentrations litter and twig components showed a trend of Ca > Mg > N > P > K. Nutrient content in the decomposing litter layer is marginally more, specially with regard to N and P of Grewia *optiva, Melia azedarach, Bauhinia purpurea* and *Morus alba*. There was no definite pattern of variation in the organic matter content in the top litter and decomposing layers due to different tree species (Table 23).

Table 23 Organic matter (kg/ha) in the top litter layer (L) and decomposing layer (F) on the plantation floor of different species growing on old river bed land in Dehradun.

Species	*Age (years)*	*L*	*F*	*Total*
G. optiva	11	200	3100	4300
Eucalyptus hybrid	14	965	455	4420
Leucaena	15	1000	2700	3700
M. azedarach	9	600	1000	1600
Bamboo	–	2400	1600	4000
B. purpurea	11	3650	1900	5550
M. alba	23	400	2600	3000

Some soil properties from two depths (0–7.5 and 7.5–15 cm) are described in Table 24. Perusal of data indicates that soil properties have undergone modifications, specially in the surface layer, these changes could not be attributed to the presence of species alone. Changes occurred with regard to water holding capacity, pH and organic carbon content. Microbial response to soil moisture is parabolic which affects mineralization, hence changes were not very significant.

Table 24 Effect of various tree species on soil properties at two soil depths in Dehradun.

Species	*Depth (cm)*	*pH*	*WHC (%)*	*Organic carbon %*
Eualyptus hybrid	0 – 7.5	6.5	50.36	3.10
	7.5 – 15	6.0	48.36	0.86
B. purpurea	0.5 – 7.5	6.8	42.23	1.23
	7.5 – 15	6.5	44.47	0.62
G. optiva	0 – 7.5	7.0	51.55	1.35
	7.5 – 15	6.5	48.31	0.73
M. azedarach	0 – 7.5	6.7	59.56	2.16
	7.5 – 15	6.9	60.14	1.02
M. alba	0 – 7.5	6.7	59.56	2.16
	7.5 – 15	6.5	46.70	0.80
L. leucocephala	0 – 7.5	6.5	59.34	1.16
	7.5 – 15	6.5	52.54	0.86
Bamboo	0 – 7.5	7.0	46.97	1.40
	7.5 – 15	7.0	36.38	0.90

WHC%—water holding capacity %

HIMALAYAN REGION

Inspite of the large areas under conifer plantations in the Himalayas, scanty information is available on the litter production, cycling of nutrients and their impact on soil properties. The litterfall from woody vegetation occurs mostly during the summer which accounts for more than 80 per cent of the years total. In deciduous species, the litter fall occurs in a single spell during autumn. Pandey and Singh (1981) developed allometric relationships between leaf fall and temperature for *Quercus floribunda, Q. leucotrichophora* and *Cupressus torulosa,* that supported this relationship. In contrast, 50-72 per cent of the total annual woody litterfall occurs during the rainy season. The contribution of rainy season to woody litter production declines with an increase in altitude (Mehra and Singh, 1985).

In the Darjeeling hills, litter production from *Cryptomeria japonica* was 4856 kg/ha/yr, with needles constituting 81.3 per cent of the total (Ram *et al.,* 1986). In Himachal Pradesh, the litter production from *P. roxburghii* increased (Table 25) with an increase in age (Rawat and Tandon, 1993). Similarly the annual return of nutrients from needles also increased (Table 26).

Table 25 Litter production (kg/ha/yr) in *P. roxburghii* plantations in H.P.

Age (yrs.)	*Needle*	*Twig*	*Total*
6	2309	92	2401
10	2753	314	3067
14	3063	148	3211
16	2875	682	3557
18	4558	441	4999

Table 26 Return (kg/ha/yr) of various nutrients from needles of *P. roxburghii* plantations in H.P.

Age (yrs)	*Nutrients*				
	N	*P*	*K*	*Ca*	*Mg*
6	37.8	0.31	8.7	8.0	8.0
10	46.0	0.43	11.0	9.8	10.0
14	50.3	0.41	11.6	10.6	10.6
16	49.7	0.50	12.6	10.8	11.1
18	75.8	0.64	18.0	16.1	16.2

Degradation of oak forests in the Himalayan region due to soil erosion, excessive lopping and biomass removal, diseases and rise in temperature has led to their gradual replacement with species like *P. roxburghii*, which are well adapted to poor, infertile sites. The ecological implications of these changes are a threat to the ecosystem in general and loss of plant biodiversity in particular. Replacing *Quercus leucotrichophora* (banj oak) with pine forests causes significant variations in soil nutrient status. Chir pine (*P. roxburghii*), adapted to infertile habitats retranslocates greater quantities of nutrients from senescing needles than do the oaks, which are adapted to fertile sites. Species adapted to infertile sites, such as *P. roxburghii*, retranslocate a greater fraction of nutrient mass from senescing leaves than those species, such as oaks, which are adapted to fertile sites (Thomas and Grigal, 1976). The mature needles (leaves) of chir pine have lower N concentrations (C : N ratio 71 : 1) than *Q. leucotrichophora* (41 : 1). Chir pine needles decompose slowly resulting in a build up of inflammable needles on the forest floor. Being fire resistant, chir pine replaces oak (which is fire susceptible) with increased incidents of burning. The high C : N ratio leads to immobilization of available N, rendering the site nutrient poor (Upadhyay and Singh, 1985a). Soil fertility increases to such an extent that oaks are rarely able to establish again on a site by chir pine (Singh *et al.*, 1984). Therefore, caution is required to be taken when revegetating barren/degraded hill slopes by planting quick growing species like *P. roxburghii*, which permanently replace broad leaved species of *Quercus*. This replacement ultimately leads to loss of plant biodiversity and adversely affects soil quality, forest hydrology and water availability.

Pinus patula, an exotic species of humid temperate Mexico was introduced and planted in considerable areas in the Himalayas. This species has been reported to be a fast growing conifer in India (Ghosh and Guhathakurta, 1973). There was a decline (Table 27) in litter production in 10 year old plantations of *P. patula* plantations in West Bengal due to poor site quality (Bhartari, 1986) and consequent slow growth. Nutrient concentrations did not vary with increasing stand age and the general pattern was in order: N > Ca > K > Mg > P. The annual return of nutrients from different ages are given in Table 28.

Table 27 Litter production (kg/ha/yr) from *P. patula* plantations in Darjeeling (West Bengal).

Age (yrs.)	*Needle*	*Branch*	*Total*
4	3120	25	3145
6	3864	13	3877
8	3971	18	3989
10	1472	–	1472
12	2651	259	2910
14	2536	125	2660

Table 28 Quantities (kg/ha) of nutrients returned by litterfall from *P. patula* plantations in Darjeeling (West Bengal).

Age	*N*	*P*	*K*	*Ca*	*Mg*
4	32.14	0.31	9.67	7.50	4.7
6	31.30	1.54	10.05	11.6	8.5
8	52.02	0.80	10.32	9.93	8.34
10	4.12	0.60	2.21	3.24	2.06
12	22.27	0.26	6.63	8.50	5.60
14	33.0	0.25	5.83	6.10	5.32

In a similar study, the annual litterfall was maximum from 25 year old stand of *P. patula,* while the forest floor accumulated the maximum quantity of litter and duff in the 34 year old stand (Table 29), (Singh, 1982).

Table 29 Quantities of litterfall, litter and duff layer in *P. patula* plantations in West Bengal.

Parameter	*Age (years)*					
	8	10	12	17	25	34
Litterfall (t/ha/yr)	–	8.75	4.15	4.67	11.42	8.75
Litter layer (t/ha)	2.99	8.82	7.22	7.91	18.84	23.24
Duff layer (t/ha)	9.69	15.14	13.82	12.00	18.00	46.90

Needle litter is rich in N and poor in P and Na. The litter and duff layers contain higher amounts of N than of Ca. Needle litter decomposition is slow in *P. patula,* as in other conifers, because the climatic conditions do not permit rapid decay. The plantation floor retains a very rich nutrient store over the mineral soil layer. Turnover rates and time of nutrient fluxes show a higher efficiency of the fluxes from the litter (nutrient) return and a greater stability of the soil pool.

RECLAMATION OF PROBLEM SOILS

Salt affected soils including sodic alkali soils cover an area of 5.88 m ha in India (Anonymous, 1994). These soils can be used for fuelwood production by planting suitable salt-tolerant species. The use of perennial vegetation for the reclamation of these soils is well known. The soil pH decreased while the soluble salt content, organic matter and N increased in the 0–15 cm soil layer, under a *P. juliflora* plantation (Yadav and

Singh, 1970). A mixture of species like *Acacia nilotica, Syzygium cuminii, F. bengalensis, E. officinalis* and *D. sissoo* was found most effective (Verma *et al.*, 1982) in reducing pH and increasing organic carbon in 'usar' soils (Table 30).

Table 30 Changes in pH and organic carbon of a sodic soil (0–30 cm depth) under various tree species, after twenty years of growth at Banthra (Uttar Pradesh).

Site	*pH*	*Organic carbon %*
Open (no trees)	10.2	0.40
T. arjuna	8.5	0.90
F. bengalensis	8.7	0.60
S. cuminii	8.3	0.84
D. sissoo	8.8	0.81
A. procera	8.0	0.90
B. variegata	8.4	0.90
A. nilotica var. indica	8.3	0.75
T. arjuna + S. cumnii	8.1	0.81
A. nilotica + S. cumnii + F. bengalensis + E. officinalis + D. sissoo	7.2	1.17
F. bengalensis + S. cuminii	8.0	0.90

The pattern of litter production and annual cycling of nutrients in *A. nilotica* and *E. tereticornis* plantations growing in highly alkaline soils (pH 1 0.1; calcium carbonate 3.13; ESP 92, EC 1.58) was investigated (Gill *et al.*, 1987) in Karnal, Haryana. Annual litter production was higher and unimodal in acacia than in *Eucalyptus* where it was bimodal. More than 55 and 40 per cent of the annual litter production was obtained during the winter season in acacia and *Eucalyptus*, respectively. Concentration of N, P, K and S and annual recycled nutrients were significantly higher in acacia litter than in *Eucalyptus*. However the reverse was true for Ca, Mg, Na, Fe and Mn accumulation, but annually recycled nutrient quantity was more in acacia than *Eucalyptus*.

Soil amelioration was better under acacia than under *Eucalyptus* due to rapid decomposition (narrow C : N ratio), nitrogen fixing properties and low Na accumulation in acacia. Therefore plantations of acacia on highly alkaline sites are beneficial for their reclamation. In the alkaline wastelands in Aligarh (UP) there was a reduction in pH, EC, ESP values and increase in organic carbon content (Singh, 1989) five years after planting various trees and shrubs (Table 31).

Table 31 Changes in soil properties after five years of afforestation on an alkaline site at Aligarh (Uttar Pradesh).

Character	*Depth (cm)*	*1984*	*1989*
pH	0 – 15	9.73	8.83
	15 – 30	9.81	9.00
EC (in mhos/cm)	0 – 15	10.7	0.23
	15 – 30	3.73	0.27
Organic carbon %	0 – 15	0.21	0.60
	15 – 30	0.16	0.55
ESP	0 – 15	86.00	30.00
	15 – 30	89.00	34.00

In the alkali 'usar' soils in the alluvial plains of U.P. Shukla and Misra (1993) investigated the impact of *Eucalyptus* hybrid, *D. sissoo, A. nilotica* and *P. juliflora* plantations, on soil properties. Soil pH decreased in the surface soil, with the maximum reduction occuring in *P. juliflora* (9.5 to 8.9), followed by that in *D. sissoo* (9.8 to 9.2), *A. nilotica* (9.6 to 9.2) and *Eucalyptus* (9.5 to 9.1). The electrical conductivity showed greatest reduction occurred with *P. juliflora* (0.45 to 0.20) followed by *A. nilotica* (0.68 to 0.43), *Eucalyptus* (0.50 to 0.28) and *D. sissoo* (0.35 to 0. 17), while in the control plots the EC values remained quite high (2.0 to 1.88).

The increase in organic carbon was maximum with *A. nilotica* (0. 18 to 0. 31%), followed by *P. juliflora* (0.27 to 0.31%), *Eucalyptus* (0.29 to 0.35%) and *D. sissoo* (0.27 to 0.34%). Improvements were also recorded in the surface soil N content in the order: *P. juliflora* > *D. sissoo* > *A. nilotica* > *Eucalyptus*. *P. juliflora* foliage also reduced exchangeable Na and Ca in the surface layer, with calcium content increasing by two times its initial value under all canopy covers. These increases occurred due to CO_2 produced from respiring roots which on desolution with water produced carbonic acid. The acid consequently increased the solubility of $CaCO_3$ minerals by lowering the pH and forming a host of complex calcium ion pairs, thus increasing Ca concentration in the soil solution (Nakayama, 1969). Bulk density decreased and water holding capacity increased in the surface layer in all species. The increase was more pronounced under *P. juliflora,* followed by *D. sissoo, Eucalyptus* hybrid and *A. nilotica.*

CONCLUSIONS

A large portion of the above ground production of trees in tropical forest ecosystems is recycled to the soil layer in the form of detritus or litterfall that may be unimodal as in deciduous species or remain steady as in evergreen species. Nutrient cycling occurs through physical and chemical breakdown of detritus, resulting in release of nutrients into the soil layer and formation of humus. In many cases soil properties are also ameliorated.

The amount of nutrients returned to the soil layer is directly related to the amount of litterfall and concentration of nutrients. Nutrients like N and P are retranslocated into woody tissues prior to senescence in deciduous species, while this nutrient 'conservation' efficiency is lower in evergreen species. The production of new foliage every year in deciduous species makes it necessary that the tree retranslocate nutrients back from the leaves, since acquisition of nutrients by active absorption is energy consumptive (Bloom *et al.*, 1985). Distinct differences in the concentration of N between evergreen and deciduous species are attributed to the inherent differences in their growth forms and physiology (Chapin 1980; Miller, 1984).

The rate at which organic matter is recycled from trees and the amount of nutrients returned, provides an index of nutrient use within the vegetation (Hirose, 1975) Forests and plantations raised on moderately fertile soil return more litter with higher nutrient concentration than do forests on other sites (Vitousek and Sanford, 1986).

The nutrient use efficiency (NUE) is the quantity of dry matter produced per unit weight of nutrients. In most forest ecosystems, the NUE for N, K, Ca and Mg is almost the same, except in tropical forests where the year round growing season leads to high rate of production and nutrient circulation. The high demand of nutrients for growth is adjusted by efficient use of nutrients from the soil. A high NUE for P in the tropical region has been observed, especially where native soil P is low (Vitousek, 1984). By internal conservation and mycorrhizal associations, it appears that tropical tree species are adapted to grow in P deficient soils.

Sal and teak are more efficient in nutrient conservation and return less amounts of nutrients annually. The nutrient conservation is poor in pine as it returns a greater proportion of nutrients through its leaf fall and a major part of N and P gets locked up in microbial tissues. In *Eucalyptus*, low conservation and high use of nutrients lead to a higher loss of nutrients, which increases further by leaching in well-drained soils (Lamb, 1985). These differences between evergreen and deciduous species result in variable levels of soil organic matter and nutrients. It is therefore advisable to introduce deciduous and/or nitrogen fixing species for enhanced N fixation and uptake along with fast growing evergreens like *Eucalyptus* and pine species.

This review reveals that soil amelioration is more vivid in the surface layer in certain drier localities to increased depths in the sub-humid tropical areas. Positive impact of trees canopy on nutrient recycling are evident in wastelands, old riverbed lands and in saline-alkaline soils.

Forest plantations strongly influence the rate of rehabilitation of degraded tropical forests. These plantations create microclimatic conditions that support recruitment of species native to the region and help them to survive through temporary harsh situations. This calls for designing and managing plantations in a way that would make them serve as 'foster' ecosystem to support the succession of native vegetation. Various extraneous factors have influenced afforestation activities by exotic species in the past, it is time that serious thought is given to the raising of plantation of native species to initiate the process of migration of species. A synthesis of silvicultural and ecological knowledge is required and can be of immense value in the process of hastening rehabilitation of degraded tropical forests in India.

REFERENCES

Aina, P.O., 1984. Contribution of earthworms to porosity and water filtration in a tropical soil under forest and long term cultivation, *Pedobiologia*, **26:** 131–136.

Anderson, J.M., and Swift, M.J., 1983. Decomposition in tropical forests, In: *Ecology and Management*, (S. Whitmore, and Chadwick, eds.), Blackwell Science Publications, Oxford, 287–309 pp.

Anonymous, 1994. *Status of land degradation in India*, Ministry of Agriculture and Co-operation, New Delhi.

Agarwal, R.K., Praveen K., and Raina P., 1993. Nutrient availability from sandy soils underneath *P. cineraria*, (Linn. Macbride) compared to adjacent open site in an arid enviromnent, *Ind. For.*, **119**(4): 321–325.

Bahuguna, V.K., Negi, J.D.S., Joshi, S.R., and Naithani, K.C., 1990. Leaf litter decomposition and nutrient release in *Shorea robusta* and *E. camaldulensis* plantation, *Ind. For.*, **116**(2): 103–114.

Berg, B., and Staaf, H., 1981. Leaching, accumulation and release of nitrogen in decomposing forest litter, In: *Terrestrial Nitrog Cycles*, (F.E. Clark, and T. Rosswall, eds.), Swed. Nat. Sci. Res. Counc., Stockholm, 163–178 pp.

Bhartari, S.K., 1986. Biological productivity and nutrient cycling in *Pinus patula* plantations in Darjeeling hills, *Ind. For.*, **112**(3): 187–210.

Bloom, A.J., Chapin, F.S., and Mooney, H.A., 1985. Resource limitations in plants—an economic analogy, *Ann. Rev. Ecol. Syst.*, **16:** 363–393.

Box, E.O., 1978. Geographical dimensions of terrestrial net gross productivity, *Radiation and Envr. Bioiphysics*, **15:** 305–322.

Chakraborty, R.N., and Chakraborty, D., 1989. Changes in soil properties under *Acacia auriculiformis* plantations in Tripura, *Ind. For.*, **115**(4): 272–273.

Chapin, F. S. III., 1980. Mineral nutrition of wild plants, *Ann. Rev. Ecol. Syst.*, **11:** 233–260.

Chowdhary, R.L., Asbdulbari, P.A., and Prasad, K.G., 1982. Performance of various pine species and their provenances in Tamil Nadu, *Seminar-cum-workshop on Tropical Pines*, Koraput (Orissa).

George, M., 1982. Litter production and nutrient return in *Eucalyptus* hybrid plantations, *Ind. For.*, **108**(4): 253–260.

Gentry, J.B., and Whitford, W.G., 1982. The relationship between wood litterfall and relative abundance and feeding activity of subterranean termites *Reticulitermes* spp. in three south eastern coastal plain habitats, *Oecologia*, **54:** 63–67.

Ghosh, R.C., and Guhathakurta, P., 1973. *Pinus patula*, Schl. & Chain.—its problems and prospects in West Bengal, *Ind. For.*, **99:** 337–348.

Gill, H.S., Abrol, I.P., and Samra, J.S.,1987. Nutrient recycling through litter production in young plantations of *Acacia nilotica* and *Eucalyptus tereticornis* in a highly alkaline soil, *For. Ecol. Mngmt.*, **22:** 57–69.

Hirose, T., 1995. Relations between turnover rate, resource utility and structure of some plant populations: A study of the matter bugets, *J. Fac Sci.*, Univ. Tokyo, III XI(II), 355–407.

Hole, F.D., 1981. Effects of animals on soils, *Geoderma*, **25:** 75–112.

Jeeva,V., and Ramakrishnan, P.S.,1997. Studies on impact of plantation forestry in Nilgiris hills of the Ghats on soil quality and nutrient cycling, *Trop. Ecol.*, **38**(2): 215–235.

Jha, M.N., and Pande, P., 1984. Impact of growing *Eucalyptus* and sal monocultures on soils in natural sal area of Doon valley, *Ind. For.*, **110**(1): 16–22.

Joshi, C.S., Rao, P.B., and Singh, R.P., 1997. Comparative study of litterfall and nutrient return in some plantations of Central Himalaya, *Proc. Ind. Natl. Sci. Acad.*, **B 63**(6): 617–624.

Jordan, C.F., 1985. *Nutrient Cycling in Tropical Forest Ecosystems*, John Wiley and Sons, New York.

Kozlowski, T.T., 1973. Extent and significance of shedding of plant parts, In: *Physiological Ecology Shedding of Plant Parts*, (T.T. Kozlowski, ed.), Academic Press, New York and London, 1–37 pp.

Lamb, R.J., 1985. Litterfall and nutrient turnover in two eucalyptus woodlands, *Aust. J. Bot.*, **33:** 1–4.

Lugo, A.E., 1988. The future of the forest ecosystem rehabilitation in the tropics, *Environment*, **30:** 16–20, 41–45.

Mallick, K.P. S. and Prakash, O., 1996. Changes in some physical and chemical properties of soil in an agri-silvicultural system with some multi-purpose tree species in Tarai region of U.P., In: *Agroforestry and Environment*, (R.K. Kohli, and K.S. Arya, eds.), Dayanand National Academy of Environmental Sciences, Chandigarh, India, 249–254 pp.

Malhotra, P.P., Tandon, V.N., and Shanker, P.P., 1987. Distribution of nutrients and their return through litterfall in an age series of *Pinus patula* Schl. & Cham. Plantations in Nilgiris, *Ind. For.*, **113** (5): 323–332.

Miller, H.G., 1984. Dynamics of nutrient cycling in plantation ecosystems, In: *Nutrition in Plantation Forests*, (G.D. Bowen, and E.K.S. Nambiar, eds.), Academic Press, London, 53–78 pp.

Meentemeyer, V., 1977. An approach to the biometeorology of decomposer organisms, *Int. J. of Biometeorology*, **22:** 94–102.

Meentemeyer, V., Box, E.O., and Thompson, N.R., 1982. World patterns and amounts of terrestrial plant litter production, *Bioscience*, **32**: 125–128.

Mehra, M.S., and Singh, J.S., 1985. Pattern of wood litterfall in five forests located along an elevational gradient in Central Himalaya, *Vegetatio*, **63:** 3–11.

Nakayama, F.S., 1969. Theoretical consideration of the calcium sulphate-bicarbonate, carbonate interaction in soil solution, *Soil Sci. Soc. An. Proc.*, **33:** 668–672.

Nandi, A., Basu, P.K., and Banerjee, S.K., 1991. Modification of soil properties under *Eucalyputs* sp. *Ind. For.*, **117**(1): 53–57.

Negi, J.D.S., Sharma, S.C., and Bist, A.P.S., 1988. Forest floor and soil nutrient inventories in an old growth *Eucalyptus* plantation, *Ind. For.*, **114**(8): 453–461.

Pandey, U., and Singh, J.S., 1981. Leaf litter decomposition in an oak-conifer forest in Himalayas: The effect of climate and chemical composition, *Forestry,* **55**(1): 47–59.

Pande, P.K., and Sharma, S.C., 1993. Biochemical cycling and nutrient conservation strategy in some plantations, *Ind. For.,* **119** (14): 299–305.

Parrotta, J.A., 1993. Secondary forest regeneration on degraded tropical lands. The role of plantations as "foster ecosystem", In: *Restoration of Tropical Forest Ecosystem,* (H. Lieth, and M. Lohman, eds.), 63–73 pp.

Prasad, A., 1994. Alley cropping in *Leucaena leucocephala* with field crops for sustained productivity, In: *Agroforestry Systems for Degraded Lands,* (P. Singh, P.S. Pathak, and M.M. Roy, eds.), 471–478 pp.

Rai, S.N., and Chakrabarti, S.K., 1996. *Fuelwood, timber and fodder from forests of India,* Forest Survey of India, Min. of Environment and Forests, Govt. of India.

Ralzada, A., and Srivastava, M.M., 1986. Litter production in a *Populus deltoides* Marsh. Plantation, *For. Ecol. Mngmt.,* **15:** 215–218.

Ralzada, A., and Srivastava, M.M., 1991a. Nutrient accumulation in standing crop and returns through litterfall in a *Populus deltoides* Marsh. Plantation, *Van Vigyan,* **29**(4): 192–199.

Raizada, A., and Srivastava, M.M., 1991b. Leaf litter decomposition and pattern of nutrient release in a *P. deltoides* Marsh. Plantation, *Van Vigyan,* **29**(3): 175–179.

Raizada, A., and Jayaram, N.S., 1995. Modification of soil properties by *Eucalyptus* hybrid plantations growing on semi-arid ferruginous red soils in India, *J. Trop. For.,* **11**(II) 227–233.

Raizada, A., Arora, Y.K., and Agarwal, M.C., 1996. Nutrient inventories underneath several tree species in old river bed land of Doon valley, In: *Agroforestry and Environment* (R.K. Kohli, and K.S Arya, eds.), Dayanand National Academy of Environmental Sciences, Chandigarh, India, 256–271 pp.

Ralhan, P.K., Rasool, S.Z.A., and Singh, A., 1996. Return of nutrients through leaf litter of san age series of poplar plantations in an agrisilvicultural system in certain parts of Punjab, In: *Agroforestry and Environment,* (R.K. Kohli, and K.S. Arya, eds.), Dayanand National Academy of Environmental Sciences, Chandigarh, India, 159–163 pp.

Ram, N., Yadav, Y.P., Vasisth, H.B., and Chauhan, J.M.S., 1986. Litter production from protected and unprotected watershed under *Cryptomerisa japonica* plantation, *Van Vigyan,* **24** (3 and 4): 56–60.

Rawat, J.K., and Tandon, V.N., 1993. Biomass production and nutrient cycling in young chir plantations in H.P. *Ind. For.,* **119**(12): 977–985.

Schlesinger, W.H., 1977. Carbon balance in terrestrial detritus, *Ann. Rev. Ecol. Syst.,* **8:** 51–81.

Sharma, D.C., Taneja, P.L., and Bist, A.P.S., 1988. Biomass productivity and nutrient cycling in a *Dalbergia sissoo* plantation, *Ind. For.,* **114**(5): 261–268

Sharma, N.K., 1999. *Populus deltoides based tree/crop interface studies on an alluvial soil under irrigated conditions,* Ph.D. thesis, Dept. of Agronomy Janta Vedic College, Baraut, C.C.S. University, Meerut.

Shukla, A.K., and Misra, P.N., 1993. Improvement of sodic soil under tree cover, *Ind. For.,* **119**(1): 43–52.

Singhal, R.M., Banerjee, S.P., and Pathak, T.C., 1975. Effect of *Eucalyptus* monoculture on the status organic matter in natural sal (*Shorea robusta*) zone in Doon valley, *Ind. For.,* **101**(12): 730–737.

Singh, B., 1982. Nutrient content of standing crop and biological cycling in *Pinus patula* ecosystem, *For. Ecol. Mngmt.,* **4:** 317–332.

Singh, B., 1989. Rehabilitation of alkaline wasteland on the Gangetic alluvial plains of Uttar Pradesh, India, through afforestation, *Land Degradation and Rehabilitation,* **1**(4): 305–310.

Singh, R.R., 1984. Nutrient cycle in *E. tereticornis* Smith. Plantations, *Ind. For.*, **110**(1): 76–85.

Singh, J.S., and Gupta, S.R., 1977. Plant decomposition and soil respiration in terrestrial ecosystems, *Bot. Rev.*, **43:** 449–528.

Singh, J.S., Pandey, U. and Tiwari, A.K., 1984. Man and forests: A Central Himalayan case study, *Ambio*, **13**(2): 86–87.

Singh, S.B., Prasad, K.G., and Banerjee, S.P., 1988. Distribution of nutrients in lateritic soils with special reference to vegetation, *Van Vigyan*, **26** (1 and 2): 24–29.

Singh, R., Singh, R.K., and Singh, Kalyan, 1990. Effect of different plant covers on soil characteristics, *Ind. For.*, **116**(10): 795–802.

Srivastava, A.K., 1993. Change in physical and chemical properties of soil in irrigated *Eucalyptus* plantation in Gujarat, *Ind. For.*, **119**(3): 227–231.

Swift, M.J., Heal, O.W., and Anderson, J.M., 1979. *Decomposition in Terrestrial Ecosystems*, Univ. of California Press, Berkeley.

Tandon, V.N., Pandey, M.C., Rawat, H.S., and Sharma, D.C., 1991. Organic productivity and mineral cycling in plantations of *Populus deltoides* in Tarai regions of Uttar Pradesh, *Ind. For.*, **117**(8): 596–608.

Thomas, W.A., and Grigal, D.F., 1976. Phosphorus conservation by evergreens of mountain laurel, *Oikos*, **27:** 19–26.

Upadhyay, V.P., and Singh, J.S., 1985a. Nitrogen dynamics of decomposing hardwood leaf litter in a Central Himalayan forest, *Soil Biol. Biochem.*, **17**: 827–830.

Venkatraman, C., Haldorai, B., Samraj, P., Nalatwadmath, S.K., and Henry, C., 1983. Return of nutrients by the leaf litter of bluegum (*E. globulus*) and black wattle (*Acacia mearnsii*) in plantations of Nilgiris in Tamil Nadu, *Ind. For.*, **100**(6): 370–383.

Verma, S.C., Jain, R.K., Rao, M.V., Misra, P.N., and Murthy, A.S., 1982. Influence of canopy on soil composition of man-made forest on alkali soil of Banthra (Lucknow), *Ind. For.*, **108**(6): 431–437.

Vitousek, P.M., 1984. Litterfall, nutrient cycling and nutrient limitation in tropical forests, *Ecology*, **65:** 285–298.

Vitousek, P.M., and Sanford, R.L. Jr., 1986. Nutrient cycling in moist, tropical forest, *Ann. Rev. Ecol. Syst.*, **17:** 137–167.

Waring, R.H., and Schlesinger, W.H., 1985. *Forest Ecosystems—Concepts and Management*, Academic Press, New York.

Yadav, J.S.P., and Singh, K., 1970. Tolerance of certain forest species to varying degrees of salinity and alkali, *Ind. For.*, **96**(8): 587–599.

CHAPTER 4

TROPICAL PLANTATIONS WITH NATIVE TREES: THEIR FUNCTION IN ECOSYSTEM RESTORATION

Florencia Montagnini

Area of Management and Conservation of Forests and Biodiversity
Centro Agronomico Tropical de Investigación y Enseñanza (CATIE)
7170 Turrialba, Costa Rica.
Present Address: School of Forestry & Environ.Studies,
Yale University, 310 Prospect St., New Hawen ET 06511, U.S.A.

ABSTRACT

Tree planting is being promoted in a number of government and private efforts towards rural development throughout the tropics. As the area in degraded lands increases, emphasis is placed on the use of species which can grow in such conditions and yield economic products as well as environmental benefits (soil conservation, watershed protection etc.).

This chapter focuses on the impacts of native trees on soil fertility and nutrient cycling, using examples from land rehabilitation programmes from tropical humid regions of Latin America. In Costa Rica, increased level of soil nitrogen, carbon Ca and Mg were found under tree plantations than under grass, with values close to those found in 20-year-old forests. Among tree species, the highest soil C, N and cations were found under *Vochysia ferruginea*, while results of nutrient cycling studies revealed important effects of other species on soil N and P availability and Ca and Mg recycling. In Bahia, Brazil positive effects were found in at least five soil parameters under 15 out of 20 tree species, in comparison with primary and secondary forests. Several species contributed to increased C and N concentrations in the top-soil: *Inga affinis, Parapiptadenia pterosperina, Plathymenia foliolosa* (leguminous N-fixing species), *Caesalpinia echinata, Copaifera luscens* (leguminous, non-N-fixing), *Eschweilera ovata, Pradosia lactescens* (of other families). In Misiones, NE Argentina, soils under *Bastardiopsis densiflora* had the highest levels of organic matter, N, cations and pH.

Nutrient use efficiencies (biomass increments per unit of nutrients in annual litterfall) were calculated for each species to help integrate results and design management recommendations geared to sustaining productivity of plantation systems including these species. In addition, analysis of nutrient stocks in soils and in forest litter, and information on nutrient release from litter, is brought forward to stress the importance of the soil-litter system on plantation nutrient conservation.

Key Words: soil rehabilitation, native trees, nutrient use efficiency, lowland humid tropics

Tropical plantations can serve diverse economic, social, political and ecological functions. With considerably higher yields than managed native forests, tropical and subtropical plantations can make substantial contributions to world timber and pulp production (Wadsworth, 1983, Evans 1992). Tree plantations can also be a source of cash, savings and insurance for individual farmers. Plantations may help stabilize rural populations in regions where shifting agriculture is the predominant land use. In combination with subsistence and commercial crops (agroforestry) or cattle (agrosilvopastoral systems), plantations have been used as tools in rural development projects worldwide. Plantations are often seen as alternatives to deforestation as they can provide products that otherwise would be taken from natural forests (Fearnside, 1990, McNabb *et al.*, 1994).

If plantation species are chosen with knowledge of their nutrient-use efficiencies and recycling capacities, they can be highly productive and also serve a function in ecosystem restoration projects. Particularly, tree plantations and tree-crop combinations represent productive land use alternatives for deforested lands with poor natural forest regeneration due to long distance to sources of propagules or intense site degradation. Among the latter, low soil fertility, soil compaction after abandonment from cattle grazing, and invasion by grasses and other aggressive vegetation can be serious obstacles to both forest regeneration and conventional agriculture. As the area in degraded lands spreads out, emphasis is increasing on the use of tree species which can grow in such conditions and yield economic products (timber, fuelwood and other) as well as prove beneficial to the environment (soil conservation, watershed protection etc.) (Evans, 1992).

The initial step in ecosystem rehabilitation projects is to identify the most important constraints to crop or tree productivity, as well as defining the specific land restoration objectives. Some soils can be recovered through the use of fertilizers, others need more drastic rehabilitation techniques, and there are situations of extreme degradation where soils cannot be recovered at all (Dedecek, 1992). The recovery of the soils' productive capacity is frequently very expensive, thus the techniques involved must produce financial returns to ensure their adoptability by the local farmers.

The choice of appropriate tree species for plantation forestry or agroforestry is influenced by knowledge on the species' performance and their economic and environmental benefits. Locally, choice of tree species is determined by seed or seedling availability and information on silvicultural characteristics and management, including fast growth and the possibility of intercropping during early establishment. Most reforestation or tree planting programmes and subsidies promote the use of well-known, often exotic species. About 85 per cent of plantation forestry in the tropics is dominated by three genera: *Pinus, Eucalyptus* and *Tectona,* while there may be thousands of indigenous species suitable for similar purposes (Evans, 1992). Native trees can be more appropriate than exotics, because (1) they are better adapted to local environmental conditions, (2) seeds and other propagules are locally available, and (3) farmers are familiar with them and their uses. Besides, the use of indigenous trees in productive systems helps preserve genetic diversity and is in better balance with the local flora and fauna.

In the present chapter, examples are presented from experimental plantations with native trees that had been established with the purpose of rehabilitating lands abandoned from extensive agriculture or cattle grazing in the humid tropics. A total of 30 indigenous tree species were evaluated in three forest regions of Latin America. The principal focus of this chapter is on the impact of trees on soils, their nutrient recycling abilities and the lessons that can be learnt for the implementation and sustainable management of plantation systems including these species. This information can be used to design management strategies to take advantage of the trees' ameliorating effects on soils and avoid site nutrient depletion with each harvest. These strategies should be useful for promoting the use of systems (mixed or pure plantations, agroforestry) including these fast growing timber species in the area and in other tropical lowland regions with similar ecological characteristics.

REHABILITATING ABANDONED LANDS IN THE LATIN AMERICAN HUMID TROPICS

A research programme to develop alternatives for rehabilitation and use of abandoned lands was conducted from 1987 to 1998 in three humid forest regions of Latin America: the Atlantic lowlands of Costa Rica, the Atlantic rainforest of Bahia in NE Brazil, and the subtropical forest of Misiones, NE Argentina (Table 1). In these regions, common situations of rapid deforestation, loss of biodiversity, resource misuse and land degradation persist. Similar methods were used in the three locations: soil chemistry and nutrient cycling parameters were measured in pure stand of selected indigenous species, using adjacent areas free of trees (abandoned agricultural field or pasture, secondary/primary forests) for comparison. The size of each project varied with the sites: in Costa Rica the studies were most complete (soils, above and below-ground tree biomass, litterfall and forest-floor litter biomass and tissue chemistry), while in Bahia and Misiones the trees were part of a forest reserve or were in private farms and thus destructive sampling was not possible.

Sampling procedures and chemical methods are described in Montagnini (1994), Montagnini and Sancho (1990a, 1990b, 1994a, 1994b), and Montagnini *et al.* (1991, 1993, 1994, 1995a, b, c). The soils under the species, grassy areas free of trees and adjacent young secondary forests were sampled for soil fertility and nitrogen availability measurements at the three research sites. Litterfall was measured biweekly with litter traps, and forest-floor litter accumulation was sampled every three months at La Selva, while at Bahia data from existing studies was used. No litter data was available from the Misiones site. For both La Selva and Bahia, nutrient use efficiencies (NUE) were calculated by dividing annual biomass increments by nutrients in annual litterfall for each species.

IMPACTS OF TREES ON SOIL AND NUTRIENT RECYCLING FROM LITTER

LA SELVA, COSTA RICA

Soil Fertility

Observations at La Selva showed that soil conditions improved in the tree plantations in only 2.5 years compared to abandoned pastures. In the top 15 cm, soil nitrogen and organic matter were higher under the trees than in nearby pastures, with values close to those found in adjacent 20-year-old forests (Table 2). The highest values for soil organic matter, total N, Ca and P were found under *Vochysia ferruginea,* a species common in mature and secondary forests in the region (Montagnini and Sancho, 1990a, 1990b). Subsequent measurements revealed similar trends in the soil parameters in the three following years.

Based on the standards determined by the Ministry of Agriculture of Costa Rica for soil fertility assessments (Bertsch, 1986), the cation levels (Ca, Mg and K) under most of the tree species were at or above the critical values for agriculture. In contrast, the cation levels in the adjacent abandoned pasture soils were low even for subsistence crops

Table 1 Characteristics of the study sites mentioned in his chapter.

Location	*Mean annual rainfall and temperature*	*Soil type*	*Previous land use*	*Experimental setting*
La Selva Biological Station, Atlantic Lowlands of Costa Rica	4,000 mm, 24°C	Fluventic Dystropepts PH 4.3–4.6	1–3 yr. agriculture, cattle	Pure plantation, 2m x 2m, 2–6 yr. old
Pau Brasil Ecological Station Porto Seguro, Bahia, Brazil	1,700 mm, 23°C	Oxisols (Haplorthox) pH 4.5–5.0	Shifting agriculture	Pure plantation, 2m x 2m, 14–15 yr. old
Eldorado, Misiones, NE Argentina (private farms)	1,700–2,400 mm, 22 °C	Acid, clayey Ultisols pH 4.5–5.0	50 yr. of agriculture, pine plantations	Pure stands from natural regeneration, 10–20 yr. old

Table 2 Topsoil chemical characteristics in pure stands of indigenous tree species at La Selva, Costa Rica; Porto seguro, Bahia, Brazil; and Misiones, Argentina.

	pH	*C* (%)	*N* (%)	*P* (*cmol. kg*$^{-1}$)	*K*	*Ca*	*Mg*
a. La Selva, Costa Rica[1]							
Stryphnodendron microstachyum	5.4ab	3.42ab	0.29b	5.6a	0.27a	0.45a	0.63ab
Vochysia ferruginea	5.4ab	3.76a	0.32a	7.la	0.22a	0.73a	0.61ab
Vochysia guatematensis	5.3ab	3.13ab	0.29b	5.2a	0.11a	0.25a	0.37ab
Hieronima alchorneoides	5.lb	2.96c	0.22b	1.5b	0.09a	0.31b	0.21b
Abandoned pasture	5.3ab	2.73c	0.22b	4.9a	0.19a	0.32b	0.27b
Secondary forest	5.3ab	4.33a	0.33a	3.6b	0.17a	0.68a	0.55ab
b. Porto Seguro, Bahia, Brazil[2]							
N-fixing leguminous species:							
Bowdichia virgilioides	4.9	1.98def	0.16def	1.32def	0.06bcd	1.35bc	0.39de
Centrolobium minus	4.6	1.87efg	0.16def	1.19efg	0.05fgh	0.53hi	0.21i
Centrolobium robustum	4.5	1.65ij	0.13f	1.07fgh	0.05fgh	0.40i	0.16i
Inga affinis	4.9	2.10cde	0.18cd	3.64a	0.07bcd	0.76gh	0.49bc
Parapiptadenia pterosperma	4.9	2.38ab	0.20bc	0.78ij	0.08gb	1.40bc	o.60a
Pithecellobium elegans	4.8	1.67hij	0.15ef	0.59kl	0.05efg	0.79gh	0.40de
Platymenia folioloso	4.7	2.08cde	0.18bcd	0.13m	0.05efg	1.05cde	0.42cd
Non N-fixing leguminous species:							
Arapatiella psilophylla	4.7	1.94def	0.18bcd	1.45de	0.06bcd	0.38i	0.37doc
Caesalpinia echimata	5.1	2.41a	0.17cde	1.54de	0.07bcd	1.17bcd	0.39de
Cassia spp.	4.7	1.94def	0.16def	1.40def	0.07bcd	0.56hi	0.34de
Copaifera luscens	5.0	2.02cde	0.170de	0.63jk	0.06cde	1.15bcd	0.34de
Dimorphandra jorgei	4.9	1.97def	0.19bc	0.97ghi	0.03j	0.98df	0.32efg
Hymenaea aurea	4.4	2.00def	0.16def	2.03c	0.06bcd	0.26i	0.24hi

Contd.

Table 2 Contd.

	pH	*C* (%)	*N* (%)	*P* (*cmol. kg*$^{-1}$)	*K*	*Ca*	*Mg*
Macrolobium latifolium	4.7	1.90efg	0.16def	0.67jk	0.04hij	0.36i	0.25fg
Of other families:							
Bombax macrophyllum	4.8	1.78ghi	0.13f	1.42de	0.06bcd	0.84efg	0.33ef
Buchenavia grandis	4.6	2.06cde	0.14f	2.09c	0.06bcd	0.80fg	0.33ef
Eschweilera ovata	5.3	1.82fgh	0.31a	0.58kl	0.11a	1.38bc	0.53ab
Lecythis pisonis	5.3	1.99def	0.18bed	0.23lm	0.04ghi	1.46b	0.32ef
Licania hypoleuca	5.0	1.63j	0.14f	1.61d	0.07bcd	1.31bcd	0.35de
Pradosia lactescens	4.9	2.15bcd	0.18bcd	0.81ij	0.05fgh	0.84efg	0.24gh
Primary forest	4.9	1.99def	0.15ef	0.96hi	0.08bc	1.23bcd	0.36de
Secondary forest	5.1	2.15abc	0.22b	2.46b	0.07bcd	2.20a	0.62a
c. Misiones Argentina[3]							
Balfourodendron riedelianum	5.8	2.6b	0.34ab	n.d.	0.55bc	7.1bc	1.7c
Bastardiopsis densiflora	7.1	6.3a	0.65a	n.d.	1.28a	20.4a	3.4ab
Cordia trichotoma	6.4	4.oab	0.46ab	n.d.	0.79b	13.6ab	2.6abc
Enterolobium contortisiliquum	6.1	3.4ab	0.39ab	n.d.	0.67b	8.7bc	3.5&
Ocotea puberula	6.1	4.4ab	0.59a	6.09a	1.11a	17.3a	4.7a
Grass control	5.8	2.2b	0.0.27b	n.d.	0.26c	6.3c	2.4bc

Sources: [1]Montagnini and Mendelsohn (1996), [2]Montagnini *et al.* (1994), [3]Fernãndez et al. (1995).
Note: For each differences among means are statistically significant when followed by different letters ($p<0.05$).
n.d.: not detected.

preferred in the region (rice, beans). The standards set by the Ministry of Agriculture do not include N or organic matter. However, an indication of the importance of the improvement of the soil organic matter levels is given by the close relation found between organic matter content and the sum of bases (Ca+Mg+K), showing that the organic matter was responsible for much of the cation retention capacity (Montagnini and Sancho, 1990a). For example, based on this relationship, a 1–2% increase in soil organic matter (in the 4–6% range) would more than double the base content, reaching values in the range recommended for agriculture (Bertsch, 1986).

Low crop yields in the humid tropics are often a result in part of unfavorable physical properties such as soil compaction (Cassel and Lal, 1992). In our site at La Selva, soil organic matter also had positive influences on soil physical properties: the soil bulk density was lower (i.e., lower compaction) while soil moisture was higher under trees than in abandoned pasturs (Montagnini and Mendelsohn, 1996).

The results of standard soil fertility tests used in agriculture (such as those shown in Table 2) may not always reveal the soil's productive potential, because they do not include all chemical forms of nutrients available for plant uptake. For example, although they make up less than 10 per cent of the total soil N pool, mineral N ($NO_3^- + NH_4^+$), are the forms of N available to plants. Nitrogen fertilizers are heavily used in the La Selva region, especially for the most demanding commercial crops such as bananas, in which case capital is available for fertilizer in a more extensive land use system. From the results shown in Table 2, *S. microstachyum,* a N-fixing tree, did not have an important effect on total N, but its litter decomposes faster than the other species, resulting in increased soil mineral nitrogen under its canopy (Montagnini and Sancho, 1994a). Evaluating the effects of trees on soil P availability is even more difficult, although experiments with test crops can determine soil impacts. For example, in other experimental research, maize seedlings, grown in plots mulched with *S. microstachyum* and *H. alchorneoides* versus the other species' litter, showed the greatest initial growth and the highest N and P plant uptake (Montagnini *et al.*, 1993). In these and in other related research at La Selva, the maize seedlings grown without mulch or fertilizer on soils from abandoned shifting agriculture fields grew very poorly, reaffirming the need for soil improvement techniques for growing conventional crops in the impoverished abandoned lands.

Nutrient Recycling from Litter

The impacts of trees on soil fertility depends on their nutrient recycling characteristics such as litter chemistry and decomposition. Tree litter can act as mulch with differing objectives: a fast mulch decomposition rate may accelerate the growth of associated crops on poor soils, while in other cases a more persistent litter may be desired. For example, high rates of litterfall and slower decomposition results in high litter accumulation and high soil organic matter under *V. ferruginea*, making this species well suited for protecting soils against erosion. In contrast, *H. alchornoides'* litter, although less abundant than the other three species, with its relatively faster decomposition and high nutrient content, promotes fast nutrient recycling, especially of N, Ca, Mg, K and P, while *V. guatemalensis* litter may be especially important for Ca and Mg recycling (Montagnini *et al.*, 1993).

Apart from their beneficial effects on soils, the tree species with their rapid canopy closure decreased the growth of weeds after 2–3 years, however with differences among species: the growth of understory vegetation was less in *V. ferruginea* and *V. guatemalensis* than in *H. alchorneoides* or *S. microstachyum* plots (Montagnini and Sancho, 1994a, 1994b). The canopy characteristics of the tree species will affect their suitability for interplanting with annual crops and the management required when used in agroforestry systems.

BAHIA, BRAZIL

SOIL FERTILITY

In Bahia, positive effects on at least five soil parameters were found under 15 out of the 20 species of the plantations, in comparison with primary and secondary forests (Table 2). Several species contributed to increased C and N, among others: *I. affinis, P. pterosperma, P. foliolosa* (leguminous, N-fixing species), *C. echinata, C. luscens* (leguminous, non N-fixing), *E. ovata, P. lactescens* (of other families) (Table 2) Others increased soil pH and/or some cations, such as *C. luscens, E. ovata, L. pisonis and L. hypoleuca* (Table 2, Montagnini *et al.*, 1994; Montagnini *et al.*, 1995a).

NUTRIENT RECYCLING FROM LITTER

Total litter accumulation (whole leaves, branches and fragments) was higher under secondary than under primary forest and mixed plantation (Montagnini *et al.*, 1994, 1995a). This is similar to observations made by Silva (1990) in Barrolandia, another experimental site located near Porto Seguro, also in Bahia. Litter nutrients were also higher than in primary forest, suggesting that in secondary forests, forest-floor litter is an important source of nutrients to the soil.

Among the 20 species in the arboretum, the highest amounts of forest-floor litter were found under *A. psilophylla, B. macrophyllum, I. affinis L. hypoleuca* and *P. pedicellare;* positive effects on soils were found under all these species, with *P. pedicellare* having the least influence. This suggests that forest-floor nutrients were incorporated into the soil via decomposition under these species. da Vinha and Pereira (1983) reported high rates of litterfall under *B. macrophyllum, B. grandis, C. echinata, H. aurea, L. pisonis* and *P. foliolosa,* in the same arboretum. Of this group of species, leaf-litter decomposition was faster under *B. grandis, H. aurea* and *B. macrophyllum,* in that order (da Vinha *et al.,* 1985).

The total N stocks (kg/ha) in the top soil (0–5 cm depth) were higher than the total N stocks in forest-floor litter in all the tree species and forest sites (Table 3). The highest top soil N stocks were in *E. ovata,* a non N-fixing species, where they were even higher than in secondary forest. Other species with high top soil N stores were *A. psilophylla, D. jorgei, L. pisonis* and *P. lactescens,* all non N-fixing species; and *P. pterosperma* and *P. foliolosa* among the leguminous N-fixing trees. Several species had higher forest-floor N stores than secondary forest: *A. psillophyla, B. macrophyllum, B. grandis, H. aurea, I. affinis, L. hypoleuca, M. latifolium* and *P. pedicellare.*

In contrast to N, P stocks were higher in the forest floor than in the top soil at all sites except for *D. jorgei, Cassia* spp. and *C. robustum.* Also different than for N, only a few species had higher topsoil P stores than secondary forest: these were *B. grandis, I. affinis* and *H. aurea* (Table 3). The highest forest-floor P stocks were under *I. affinis,* a leguminous N-fixing species, however, most of the species with highest forest-floor P reserves belonged to other families (Table 3).

The comparison of top soil and forest-floor nutrient stores presented in Table 3 contributes further to interpretation of results on impacts of tree species in land rehabilitation. For example, litter under *L. hypoleuca* had low nutrient concentrations but due to its relatively high mass it resulted in large nutrient stores, and positive effects on soils. Under the species with high litter nutrient concentrations and slower decomposition, forest-floor litter may serve as a nutrient reservoir. Forest-floor litter with slower decomposition may also serve to protect the soil against erosion.

MISIONES, NE ARGENTINA

SOIL FERTILITY

The greatest differences in soil C and N levels under tree species and grass were found under *B. densiflora,* where they were twice those in

Table 3 Soil and forest floor nutrient stocks of N and P under plantations and forests at Bahia (kg/ha).

Species	N		P	
	Soil	*Forest-floor*	*Soil*	*Forest-floor*
Leguminous, N-flxing species				
Bowdichia virgilioides	952	147.6	0.79	1.24
Centrolobium minus	976	97.0	0.73	1.59
Centrolobium robustum	799.5	23.4	0.66	0.60
Inga affinis	981	388.7	1.98	8.96
Parapiptadenia pterosperma	1060	210.6	0.41	3.50
Pithecellobium pedicellare	930	339.4	0.37	3.79
Platymenia foliolosa	1044	140.6	0.08	1.98
Leguminous, non N-fixing species				
Arapatiella psilophylla	1044	307.5	0.84	7.08
Caesalpinia echinata	960.5	128.6	0.87	2.88
Cassia spp.	992	9.3	0.87	0.15
Copaifera luscens	935	102.4	0.35	2.83
Dimorphandra jorgei	1178	60.3	0.60	0.60
Hymenaea aurea	856	212.2	1.09	5.67
Macrolobium latifolium	744	192.2	0.31	3.35
Species of other families				
Bombax macrophyllum	689	233.1	0.75	4.05
Buchenavia grandis	735	192.8	1.10	5.44
Eschweilera ovata	1798	156.7	0.34	3.72
Lecythis pisonis	1116	130.1	0.14	3.63
Licania hypoleuca	798	277.9	0.92	8.21
Pradosia lactescens	1107	137.5	0.50	3.62
Mixed plantation	896	87.1	1.10	1.30
Primary forest	802.5	182.3	0.51	3.07
Secondary forest	1078	137.5	1.21	2.06

areas beyond the canopy influence (Table 2) (Fernàndez *et al.*, 1997). The pH was higher under *B. densiflora* and *C. trichotoma*, while the sum of bases (Ca+Mg+K) was highest under *C. trichotoma, B. densiflora* and *E. contortisiliquum.*

Data from biomass chemical analyses helped to explain the trends shown above through nutrient recycling mechanisms (Montagnini *et al.*, 1994b). *B. densiflora* has interesting possibilities because it is a forest pioneer which colonizes areas abandoned from agriculture, forming dense even-aged stands which can reach commercial size in 10–14 years. *C. trichotoma* with a slower growth rate is however a valuable timber species known to participate successfully in plantation and agroforestry systems. The potential of *E. contortisiliquum* as a soil improver with its N-fixing capacity needs to be verified.

ASSESSING POTENTIAL LONG-TERM IMPACTS: NUTRIENT USE EFFICIENCIES

As shown above, a number of native timber species which combine good economic value with important soil rehabilitation properties were found in this programme. The potential of the species for rehabilitation of degraded areas sometimes was obvious through their direct impacts on soil fertility parameters such as organic matter or cations. In other cases their effects only became evident through more detailed studies such as litter decomposition, N mineralization, or by studying the response of test plants to the tree species mulch or litter.

The calculation of nutrient use efficiencies is a useful means of integrating results for drawing recommendations for system design and management. The nutrient use efficiency (NUE) concept has been employed to describe the differential ability of tree species to accumulate organic matter in relation to nutrients taken up from the soil. It has been defined on different scales of space and time. At the plant population and community levels, NUE is generally defined as the amount of biomass produced per unit of nutrient taken up (Grubb, 1989; Binkley *et al.*, 1992; Medina, 1995). Since direct measurements of nutrient absorption are not usually made for mature tree stands, leaf litter fall nutrients are generally used as an estimate of annual nutrient uptake (Vitousek, 1984; Grubb, 1989; Binkley *et al.*, 1992). Ideally, efforts should be made to measure total production by stems, branches and roots and to consider nutrient uptake plus losses to herbivory and leaching of nutrients (Grubb, 1989). Differences in nutrient cycling at efficiency of use may result from an ability to use various nutrient conserving mechanisms, from physiological to mutualistic interactions (Chapin, 1980, 1983).

It is also important to consider the relationship between the recycling ability of the species and their potential short- and long-term impacts on soil nutrient amelioration. A "nutrient cycling index" (NCI), taken as the inverse of NUE, i.e., the amount of nutrients in annual litterfall/ annual tree biomass production, has been used to assess suitability of tree species for agroforestry combinations. For example, Fassbender *et al.* (1991) found that the P recycling index was about six times higher in combinations of cacao with *Erythrina poeppigiana* than with *C. alliodora* in agroforestry systems in Turrialba, Costa Rica. *Cordia,* a timber species, accumulated much P in stem biomass, while *Erythrina,* a shade tree with good nutrient cycling properties, produced large amounts of leaves and branches resulting in greater P recycling.

When put in context with nutrient recycling characteristics of a species, NUE can indicate appropriate system design and management to maintain productivity and recover or conserve nutrients over the long term. The ability of a species to produce large amounts of biomass with less nutrients may be an important consideration in choosing species for degraded, nutrient-poor sites.

APPLICABILITY OF NUE IN PLANTATION DESIGN AND MANAGEMENT

From the 24 tree species shown in Table 2, data on tree productivity, leaf litter fall and litter chemistry of eight species in Bahia, Brazil and four species at La Selva, Costa Rica were used to calculate NUE values (Table 4). Because the species were part of a forestry project with fast-growing timber trees, NUE values were calculated as the annual stem biomass increments/nutrients in annual leaf litter fall (Montagnini, 1994). There were not enough data on litter fall or productivity for the other species shown in Table 2.

At both Bahia and La Selva, the highest efficiencies were for K and P, and the lowest were for N, Ca and Mg (Table 4). Results from Bahia suggest that *B. macrophyllum* and *P. foliolosa,* with overall high NUE values, would grow well on relatively nutrient-poor soils, and thus could be good alternatives for reforestation of degraded sites following the abandonment from agriculture and pasture that is frequent in the region. *B. macrophyllum* tended to accumulate high amounts of litter under its canopy while *P. foliolosa* stands had relatively high amounts of organic matter and total N in the topsoil in comparison with adjacent areas of secondary forest (Montagnini *et al.*, 1995a). These features indicate that these species may be well suited for soil rehabilitation, including increasing soil organic matter content and protecting against soil erosion. Also at Bahia, species such as *B. grandis* and *H. aurea,* with overall lower NUE values, would be most appropriate for agroforestry combinations where crops could benefit from nutrient recycling from litter.

At La Selva, *V. ferruginea* showed comparatively low efficiency values for all the nutrients considered, confirming the beneficial role of this species in recycling organic matter and positively impacting soil fertility as shown in Table 2. The relatively low efficiency (high recycling) of N and P found for *S. microstachyum* and *H. alchorneoides* was also shown in experiments where maize grown with mulch of these species grew better and absorbed more N and P than with mulch of other species (Montagnini *et al.*, 1993).

Table 4 Nutrient use efficiencies (NUE) (Megagrams of annual stem biomass increment/kg nutrient in annual litter fall).

Porto Seguro, Bahia, Brazil	*N*	*Ca*	*Mg*	*K*	*P*
N-fixing leguminous species:					
Centrolobium rohustum	0.05	0.05	0.37	0.87	1.74
Platymenia foliolosa	0.08	0.34	7.00	3.51	4.66
Non N-fixing leguminous species:					
Caesalpinia echinata	0.04	0.03	0.54	0.78	1.81
Hymenaea aurea	0.06	0.09	0.45	0.64	1.91
Species of other families:					
Bombax macrophyllum	0.36	0.20	0.88	20.70	20.70
Buchenavia grandis	0.05	0.04	0.38	0.68	1.70
Eschweilera ovata	0.12	0.10	0.44	0.96	5.73
Lecythis pisonis	0.12	0.11	0.89	1.58	3.56
La Selva, Costa Rica	*N*	*Ca*	*Mg*	*K*	*P*
Stryphnodenciron mycrostachyum	0.08	0.06	0.35	0.63	0.53
Vochysia ferruginea	0.08	0.03	0.26	0.38	0.56
Vochysia guatemalensis	0.16	0.05	0.29	0.84	1.07
Hieronima alchorneoides	0.09	0.03	0.15	0.22	0.45

In projects that aim to recover soil nutrients in degraded sites, species with high nutrient use efficiency should be combined in time or space with species with low use efficiency. However, NUE values alone may not be enough to assess the role of a tree species on ecosystem nutrients. For example, inspite of high NUE values, a tree species may pose high demands on soil nutrients over the long term. In humid tropical regions nutrients are expected to be critical factors influencing tree productivity, while in regions with a marked dry season water use efficiency rather than NUE would be a more important factor influencing species choice and system design. Still other ecological adaptations of the species (e.g., light use efficiency, root architecture, resistance to pests and diseases) could be more important in selecting species for agroforestry combinations.

TREE PRODUCTIVITY IN PLANTATIONS USED FOR LAND REHABILITATION

As stated earlier in this chapter, land rehabilitation programmes must use species and designs that not only help in soil recovery, but are also productive. If a land rehabilitation system is productive it may represent an economic incentive for the local farmers. At La Selva, the values of whole tree biomass for the plantations (Table 5) were greater than those reported for 4-year-old *Albizia lebbek* (Parrota, 1989), and for 5.5 year-old *Leucaena leucocephala* (Wang *et al.*, 1991), both growing in dense plantations for biomass production in Puerto Rico. Values of total tree above-ground net primary productivity (calculated by dividing whole-tree biomass by tree age) lie within the ranges reported elsewhere for monospecific plantations in the humid tropics. The value for *V. guatemalensis* is close to that reported for *Gmelina arborea* (12.8 tons per ha/yr) in the Brazilian Amazon (Russell, 1987), and to *G. arborea* (12.7 tons/ha) and *A. falcataria* in the Philippines (11.3) (Kawajara *et al.* (1981), in Young (1989). The increments shown here are however lower than those reported for some of the fastest growing trees in the humid tropics such as *Acacia mangium* (15.5 to 18.0 tons/ha in Malaysia), *L. leucocephala* (20.0–30.0 and even up to 80.0 tons/ha in Hawaii and other tropical sites (Young, 1989).

Stemwood biomass annual increment for broadleafs ranges from 1 to 28 tons/ha/yr. Fast growing species such as *G. arborea* and *E. saligna* range from 10–20 and 8–28 tons/ha, respectively, and relatively slower growing species such as *Swietenia* sp. and *Tectona grandis* range from 1–4 and 3–12 tons/ha, respectively (Wadsworth, 1983). Other values for fast-growing trees in tropical humid regions include various eucalyptus species grown in the Americas and Asia (7.2 to 11.9 tons/ha), *G. arborea*

Table 5 Means, aboveground biomass and annual increments for the four native species in plantation at La Selva, Costa Rica[1].

Tree species	*Aboveground live biomass, kg/ha*				*Mean annual Increment, t ha^{-1} yr^{-1}*	
	Stem	*Branches*	*Leaves*	*Total*	*Total*	*Stems*
Stryphnodendron microstachyum	35,250a	15,250a	4,325a	54,825a	13.7a	8.8a
Vochysia *ferruginea*	24,750b	14,250a	5,925a	44,925	11.2b	6.2b
Vochysia guatemalensis	41,750a	6,500b	7,250a	55,500	13.9a	10.4a
Hieronima alchorneoides	26,250a	12,250a	5350a	43,850	12.0b	6.5b

[1]Differences between sites for a given parameter are statistically significant ($P<0.05$) when means are followed by different letters.

in Costa Rica (11.8) (Lugo *et al.*, 1988); 1.3 to 5.3 year-old *L. leuocephala* growing in pre-montane and lowland humid sites (2.8 to 15.9), *Prosopis juliflora*, moist site in India (9.4), *Populus deltoides*, subtropical site in India 6.4 (Lugo *et al.*, 1990a). Thus the mean annual stemwood biomass increments for the species of this land rehabilitation program also fall within the ranges reported for other fast-growing tree species in the humid tropics.

NUTRIENTS IN PLANTATION BIOMASS: IMPLICATIONS FOR MANAGEMENT

In the long-term, the impacts of plantation trees on soil nutrients depend on: (1) the annual nutrient uptake by the trees in relation to the nutrient supplying capacity of the soil, (2) amounts of leaf litterfall and decomposition, (3) differential nutrient allocation in tree parts: stems, foliage, roots, and (4) the parts of the tree removed at harvest whether the whole tree or stemwood, and their biomass and nutrient content. To help understand these relationships, biomass nutrients were compared among tree parts for the four species of the La Selva site. Consistent with trends found in N tissue concentrations (Montagnini *et al.*, 1991), *S. microstachyum* had the highest stem, branch and whole-tree biomass N (Fig. 1). *V. guatemalensis* had a similar proportion of N to *S. microstachyum* in its leaves+branches portion; over 50 per cent of whole-tree biomass N could be recycled if left at the site at the time of harvest. *V. ferruginea*, with relatively lower stem biomass, had proportionally more N in leaves (53%) and branches (42%); *H. alchorneoides* showed a more even distribution of whole-biomass N in stems (39%) branches (27%), and leaves (34%). *V. guatemalensis* and *H. alchorneoides* had the highest proportions of stemwood P (72% and 62%, respectively) (Fig. 2).

V. guatemalensis, with the highest stem biomass and Ca concentration, also had the highest stemwood Ca (over 600 kg/ha, or 84 per cent of whole-tree biomass Ca), approximately twice as much as either *S. microstachyum* or *V. ferruginea*, and several times more than *H. alchorneoides* (Montagnini and Sancho, 1994b). Therefore, the harvest of *V. guatemalensis* trees could substantially reduce the amount of Ca in the site. However, while the *V. guatemalensis* trees are living, relative large amounts could be recycled, because although it represented only 16 per cent of whole-tree biomass, the amount of Ca in leaves and branches together was over 100 kg/ha. Likewise, *V. guatemalensis* with its high stem biomass and high Mg concentration also had the highest Mg in stemwood (55% of whole-tree biomass, or approximately

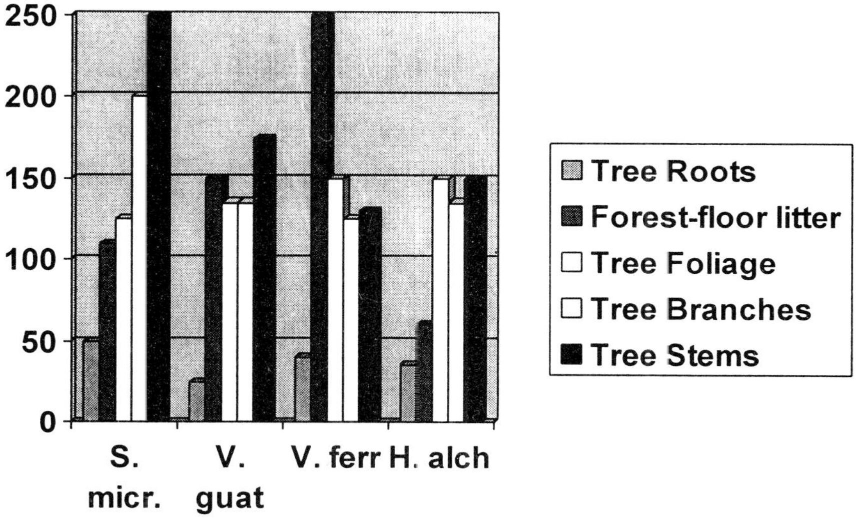

Fig. 1 Biomass nitrogen.

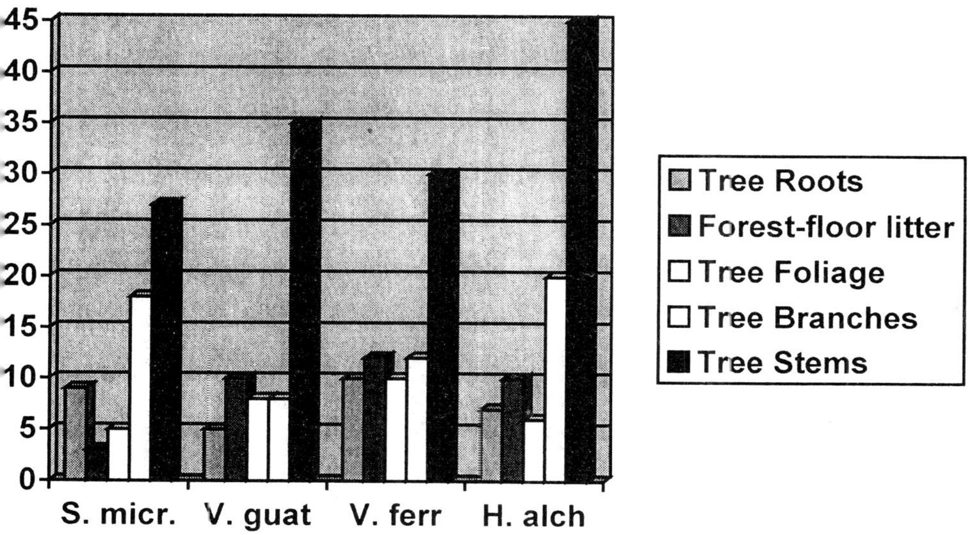

Fig. 2 Biomass phosphorus.

30 kg/ha) (Montagnini and Sancho, 1994b). Again, removal of *V. guatemalensis* stems would affect the sites's Mg budget more dramatically so than the other species, especially if the whole tree is harvested. For K the picture changed: the highest accumulation of K in stems was found in *H. alchorneoides* (252 kg/ha), followed by *V. guatemalensis* with 175 kg/ha, which represent 77% of whole-tree biomass K. Thus whole-tree harvest of *H. alchorneoides* and *V. guatemalensis* may have the greatest effects on the K budget (Montagnini and Sancho, 1994b).

For the nutrients with lowest use efficiency values (e.g. N) the forest-floor comprised a relatively high proportion of the total (Fig. 1). Therefore, the forest-floor appears as an important compartment for N accumulation and recycling, although with marked differences among the species considered (Fig. 1). If the forest-floor is burned or collected for fuelwood, a substantial loss of organic matter and N may occur, while if the litter is left on the floor after harvest, it represents a significant reservoir for the next rotation.

In contrast for P, which was relatively more scarce and showed higher use efficiency values, a higher proportion was stored in live tree biomass (Fig. 2). Although the forest-floor was still a substantial compartment it was proportionally smaller than for N. In comparison, especially for *V. ferruginea* and *S. microstachyum*, root biomass P was relatively more important than forest-floor P (Fig. 2). Under tropical humid conditions P recycling from roots could be particularly important because root turnover is expected to be relatively fast. The roots are less likely to be affected during manual timber harvest operations than the forest-floor, while mechanical site operations could seriously affect the root system and its nutrient storing and recycling capacity. This can be especially critical for P and K, often mentioned as the nutrients which are most likely to be depleted from soils with subsequent rotations (Wadsworth, 1983; Bruijnzeel, 1991).

MANAGEMENT OF PLANTATION FLOOR LITTER

The relative importance of the litter layer in plantation nutrient conservation has been shown by Lugo *et al.* (1990b), who reported that the amount of litter in 10 tropical plantations ranged from 5 to 28 Mg ha^{-1}. Results of the present research program at La Selva show that plantation floor leaf litter represented up to 50 per cent of the aboveground tree nutrient content (Fig. 1). This figure emphasizes the importance of floor leaf litter to nutrient cycling. Green mulch from decomposing indigenous leaf litter can be a low-cost and effective soil supplement in several regions, so it is likely that small farmers will remove litter from the plantation floor to use in fields or homegardens or otherwise aid in the growth of subsistence crops (Byard *et al.*, 1996). This could have a significant nutrient cost. For example, at La Selva, floor leaf litter represented between 40 per cent and 45 per cent of the aboveground N found in the plantations. Furthermore, in some of the studied plantations up to 54 per cent of the aboveground K, and 56 per cent of the Mg, were in floor leaf litter (Stanley and Montagnini, 1999). On the other hand, floor leaf litter never represented more than 28 per cent of aboveground P.

When leaf litter is not removed it decomposes and is reincoporated into the soil or is recycled. However, decomposition and litterfall rates vary along the year; for example, we observed large temporal variations in leaf litter accumulation and nutrient content. Litterfall rates, decomposition rates, and litter accumulation also varied significantly between species.

While quick decomposition may have a positive effect on productivity by making nutrients available for incorporation into the soil where it is available to plants, species that have large amounts of litter that decompose slowly may provide other benefits. For example, thick litter layers, such as those found under *V. ferruginea,* provide ground cover that reduces susceptibility to soil erosion and are an important nutrient source in subsequent rotations. *Vochysia ferruginea,* also has a dense canopy, which helps to deter the growth of understory vegetation and decrease costs of weeding (Horn and Montagnini, 1998, Montagnini and Mendelsohn, 1996). In situations where soil fertility is severely depleted and rotations are short, species characterized by quick nutrient recycling become more desirable. If more than one benefit is sought, a mixture of species may offer an attractive option.

CONCLUSIONS AND RECOMMENDATIONS

A number of economically valuable native tree plantation species were found to have positive impacts on soils in three regions of the Latin American humid tropics. The calculation of nutrient use efficiencies proved a useful tool for system design and management: to conserve nutrients in the long term, species with low use efficiency should be combined in time or space with others with high use efficiency. Evaluation of biomass nutrient allocation helped in drawing specific management recommendations: preserving the forest-floor was key for nutrient conservation, but for the most limiting nutrients, the roots were more important. For ecosystem rehabilitation, the examination of tree growth, soil chemistry and litter biomass and nutrients constitutes a minimum set of measurements needed for the design and long-term management of sustainable forest systems.

REFERENCES

Bertsch, F.H., 1986. *Manual para Interpretar la Fertilidad de los Suelos en Costa Rica*, Universidad de Costa Rica, Escuela de Fitotecnia. San José, Costa Rica, 76 p.

Binkley, D., Dunkin, K.A., DeBell, D., and Ryan, M.G., 1992. Production and nutrient cycling in mixed plantations of *Eucalyptus* and *Albizia* in Hawaii, *Forest Science*, **38:** 393–408.

Bruijnzeel, L.A., 1991. Nutrient input-output budgets of tropical forest ecosystems: a review, *J. Tropical Ecology*, **7:** 1–24.

Byard, R., Lewis, K.C., and Montagnini, F., 1996. Leaf litter decomposition and mulch performance from mixed and monospecific plantations of native tree species in Costa Rica, *Agriculture, Ecosystems and Environment*, **58:** 145–155.

Cassel, D.K., and Lal, R., 1992. Soil Physical Properties in the Tropics: Common Beliefs and Management Restraints, In: *Myths and Science of Soils in the Troics*, (R. Lal, and P.A. Sanchez, eds.), Soil Science Society of America Special Publication Number 29, SSSA and American Society of Agronomy. Madison, Wisconsin, 61–90 pp.

Chapin, F.S., 1980. The mineral nutrition of wild plants, *Annual Review of Ecology and Systematics*, **11:** 233–160.

Chapin, F.S., 1983. Patterns of nutrient absorption and use by plants from natural and man-modified enviromnents, In: *Disturbance and Ecosystems*, (H.A. Mooney, and M. Godron, eds.), Ecological Studies, Vol. 44, Springer-Verlag, Berlin, 175–187 pp.

Dedecek, R.A., 1992. A dinamica dos solos em areas degradadas, In: *Simposio Nacional Recuperacao de Areas Degradadas*, Universidad Federal do Parana. Curitiba, 2529 October 1992 44–57 pp.

Evans, J., 1992. *Plantation forestry in the tropics. Tree planting for industrial, social, environmental and agroforestry purposes*, 2nd. ed., Clarendon Press. Oxford: England, 403 pp.

Fassbender, H.W., Beer, J., Heuveldop J., Imbach, A., Enriquez, G., and Bonriemann, A., 1991. Ten year balances of organic matter and nutrients in agroforestry systems at CATIE, Costa Rica, *Forest Ecology and Management*, **45:** 173–183.

Fearnside, P.M., 1990. Predominant land uses in Brazilian Amazonia, In: *Alternative Deforestation: Steps Towards Sustainable Use of the Amazon Rain Forest*, (A.B. Amderson ed.), Columbia University Press, New York, USA, 233–251 pp.

Fernàndez, R., Montagnini, F., and Hamilton, H. 1997. The Influence of native tree species on soil chemistry in a subtropical humid forest region of Argentina, *Journal of Tropical Forest Science*, **10**(2): 188–196.

Grubb, P.J., 1989. The role of mineral nutrients in the tropics: a plant ecologist's view, Pages 417–43, In: *Mineral nutrients in tropical forest and savanna ecosystem*, (J. Proctor ed.), Blackwel Scientific Publications, Oxford, 417–439 pp.

Horn, N., and Montagnini, F., 1998. Litterfall, litter decomposition and maize bioassay of mulches from four indigenous Costa Rican tree species in mixed and monospecific plantations, *International Tree Crops Journal*, Submitted.

Lugo, A.E., Brown, S., and Chapman, J., 1988. An analytical review of production rates and stemwood biomass of tropical forest plantations, *Forest Ecology and Management*, **23:** 179–200.

Lugo, A.E., Wang, D., and Bormann, F., 1990a. A comparative analysis of biomass production in five tropical tree species, *Forest Ecology and Management*, **31:** 153–166.

Lugo, A.E., Cuevas, E., and Sánchez, M.J., 1990b. Nutrients and mass in litter and top soil of ten tropical tree plantations, *Plant and Soil*, **125:** 263–280.

McNabb, K., Borges J., and Welker, J., 1994. Jaré at 25. An investment in the Amazon, *J. of Forestry,* **92:** 21–26.

Medina, E., 1995. Physiological ecology of trees and application to forest management, In: *Tropical Forests: Management and Ecology. Ecological: Studies* 112, (A.E. Lugo, and C. Lowe., eds.), Springer-Verlag, New York, 289–367 pp.

Montagnini, F., 1994. Rehabilitating forest ecosystems in the humid tropics: Recent experiences from Latin America, In: *Proceedings from IUFRO International Symposium on Growth and Yield of Tropical Forests. Sept. 26–Oct. 1, 1994,* Tokyo University of Agriculture and Technology. Fuchu, Tokyo, Japan, 224–234 pp.

Montagnini, F., and Mendelsohn, R., 1996. Managing forest fallows: improving the economics of swidden agriculture, *Ambio,* **26**(2): 118–123.

Montagnini, F., and Sancho, F., 1990a. Impacts of native trees on tropical soils: a study in the Atlantic lowlands of Costa Rica, *Ambio,* **19**(8): 386–390.

Montagnini, F., and Sancho, F., 1990b. Influencia de seis species de árboles nativos sobre la fertilidad del suelo en una plantación experimental en la llanura del Atlántico en Costa Rica, *Yvyraretá,* (Argentina), **1**(1): 29–49.

Montagnini, F., and Sancho, F., 1994a. Net nitrogen mineralization in soils under six indigenous tree species, an abandoned pasture and a secondary forest in the Atlantic lowlands of Costa Rica, *Plant and Soil,* **162:** 117–124.

Montagnini, F., and Sancho, F., 1994b. Above-ground biomass and nutrients in young plantations of four indigenous tree species: implications for site nutrient conservation, *J. of Sustainable Forestry,* **1:** 115–139.

Montagnini, F., Sancho, F., Ramstad, K., and Stijfhoorn, E., 1991. Multipurpose trees for soil restoration in the humid lowlands of Costa Rica, pp. 41–58, In: *Research on multipurpose trees in Asia,* Proceedings of an International Symposium, (D.A. Taylor, and K.G. McDicken, eds.), November 19–23, 1990, Los Baños, Philippines, Winrock International Institutre for Agricultural development, Bangkok.

Montagnini, F., Ramstad, K. and Sancho, F., 1993. Litterfall, litter decomposition and the use of mulch of four indigenous tree species in the Atlantic lowlands of Costa Rica, *Agroforestry Systems,* **23:** 39–61.

Montagnini, F., E. Gonzádez, Rheingans, R., and C. Porra, 1995c. Mixed and pure forest plantations in the humid neotropics: a comparison of early growth, pest damage and establishment costs, *Commonwealth Forestry Review,* **74:** 306–314.

Montagnini, F., Fanzeres, A., and da Vinha, S.G., 1994. Studies on Restoration Ecology in the Atlantic Forest region of Bahia, Brazil, *Interciencia,* **19**(6): 323–330.

Montagnini, F., Fanzeres, A., and da Vinha, S.G., 1995a. The potential of twenty indigenous tree species for reforestation and soil restoration in the Atlantic Forest region of Bahia, *J. of Applied Ecology,* **32:** 841–856.

Montagnini, F., Femández, R., and Hamilton, H., 1995b. Relaciones entre especies nativas y la fertilidad de los suelos, Parte 1: Contenido de elementos en la biomasa, *Yvyraretá,* (Argentina), **6**(6): 5–12.

Parrotta, J.A., 1989. The influence of management practices on the biogeochemical stability of tropical fuelwood plantations, *Tropical Ecology,* **30:** 1–12

Russell, C.E., 1987. Plantation forestry. Case study No. 9: The Jari project, Pará, Brazil, In: *Amazonian Rain Forests: Ecosystem Disturbance and Recovery,* (C.F. Jordan, ed.), Ecological Studies 60, Springer-Verlag, New York, 76–89 pp.

Silva, L.F., 1990. *Interacao solo-vegetacao em floresta primária e capoeira do ecossistema de tabuleiro do sul da Bahia, CEPLAC,* Itabuna Bahia, Brazil.

Stanley, W., and Montagnini, F., 1999. Biomass and nutrient accumulation in pure and mixed plantations of indigenous tree species grown on poor soils in the humid tropics of Costa Rica, *Forest Ecology and Management,* **113:** 91–103.

da Vinha, S.G., and Pereira, R.C., 1983. Producao de folhedo e sua sazonalidade em 10 especies arboreas nafivas no Sul da Bahia, *Theobroma,* (Brazil), **13:** 327–341.

da Vinha, S.G., de Carbalho, A.M., and Silva, L.A.M., 1985. Taxa de decomposicao do folhedo de dez especies de arvores nativas no Sul da Bahia, Brasil, *Theobroma*, (Brazil), **15:** 207–212,

Vitousek, P.M., 1984. Litterfall, nutrient cycling, and nutrient limitation in tropical forests, *Ecology*, **65:** 285–298.

Wadsworth, F.H., 1983. Production of usable wood from tropical forests, In: *Tropical Rain Forest Ecosystems. Structure and Function. Ecosystems of the World 14A*, (F.B. Golley, ed.), Alsevier, New York, 279–288 pp.

Wang, D., Bormann, F.H., Lugo, A.E., and Bowden, R.D., 1991. Comparison of nutrient-use efficiency and biomass production in five tropical tree taxa, *Forest Ecology and Management*, **46:** 1–21.

Young, A., 1989. *Agroforestry for Soil Conservation*, C.A.B. International and International Council for research in Agroforestry, Wallingford, UK.

CHAPTER 5

COMPARATIVE NUTRIENT DYNAMICS WITH PARTICULAR REFERENCE TO LITTERFALL IN *HEVEA* RUBBER PLANTATIONS AND LOWLAND TROPICAL RAIN FORESTS IN ASIA

I.A.U.N. Gunatilleke

Department of Botany, University of Peradeniya, Peradeniya 20400, Sri Lanka

ABSTRACT

Hevea rubber cultivation in the Asian region is currently dominated by small-holdings and their management is strongly influenced by the socio-cultural outlook of these farmers. An ecosystem level perspective in understanding the mechanistic processes regulating nutrient dynamics and soil fertility in large- and-small scale rubber holdings in relation to that of lowland rain forests which they replaced is of critical importance in sustainable management of these rubber holdings.

A comparative survey of major nutrient pools and fluxes in different subsystems of forest and plantation ecosystems was made so as to get an insight into the contribution of each subsystem to the ecosystem functioning through the biogeochemical cycling of nutrients. Macronutrient levels in the live vegetation of rubber and its herbaceous ground cover was substantially high yet considerable losses result from extraction of latex and eventual removal of rubber trees at the end of their cultivation cycle. Leguminous cover crops immobilize higher levels of nutrients than grasses and shrubs thus contributing significantly to nutrient conservation during the early phase of rubber cultivation especially in large, well managed holdings. Soil nutrient status with the exception of extractable P was highest in lowland rain forest and least in the rubber small holding. Total above ground litter mass in the rain forest was twice as high and display an aseasonal litterfall pattern compared with that of the rubber plantation.

Decomposition of leaf litter in large rubber holdings was greatest suggesting better soil fertility management in them. Canopy interception of precipitation was highest in multi-layered rain forest while the mature rubber holdings had the least. The study reveals the potential for the regulation and synchronization of nutrient fluxes among different subsystems by simulating multi-layered forest model which is of considerable interest to the subsistence farmers managing rubber small holdings.

Key Words: Hevea brasiliensis, para rubber plantation, nutrient cycling, leguminous cover crops, litterfall, litter decomposition, latex, rain forest.

Hevea brasiliensis or Para rubber was introduced to the Asian tropics from the Amazonian rain forests as a commercial plantation crop during the latter half of the nineteenth century *via* Kew Gardens in England and Henerathgoda Botanic Gardens in Sri Lanka. Since its introduction, the *Hevea* rubber plantations have gone through at least three planting cycles each of about 30 years in the pioneer rubber growing countries in Asia viz. Sri Lanka, Malaysia and Indonesia (Samarappuli, 1996). In all these countries, the rubber plantations have replaced some of the most luxuriant lowland rain forests.

Introduction of natural rubber has played a predominant role in the economy of most of the South and South-east Asian countries for many decades as a major foreign exchange earner for the producer countries and a source of gainful employment for the rural poor as well as the urban rich communities. In recent times however, there have been major shifts in areas cultivated, ownership patterns, productivity levels, production costs, profit margins, consumer preferences, research priorities and trading systems with respect to natural rubber (Samarappuli *et al.*, 1997). While both Malaysia and Sri Lanka have shown gradual decline in natural rubber production during the period of 1970–1995, Thailand, China, India and Vietnam have shown an increase in production. For example, Malaysia's market share has declined from 36 per cent in 1960 to 18 per cent in 1995, Thailand has increased its share from 8 per cent to 31 per cent during the same period.

Large rubber plantations are managed even today primarily as monocultures for their latex with intensive management interventions by professionally trained managers and a skilled work force backed up by well established research and extension services. However, high value prevailed during and after the Second World War for natural rubber prompted local villagers also to cultivate rubber as small holdings ranging from 0.5–25 ha taking the cue from large plantations. One of the significant trends common to all the rubber growing countries is the expansion of this small holding sector. In Malaysia this increase was 88 per cent followed by Indonesia and India with increases of 28 per cent and 14 per cent from 1960 to 1995. Currently 85 per cent of the world's rubber growing area belongs to the small-holders—a 39 per cent increase since 1960 (Samarappuli *et al.*, 1997).

The intensively managed large-scale monoculture *Hevea* plantations have been geared to achieve sustainability in terms of yield, yield stability and economic viability.

Paradoxically however, these management skills have not been infused into the small holders to the same level in most of the rubber growing countries. Although in most instances, the small-holders receive government assistance in the form of replanting subsidies for the control of soil erosion, and advice on planting high yielding clones and application of recommended inorganic fertilizer inputs, their productive capacity has been far below that of the large estates managed by professional planters. It is a common sight in rubber growing areas in most rural landscapes that *Hevea* small-holdings contain a number of other utility tree species growing amongst them as an extension of the traditional home garden concept. As such, the management packages geared to meet the specific needs of the small holders need to be developed with a proper understanding of their socio-economic and ecological milieu. Research and extension work has not given the due attention these small-holders need in developing management systems characterized by a relatively inexpensive and affordable levels of input, a higher efficiency of internal resource use and consequently a more sustainable production in both economic and ecological terms.

With increased global attention on environmental and biodiversity conservation, especially in the humid tropics, both large and small-holdings of *Hevea* rubber are being critically examined for their ecological sustainability as well as economic viability in a climate of rapidly fluctuating and often declining market prices. In such an ecological and economic climate, an ecosystem level perspective in understanding the mechanistic processes regulating nutrient dynamics and soil fertility in the two somewhat contrasting types of management systems in relation to that of natural forest ecosystems which they replaced would be useful from both socio-economic and biological points of view. The fertility status of many tropical soils, particularly that of farmers having smaller holdings is regulated by key biological processes and organic and inorganic inputs available to these farmers. The complexities of these biological processes affecting soil fertility have been poorly understood by agriculturists. As a result there is a lack of well integrated management strategies. Unfortunately, agricultural and plantation forest sciences have developed independent of ecology geared towards the goal of sustained yield production of food, fiber and timber. In contrast, the ecological approach has focused primarily on the functioning of natural ecosystems where sustainability is viewed in terms of the maintenance and stability of ecosystem productivity and the tightness of the nutrient cycles. Modern interdisciplinary holistic approaches in understanding natural and agro-ecosystem structure and the mechanistic processes that regulate soil fertility have brought these disciplines closer. The outcome being an ecosystem perspective in agriculture which both, addresses production and conservation aspects (Woomer and Swift, 1994). This system-and-process paradigm has become increasingly popular in resource management research in the recent past for the development of sustainable means of land management enhanced by technical developments such as simulation modelling, geographic information systems and hierarchy theory (Scholes *et al.*, 1994).

For sustainable management of both these plantation types, a better understanding of the application of biological principles to enhance stability, resource-use efficiency and productivity is essential. An understanding of the biological principles and processes that regulate soil biological activities in the corresponding natural ecosystems, in this case the lowland rain forests, will assist in determining management options for maintaining optimal levels of soil fertility under managed conditions. Among the key processes regulating the sustainability of ecosystems under natural disturbance or exploitation by humans are (i) hydrology and drainage pattern, (ii) soil organic matter dynamics, and (iii) synchrony between soil processes and plant nutrient demand (Brown *et al.*, 1994).

NUTRIENTS IN VEGETATION AND SOIL

Biomass and Nutrient Stocks in Live Vegetation

Most of the modern clones of *Hevea* rubber bred for increased latex production are highly fertilizer-responsive and have faster growth rates under intensively managed conditions than that of natural forest tree species in a rain forest. A Malaysian study concluded that nutrient levels similar to those of a humid tropical evergreen forest have been achieved by a 24 year old rubber plantation and a comparable biomass at the age of 33 years (Sivanadyan and Norhayati, 1992); (Table 1). This can be attributed to the regular application of higher levels of inorganic fertilizers to superior clones selected for faster growth and higher yield. The tropical rain forest species on the other hand, are adapted to grow in nutrient poor highly weathered soils with efficient nutrient conservation mechanisms (Jordan, 1985).

Table 1 Tree biomass and nutrient stocks in humid tropical evergreen forests and *Hevea* rubber plantations in Malaysia (From Sivanadyan and Norhayati, 1992)

Ecosystem	*Biomass* ($t\ ha^{-1}$-*DM*)	*Nutrient Content* ($kg\ ha^{-1}$ *Dry Mass*)				
		N	*P*	*K*	*Ca*	*Mg*
Humid Tropical Evergreen Forest	426	1938	52	618	759	230
Hevea rubber plantation						
5 yr old	48.6	470	56	234	360	48
11 yr old	206.1	839	107	242	466	93
24 yr old	248.6	1172	n.a.	717	779	163
33 yr old	444.9	1595	241	1091	1854	351
33 yr old (untapped)	963.8	3598	925	3919	2932	777

The distribution of dry mass among both above and below ground components of the mature rubber trees reveal that bulk of the dry mass is contributed by the main trunk and the branches (83.1%), the leaves and twigs contribute very little to the total biomass (Fig. 1).

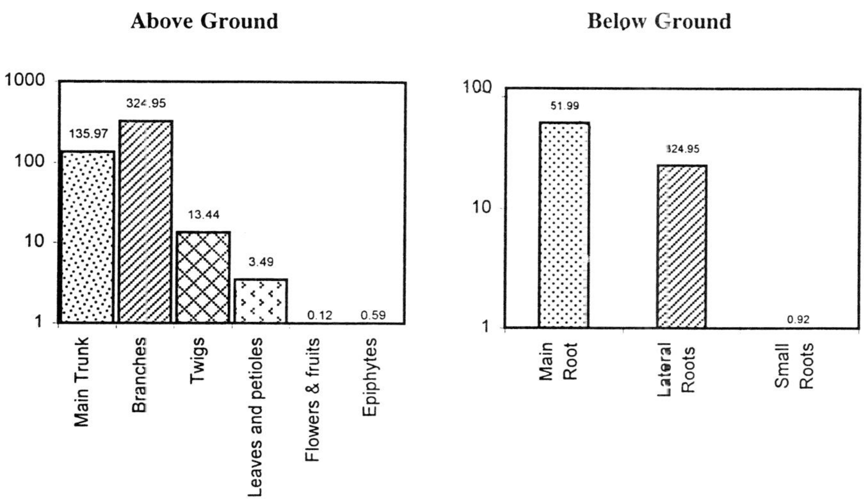

Fig. 1 Mean above-ground and below-ground dry mass (kg) of different components of mature rubber trees (n=6) of the clone PB 86.

Among the ground vegetation recorded in a mature rubber plantation (MR) in Hedigalla, Sri Lanka, *Paspalum conjugatum, Axanopus affinis, Hedyotes fruiticosa, Adiantum* species were found with a frequency of occurrence ranging from 45–90%. The more light demanding ground cover crop species, *Pueraria phaseoloides* had only 20% frequency of occurrence and showed patchy distribution in areas where the canopy was thin. In a nearby young rubber plantation (YR), most of the floor was bare with a few patches of *Puraria* ground cover still surviving in areas with less canopy closure. Exposed rocky outcrops and rubber roots were very frequent in this site indicating heavy surface erosion despite it being on a gentle slope. 50% of the ground quadrats had a gravelly texture or sandy top-soil layer further substantiating this observation. The young rubber stands of 8–12 years having thick foliage and consequently a poor ground cover are particularly vulnerable to loss of fertility through soil erosion. In a small-holder's site (SH) in the same locality, the ground vegetation was dominated by *Desmodium heterophyllum* (35% cover) and the fern *Dicranopteris linearis* (30%). Among the other species in the ground vegetation were species of *Adiantum, Pteris, Axonopus, Phyllanthus and Cyrtoccocum.*

Nutrients Stocks in Live Vegetation

Nutrient contents of the main trunk, branches, fruits, seeds and leaves have been estimated for the clone RRIM 501 in a Malaysian study (Shorrocks, 1965). While a considerable amount of nutrients are stocked in the trunk and the branches which are often removed from the land after felling, the annual turnover of leaf nutrients could be an important contributor to the nutrient dynamics especially in poorly managed rubber stands like those of the small-holders (Table 2). It has been estimated that at each replanting, about 1700 kg ha^{-1} of nitrogen is removed when felled wood is removed and the remainder burnt (Pushparajah, 1981).

Table 2 Estimated nutrient stocks (kg ha^{-1}) in trunks and branches of 33 year old rubber trees and their annual leaf and fruitfall.

	N	*P*	*K*	*Ca*	*Mg*
Trunk and branches	1389	218	989	N/A	328
Leaves	45–90	3–7	10–20	9–18	60–120
Capsules and seeds	1.7	0.2	1.8	0.3	0.2

Live vegetation (excluding the main trunk, branches and roots) of each site analyzed for five major nutrients N, P, K, Ca and Mg in Hedigalla Plantation in Sri Lanka are given in Tables 3–7. In the Rain Forest (RF) site, the nutrients of the five most dominant species are given (Table 3).

Table 3 Percentage nutrient concentration of the live vegetation of lowland rain forest (RF) site (n=10) in Hedigalla, Sri Lanka.

Component	*Total N* (%)	*Total P* (%)	*Total K* (%)	*Total Ca* (%)	*Total Mg* (%)
Leaves	1.832	0.105	1.023	0.953	0.254
Green Twigs	0.885	0.068	0.804	0.957	0.178
Brown Twigs	0.603	0.044	0.542	0.777	0.120
Woody	0.998	0.094	1.303	0.917	0.219

Table 4 Percentage nutrient concentration of the live vegetation of a Mature Rubber (MR) stand (n=10) in Hedigalla, Sri Lanka.

Component	*Total N*	*Total P*	*Total K*	*Total Ca*	*Total Mg*
Mature Leaves	1.190	0.203	1.460	0.953	0.162
Young Leaves	3.115	0.350	1.565	0.195	0.203
Rubber Petioles	0.990	0.183	1.536	0.203	0.093
Green Twigs	1.210	0.266	1.850	0.618	0.125
Brown Twigs	1.200	0.194	1.459	0.442	0.087

Nutrient levels of the grasses and other herbaceous flora in the MR site is given in Table 5. These results show that the ground vegetation which is usually removed around the trees is a good source of mulch and after weeding they are to be left on site for decomposition.

Table 5 Percentage nutrient concentration of the Ground Vegetation in MR site in Hedigalla, Sri Lanka.

	N	*P*	*K*	*Ca*	*Mg*
Ground Flora	1.936	0.171	2.256	0.472	0.286

Table 6 Percentage nutrient concentration of the live vegetation of the young rubber (YR) stand (n=10) in Hedigalla, Sri Lanka.

Component	*Total N* (%)	*Total P* (%)	*Total K* (%)	*Total Ca* (%)	*Total Mg* (%)
Rubber Leaves	3.000	0.264	1.282	0.690	0.323
Rubber Petioles	1.140	0.239	1.238	0.721	0.215
Green Twigs	1.080	0.289	0.894	1.401	0.253
Brown Twigs	0.575	0.170	0.535	1.300	0.234
Rubber Fruits (green)	0.630	0.199	1.022	0.520	0.176

Table 7 Percentage nutrient concentration of the live vegetation of the small-holder's (SH) site (n=10) in Hedigalla, Sri Lanka.

Component	*Total N* (%)	*Total P* (%)	*Total K* (%)	*Total Ca* (%)	*Total Mg* (%)
Rubber Leaves	2.78	0.181	1.100	0.667	0.374
Rubber Petioles	0.860	0.104	1.288	0.632	0.209
Green Twigs	0.685	0.107	1.079	1.121	0.206
Brown Twigs	0.630	0.167	1.198	0.906	0.237

Nutrient concentrations in live vegetation is relatively high in all the sites sampled (Tables 3–7). Even the poorly managed SH site has relatively higher levels of nutrients in its tree biomass. Leaves, particularly the young leaves contain high levels of nitrogen. However, the mixed ground cover has variable nutrient concentrations. In contrast, the legume ground cover of the YR site has a uniformly high nutrient concentration.

Similarly, high concentrations of nutrients have been recorded from Malaysia as well. Watson *et al.* (1964) have shown that after two years of growth, the legume creepers (*Pueraria phaseoloides, Calopogonium mucunoides, Centrosema pubescens*) immobilized 128 kg N ha^{-1} while covers of grasses (*Ischaemum muticum, Ottochloa nodosa*), *Mikania cordata* and 'naturals' (species of *Trema, Macaranga, Ficus, Solanum*) immobilized 45 and 84 kg N ha^{-1}, respectively (Table 8). Immobilization by grasses was the least for all the nutrients.

Table 8 Nutrient quantities (kg ha^{-1}) in different cover crop species 20-24 months after planting (Anon, 1972).

Cover plants	*Dry mass (g)*	*N*	*P*	*K*	*Mg*
Leguminous creepers	2,696	128	12	95	10
Grasses	5,247	25	5	50	5
Mikenia sp.	3,672	45	8	84	9
Naturals	6,381	84	12	143	19

However, during a period of four to five years when the ground cover prevails especially the leguminous covers, they return considerable quantities of plant nutrients, particularly nitrogen, to the soil in the form of litter during the period of canopy closure (Table 9).

Table 9 Amounts of Nutrients (kg/ha) returned to the soil by different cover crop categories over the five-year immature period of rubber growth (Anon., 1972).

Cover Crop	*N*	*P*	*K*	*Mg*
Leguminous creepers	226–353	18–27	85–131	15–27
Grasses	24–65	8–16	31–86	9–15
Mikania sp.	74–119	9–14	63–99	9–24
Naturals	13–117	3–10	46–140	3–18

Pushparajah and Tan (1979) found that in the absence of legume covers about 860 to 1000 kg N ha^{-1} had to be applied to rubber stands to obtain growth and yield comparable to those with legume covers. Furthermore, Pushparajah *et al.* (1978) showed that the residual effect of legume covers on the nitrogen status persists until about the thirteenth year of cover establishment. This long-term residual effect of legume covers could in part be due to the early build-up of nitrogen in soil and also consequently sustained in rubber trees (Table 10). Watson *et al.* (1964) showed that at the sixth year after establishment, the annual leaf fall from trees in the legume cover areas returned twice as much nitrogen as trees did in the non-legume cover areas, thus ensuring a greater annual cycling of nitrogen. Leguminous creepers have been shown to mobilize greater quantities of phosphorus and calcium as compared with other types of cover crops during the first two years (Watson *et al.*, 1964).

The use of shade tolerant legumes viz. *C. caeruleum* in combination with more conventional legumes would sustain higher residual levels of nitrogen in soils over a longer period of time after canopy closure (Tan *et al.*, 1976). Another contributory factor for this long-term residual effect is the relatively slower decomposition rate of this species compared with that of conventional legumes and hence a considerable saving in the addition of nitrogenous fertilizer (Pushparajah, 1981).

Table 10 Levels of soil nitrogen (%) under different ground cover species in rubber plantations in Malaysia (Pushparajah, 1981).

Type of cover	*Nitrogen in soil* (%)		
	2nd year	*4th year*	*8th year*
Legumes	0.140	0.144	0.115
Grasses	0.130	0.123	0.109
Naturals	0.119	0.124	0.106
Bare	0.099	0.095	0.092

Soil Physico-Chemical Properties

Soil pH of the four study sites (RF,MR,YR and SH) in Hedigalla, Sri Lanka ranged between 4.5 and 4.95. The soil pH of the RF site up to a depth of 20 cm ranged from 4.61–4.67 while that of MR site was 4.50. In the YR site it ranged from 4.53–4.68 and in the SH site it was between 4.81 and 4.95. Inspite of the regular application of inorganic fertilizer, the pH of the MR site was not different from that of the adjacent natural forest site. The mean organic carbon content of the RF soil was 1.93 and that of MR, YR and SH sites were 1.55, 1.36 and 1.25, respectively. The comparative soil nutrient status determined at the beginning of the experiment in 1985 is given in Figure 2.

The RF site had higher nutrient levels than the other sites for all nutrients except for NH4F/HCL extractable P. The higher levels of organic matter in the RF soils would have affected the more complete release of P in the NH4F/HCL extraction process. The SH site had the lowest levels of P, K and Ca while the YR site had the lowest levels of N and Mg. This indicates that the demand for N and Mg in young rubber trees of the clone PB 86 and also for the vigorously growing legume cover crop to be greater during this rapidly growing phase. These results show that the RF site is adapted to conserve nutrients more efficiently than the rubber sites.

Phosphorus levels at all three depths of the two well managed plantations (MR and YR) were higher than those in the SH site which received minimal agronomic management. Consequently, its P levels are similar to those of the natural forest (RF) site. The exchangeable Ca and Mg levels in the rubber sites declined more rapidly from the surface to the measured soil depth of 30 cm than in the natural forest. The demand for these two nutrients appear to be higher in all three rubber sites than that in the natural forest in which higher levels of soil Ca and Mg stocks are maintained in a steady state by natural recycling processes.

Nutrient status of rubber replanting soils in different soil series in Malaysia is also relatively low and ranging widely among different soil series (Table 11) . The top 45 cm of soil which has been used for nutrient analysis is considered as the most active zone of root activity in which the feeder roots are most prevalent.

Table 11 Estimates of total nitrogen in the top soil (0–45 cm) in different soil types of Malaysia (Pushparajah, 1981).

Soil series	*Soil order*	*Bulk density (g/cc)*	*Range of N (kg/ha)*
Rengam	Ultisol	1.15	4190–11,300

Contd.

Table 11 Contd.

Soil series	*Soil order*	*Bulk density (g/cc)*	*Range of N (kg/ha)*
Durian	Ultisol	0.95	2720–13,870
Malacca	Oxisol	1.07	2800–10,100
Holyrood	Entisol	1.25	5150–13,560
Selangor	Inceptisol	0.69	4650–22,180

NUTRIENT FLUXES IN THE ECOSYSTEMS

Latex Yield and Its Nutrients

The yield and the nutrient content in latex of different clones of rubber belonging to different age groups have been studied in Hedigalla Plantation (MR and YR sites) and a nearby small-holding (SH site) in Sri Lanka (Gunatilleke and Perera, 1989). The MR stand had approximately normal distribution of yield classes with a frequent mode of 21–30 g/tree yield class. The SH stand showed an approximate skew type distribution of yield classes with a mode of 11–20 g/tree during most high yielding months e.g. July–October. The SH site had a high % CV compared to that of MR due to greater variability in individual tree diameter. The clone RRIC 45 in SH site is a poorer yielder compared to RRIC 103 of the MR site. In addition, site management such as the frequency and quantity of fertilizer application also may have contributed to higher yield in the MR site. Except for Ca, all other nutrients were found in higher levels in the latex from the MR sample (Tables 12 and 13). Application of ethrel as a latex stimulant in high-yielding clones appear to drain substantial quantities of nutrients compared to that of unstimulated normal tapping (Table 14).

Table 12 Mean percentage of latex nutrients in the MR site (n = 10) in Hedigalla Plantation, Sri Lanka (Gunatilleke and Perera, 1989).

	N	*P*	*K*	*Ca*	*Mg*
Mean (%)	0.632	0.245	1.157	0.031	0.184
SE	0.054	0.013	0.107	0.002	0.021
CV %	26.78	16.68	29.23	24.13	36.40

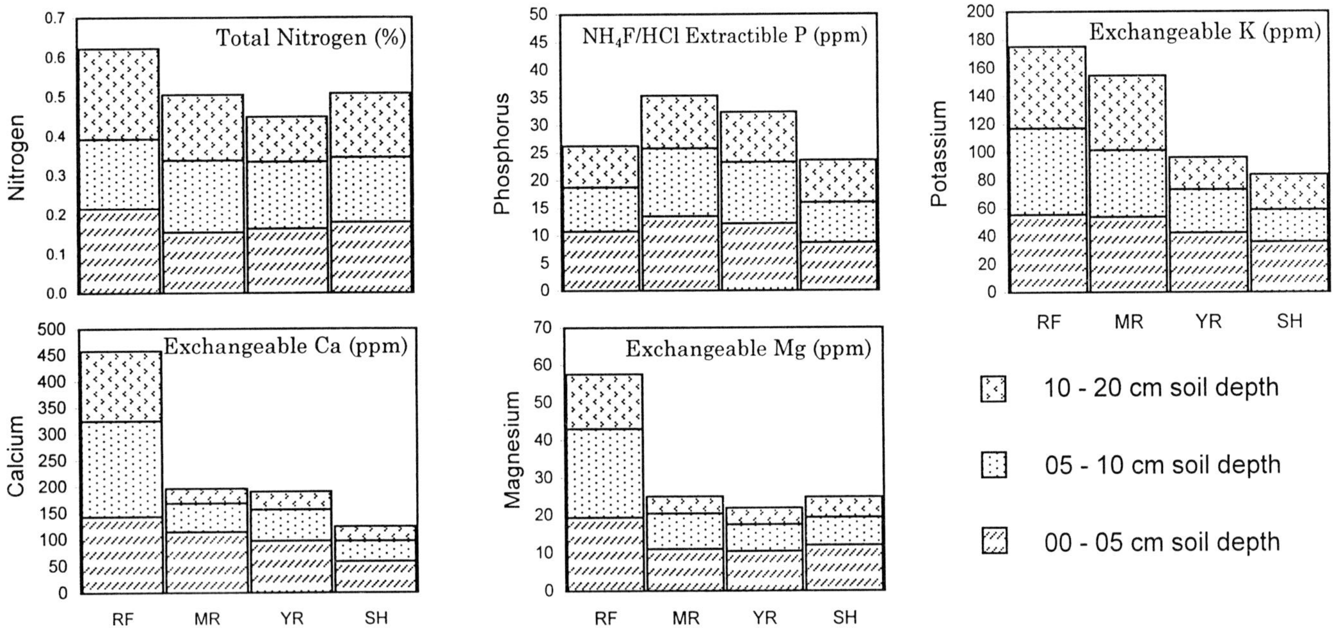

Fig. 2 Comparative soil nutrient status at three different depths in the four study sites RF, MR, YR and SH in Hedigalla, Sri Lanka (see the text for study site details).

Table 13 Mean percentage of latex nutrients in the SH site in Sri Lanka.

	N	*P*	*K*	*Ca*	*Mg*
Mean (%)	0.521	0.156	0.572	0.043	0.127
SE	0.07	0.009	0.005	0.008	0.022
CV %	42.67	18.08	30.54	60.47	53.83

Table 14 Amount of nutrients removed in latex per year under different tapping methods (Anon., 1972)

Tapping Method	*Nutrients removed kg ha^{-1}*			
	N	*P*	*K*	*Mg*
Normal tapping (Yield of 1400 kg yr^{-1})	9.4	2.3	8.3	1.7
Stimulation by 10% Ethrel (Yield of 2600 kg yr^{-1})	23.9	7.2	22.3	4.1

LITTERFALL

Total litter input (excluding large and medium sized branches) to the ground has been estimated using litter traps (1m^2) placed at 1.3 m above ground collecting litter from trees, shrubs, lianas, parasites and epiphytes and ground quadrats collecting litter from the herbaceous ground vegetation at biweekly intervals for two years from July 1985–June 1987 in Hedigalla, Sri Lanka (Gunatilleke and Perera, 1989). While the ground quadrats enabled the collection of litter input from all different life forms in the natural forest and in the rubber sites, the disadvantage had been that an unquantified amount of ground litter was turned over during the two-week period between collections due to comminution, mechanical removal by water and decomposition. Therefore, it can be assumed that the ground quadrat data provided at best, an underestimation of the total above ground litter input.

Litterfall from ground vegetation in the RF site which includes seedlings and saplings yielded a higher litter input in the ground quadrats despite its inherent limitations (Fig. 3). In the rubber sites too, except one year each in the YR and SH sites, the ground quadrats yielded higher input of litter. The RF site recorded, by far the highest total litter input which was more than twice that of rubber stands. Interestingly, all the rubber sites had more or less similar levels of above ground litter input in traps as well as in quadrats when the two year means were compared. A certain degree of year to year variation in total litterfall was also evident (Fig. 3). The non-rubber litter input

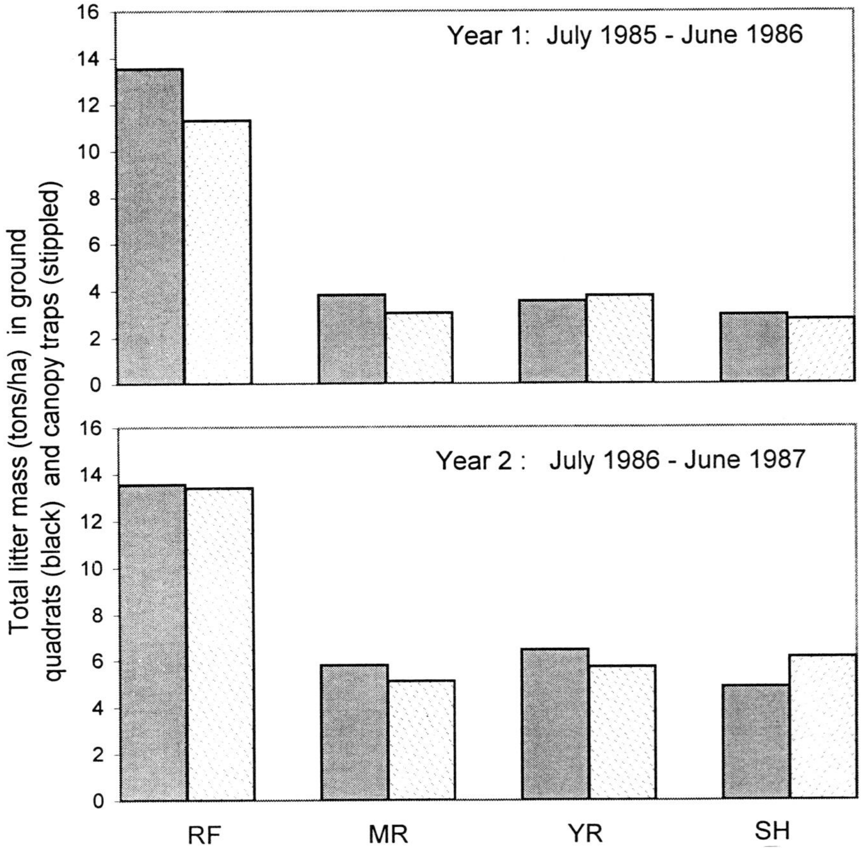

Fig. 3 Total litter input collected in canopy traps (black) and ground quadrats (dotted) in different study sites RF, MR, YR and SH (see the text for study site details) in Hedigalla, Sri Lanka.

was lowest in the YR site and dominated by the litter from the planted cover crop, *P. phaseoloides* (Fig. 4). In the MR site, it was dominated by grasses while in the SH site where the non-rubber litter input was highest among the three rubber sites, the contribution was from grasses, ferns and other weedy shrub species. In each study site, the leaves formed the major litter component ranging from 39–70% of the total litter input (Fig. 4). Small woody components, petioles, litter fragments etc. contributed in smaller proportions to the balance.

A comparative study of litterfall in rubber and teak plantations and a natural forest in India has shown that teak plantations recorded the highest and rubber plantation the lowest litter accumulation rates (Krishnakumar *et al.*, 1991). Another study carried out by Norhayati

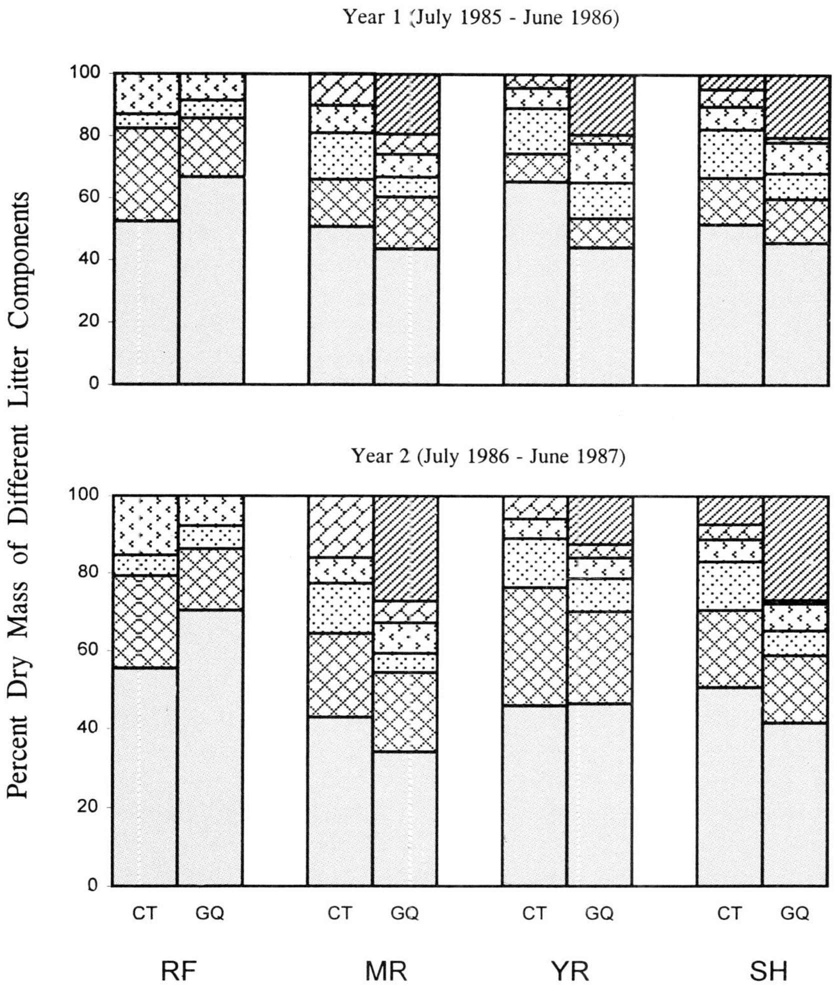

Fig. 4 Proportion of annual total dry mass of different litter components leaves (gray), woody twigs (diamond checks), petioles (dots), litter fragments (divots), rubber leaves (left slanting bricks), non-rubber and grass leaves (left slanting lines) collected biweekly in canopy traps (CT) and ground quadrats (GQ) in the four study sites RF, MR, YR and SH (see the text for study site details) in Hedigalla, Sri Lanka.

and Lau (1990) showed that natural forests recycled about 8.33 tons of dry litter as compared to about 4.43 tons per year under legumes and 3.7–7.7 tons per year under mature rubber. While litterfall in natural forests in this study was lower than that of Hedigalla in Sri Lanka, the values for rubber litterfall were comparable.

Seasonality in Above Ground Litterfall

As expected in an aseasonal species-rich rain forest, the RF site in Sri Lanka did not show any prominent peaks during any specific period of the year although slightly elevated levels were evident especially in ground quadrats during the relatively dry periods in February and July/August in both years. This implies that while the litterfall of ground vegetation of the rain forest which comprise predominantly saplings and seedlings of the species of the upper strata and shade-loving herbaceous vegetation may be higher during the dry periods, litter removal and turnover may also be less during the same periods. A highly significant ($P > 0.001$) variation was observed between biweekly leaf litter inputs in both traps and quadrats but no significant variation was evident between the higher and lower slopes of the RF site. The canopy leaf litter input was only 77.9% of the ground quadrat collections which included those from seedlings, saplings and other herbaceous ground vegetation.

Based on identification of fallen leaves in litter traps, it was possible to estimate the biweekly contribution of leaf litter from canopy individuals of each species. It was found that the litterfall peaks of different species showed some variation with respect to their timing (Fig. 5). Each of the three rubber sites, on the other hand, showed a single peak in January/February, depending on the clone, corresponding with the major litterfall characteristic of *H. brasiliensis.* Only in instances where there was a severe incidence of foliar diseases such as *Corynospora* leaf spot in RRIC 103, that a secondary premature leaf fall was observed.

In the MR site, a major peak in mature rubber leaf litterfall was observed in January while in the YR and SH sites it was in early February in both years. In the MR site, although there were fluctuations in other components of the litterfall throughout the year their contributions remained relatively low and peaks and troughs were less prominent. In the YR site, in ground quadrats that were randomly placed in *Puraria* patches, the *Puraria* leaf litter input in July and in December remained higher than during other months. The July peak was due to premature litterfall of *Corynospora* infected leaves of this highly susceptible clone. In the SH too, litter contributions from ferns and grasses peaked in July/August and again in Feb/March coinciding with the relatively dry periods.

A comparable nutrient analysis carried out in Malaysia using rubber leaves of the clone GT 1 and two leguminous cover crops revealed significantly lower C : Nutrient ratios (except for Ca) in legume litter as compared to that of rubber leaves (Table 15). The lower C : N ratios in legume litter may have contributed to faster decomposition and higher mineralization rates of legume leaves than those of rubber leaves.

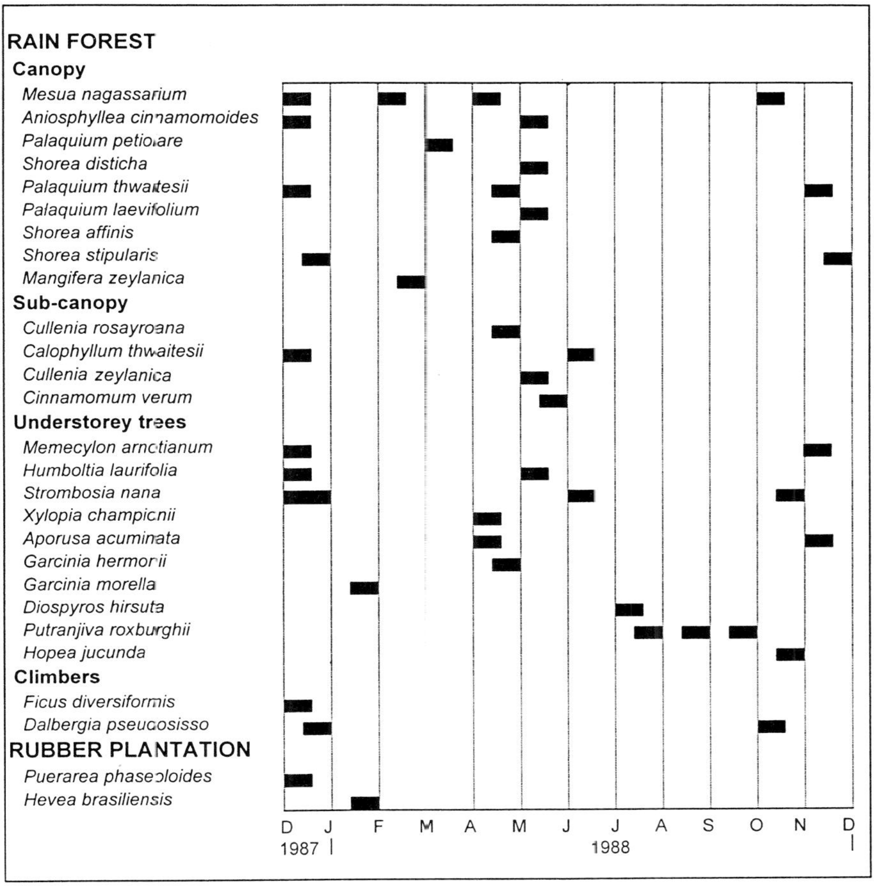

Fig. 5 Peak leaf fall times for some dominant rain forest species, *Hevea* rubber and the cover crop *Pueraria phaseoloides* in Sri Lanka.

However, having a litter mixture of high and low quality leaf litter would have enhanced the decomposition rate of rubber leaves in the legume covered stands than in the others.

Table 15 Chemical composition of rubber and leguminous leaf litter (Haut and Pushparajah, 1985).

Leaf litter	*C* (%)	*N* (%)	*C:N*	*P* (%)	*K* (%)	*Ca* (%)	*Mg* (%)
P. phaseoloides	36.4	2.56	14	0.11	1.09	0.98	0.37
C. caeruleum	37.4	2.36	16	0.11	0.92	1.20	0.24
H. brasiliensis	47.1	1.36	35	0.03	0.34	1.00	0.16

A Malaysian study has estimated that the annual litter fall of rubber leaves returns nitrogen ranging from 30 to 70 kg ha^{-1} yr^{-1} depending on the age of the tree with a mean over the 25 years being about 56 kg ha^{-1} yr^{-1} (Pushparajah, 1981).

Loss of Litter Mass During Decomposition

Both Rain Forest (RF) and Mature Rubber (MR) litter in Hedigalla, Sri Lanka has shown high initial rate of mass loss up to about three months for all litter types in all the sites examined (Figure 6). The rate of mass loss of mixed leaves of forest species (RF) as well as that of mature rubber (MR) leaves was greater in the MR site than in the forest during most part of the decomposition period. MR leaf litter decomposed faster than the forest litter in both MR and RF sites. This infers that the physico-chemical as well as biological environment of the MR site is more favourable for litter decomposition than that of RF site. The plantation management practices such as periodic addition of inorganic fertilizer (N, P, K, mixtures) and dolomite/kieserite may have adjusted the C : N, C : P and other elemental ratios and the pH of the soil in the MR site to a more favourable level for faster decomposition than that in the RF site. Faster decomposition of rubber than that of forest leaf litter could perhaps be due to the uniformly less coriaceous nature of rubber leaves belonging to clones which have been bred for higher yielding of latex as compared with the variable degree of coriaceousness in natural forest species. The forest species are also known to have efficient nutrient conservation mechanisms under natural conditions such as withdrawal of nutrients prior to litterfall (Jordan, 1985) resulting in their higher C : Nutrient ratios. Rubber trees on the other hand have been bred for higher yield of latex in response to high inorganic nutrient input. It may be likely that under such conditions, the nutrient conservation mechanisms may be less than efficient compared with those found in the naturally growing rubber trees.

While the leaf litter of well managed young rubber showed a rapid rate of decomposition in its own site, that of small-holdings showed the slowest rate of decomposition among the three rubber litter types. As no interchange of litter from the three rubber sites was made in the decomposition study, the clonal variation with respect to litter decomposition could not be determined. However, the slowest decomposition rate observed in the small-holding could be resulting from a combination of lower soil fertility levels due to poor management operations in it and possibly lower resource quality of the leaf litter. In the YR site, with regular application of inorganic fertilizer and the presence of nitrogen-fixing ground cover crop at least in some places would have helped to improve the soil fertility levels.

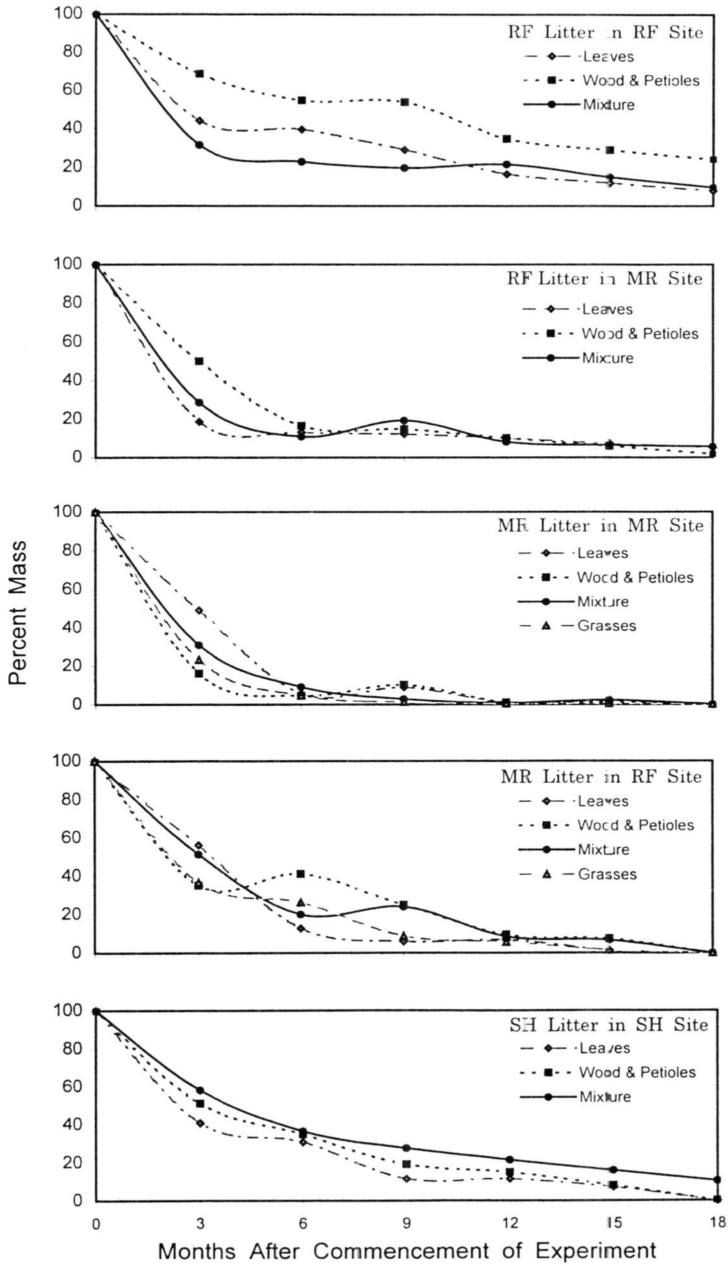

Fig. 6 Percent mass loss of different litter components from each study site during decomposition in different study sites.

Woody components of petioles as well as the mixture of major litter components of each vegetation type broadly followed the trends exhibited by their respective leaf litter during decomposition (Fig. 6). Woody components with petioles and the mixed leaf litter from the natural forest decomposed faster in the MR site than in the adjacent RF site.

Nutrient Release During Litter Decomposition

The concentrations of nitrogen and phosphorus in freshly collected litter from the different study sites in Hedigalla, Sri Lanka are given in Table 16.

Table 16 Concentrations of total N and P in different litter types from rain forest and rubber stands in Hedigalla, Sri Lanka (site description given in the text).

Litter Type	*Total N* (%)	*Total P* (%)
Forest Leaves	1.073	0.026
Forest Twigs and Petioles	0.710	0.055
Forest litter mixture	0.927	0.055
MR Leaves	2.027	0.139
MR Wood and Petioles	0.813	0.041
Grasses	1.454	0.101
MR Litter Mixture	1.325	0.054
SH Leaves	1.308	0.066
SH Wood and Petioles	0.655	0.027
SH Litter Mixture	1.160	0.044

Release of major nutrients (N and P) from decomposing litter was determined by chemical analysis of litter remaining at each sampling time for all litter types in their own as well as in other sites (Figs. 7 and 8). The rates of mass loss of different litter types appear to be closely related to their initial total N content while the initial concentration of total P did not show such a clear pattern with litter decomposition. MR leaves with the highest concentration of N showed a high rate of litter decomposition and the RF site a relatively low rate. MR leaves released both N and P at a faster rate in both sites as compared with that of mixed leaf litter (Figs. 7 and 8).

Mineralization of Leaf Nutrients in Rubber Soils and Growth Responses

A laboratory incubation method has been used to estimate how readily the organically bound nitrogen in rubber and legume leaves is made available upon biological mineralization in four soil types in Malaysia (Haut and Pushparajah, 1985). When oven-dried and finely ground

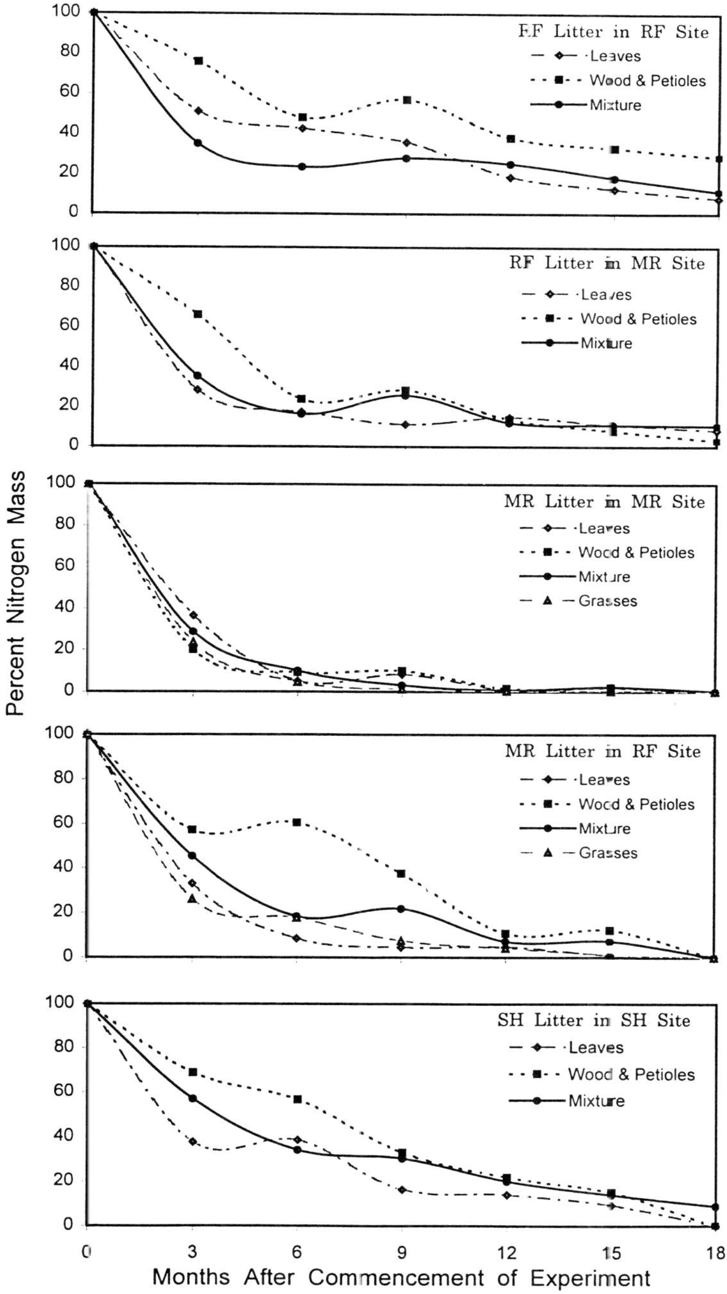

Fig. 7 Percent mass loss of nitrogen in different litter components from each study site during decomposition in different study sites.

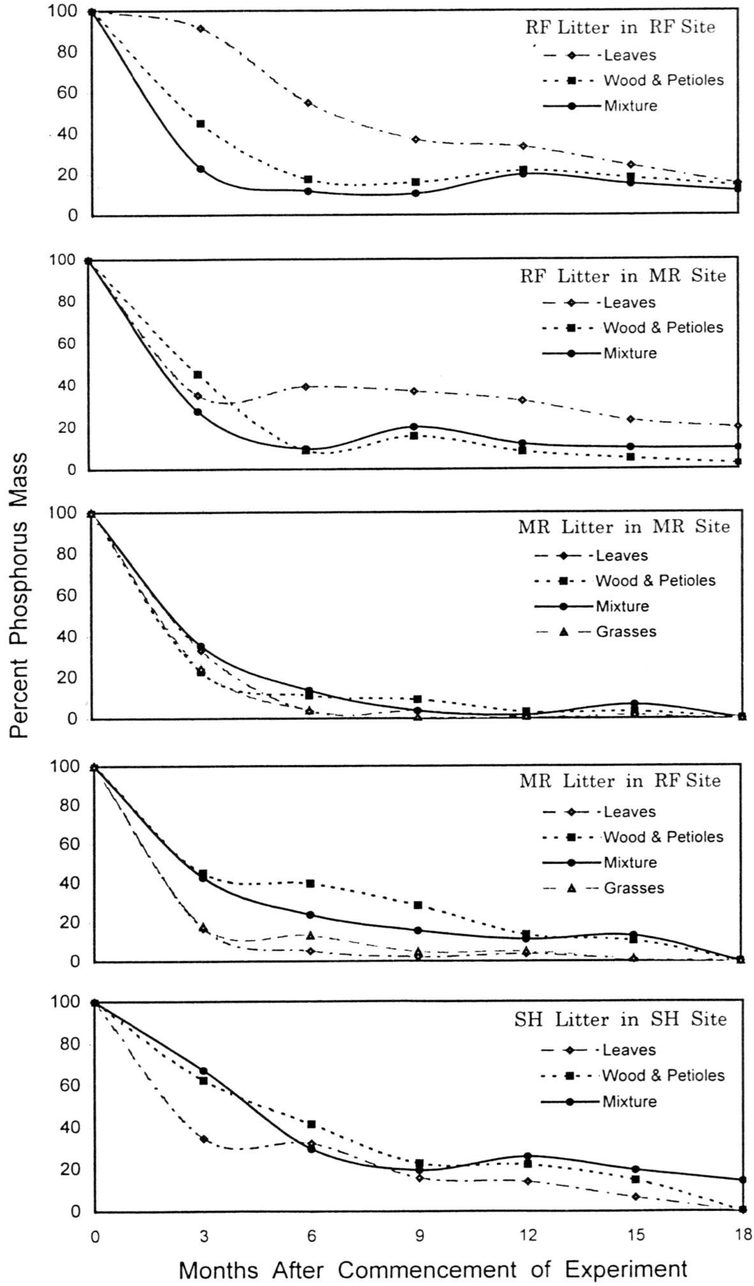

Fig. 8 Percent mass loss of phosphorus in different litter components from each study site during decomposition in different study sites.

rubber leaf samples were incorporated into laboratory incubated soil samples from four different soil series (Batu Anam, Holyrood, Munchong and Serdang), it resulted in net immobilization of mineral-N as indicated by the consistently lower levels of ammonium and nitrate-N over a period of ten weeks as compared with the corresponding untreated control samples. In contrast, incorporation of similarly prepared leaf powder from *P. phaseoloides* and *C. caeruleum* to these soils accumulated considerably more mineral-N than the control samples. Leaf powder of *P. phaseoloides* appeared to have released more mineral-N compared to that of *C. caeruleum* although they have similar C : N ratios which may be attributed to possible differences in secondary compounds such as polyphenols (Haut and Pushparajah, 1985). Net immobilization of available mineral-N by rubber leaf powder may probably be due to its lower nitrogen content and hence higher C : N ratio. These results suggest that a temporary unavailability of soil nitrogen could occur during the decomposition of rubber leaves. However, these studies carried out under laboratory conditions using powdered leaf material may not reflect the real field conditions in the rubber soils. The use of labeled nitrogen might yield more accurate results.

In another experiment, in which rubber seedlings were grown in containers to which either ammonium sulphate, leaf litter of *P. phaseoloides* or *C. caeruleum* were added as the sole source of nitrogen over a period of seven months has shown highest girth increment when the chemical fertilizer was added. However, addition of leaf litter of legumes also showed a better growth and nutrition of rubber seedlings than that of the control (Haut and Pushparajah, 1985).

Nutrient Loss from Soil

Apart from nutrient removal by tapping for latex, considerable amounts of soluble soil nutrients may be lost from the rubber plantation system by way of leaching and volatilization from the soil. While phosphorus is relatively immobile in the soil, leaching losses of other nutrients, particularly N, K, and Mg are likely to be high in light textured sandy soils with poor nutrient retention ability. Loss of nutrients such as nitrogen due to leaching in rubber soils can be around 12–28% when added as ammonium sulphate. The losses due to denitrification, volatilization and leaching could be as high as 50% of the amount added under normal field conditions when added as urea (Pushparajah, 1981).

Throughfall and Canopy Interception

Based on the readings taken at seven rain storms of varying intensities distributed over the 2 year study period in Hedigalla, Sri Lanka, it was found that the Rain Forest (RF) site experienced higher canopy interception than the mature rubber site under low rainfall conditions (Fig. 9). With a multi-layered canopy structure, a higher threshold rainfall is needed for the percentage throughfall to increase with the rainfall. Below this level, there is high canopy interception in the natural forest.

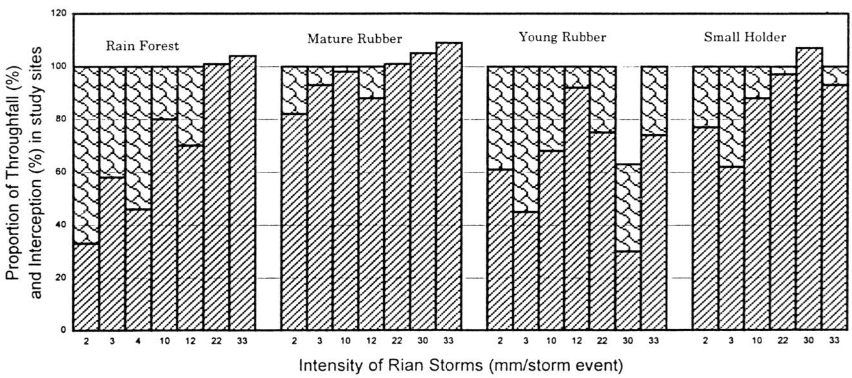

Fig. 9 Percent canopy interception (left slanting) and throughfall (shingles) during seven rain-storm events in the four study sites.

The Mature Rubber (MR) site with a relatively thin canopy had > 80 per cent throughfall even in the slightest rainfall. In heavy rainstorms, there was 100 per cent or more throughfall. "Secondary rain' rebounding off lateral branches and tree trunks into some of the randomly placed collection bottles in some study sites has resulted in having cumulative totals of over 100% compared with open area control sites in high intensity rain storms. The young rubber trees of the YR site having a dense canopy with branching also experienced higher canopy interception than mature rubber stands in all the rain storms (Fig. 9). The Small Holders' site with a canopy of rubber and a mixture of other utility tree species had intermediate levels of interception and throughfall among the rubber sites. The adjacent RF and MR sites had significantly different (t = 4.296**) canopy interception in low intensity rain storms due to their differences in canopy architecture, stratification and density and biomass of the above ground vegetation. In the rubber sites, the canopy interception and throughfall are also determined by the leafing phenology and premature leaf-fall due to leaf diseases during certain periods of the year. In the YR site, PB 86 clone shed their mature leaves in January.

The N content of the water samples were less than 1 μg/g and hence could not be detected either by the Indophenol blue or the nitroprusside method used. Potassium being the most mobile nutrient varied the most among all nutrients while Mg and Phosphorus varied slightly. The control readings showed relatively high values for all nutrients probably due to their presence in the atmosphere from dust etc.

Rain water collected in an open area in a West Malaysian study site indicates that with an annual rainfall of 2540 mm approximately 20 kg N, 12 kg K, 3 kg Mg and 0.2 kg P may be added to one ha of soil per year (Shorrocks, 1965). Although not quantified in any of the studies available to the author, the contribution by weathering of rocks could contribute varying quantities of potassium and magnesium.

In the lowland rain forest as well as in the rubber plantations, nutrients are cycled between two closely associated reservoirs, namely soil and vegetation. Species diversity in lowland rain forests and their life history variations including leaf fall phenology contribute to synchronous supply of nutrients to meet the continuous demand of growth and reproduction of incumbent species. Furthermore, it is known that in tropical rain forests, there exists a nutrient retranslocation before leaf fall as a mechanism of nutrient conservation. Such nutrient conservation mechanisms act as built-in synchrony controls or regulators (Fig. 10a). In the less-diverse *Hevea* rubber plantations, litterfall showed a more or less overlapping peak with a relatively faster decomposition than that of the mixed forest leaf litter thus adding a short pulse of nutrients annually during the early period of refoliation. The extent of retranslocation is expected to be less in high yielding fertilizer-responsive hybrid clones of rubber. Nutrient conservation mechanisms prevalent in rain forests are less well developed in well managed rubber holdings and consequently, only a partial synchrony between nutrient demand and supply seems to be in operation in the latter (Fig. 10b).

COMPARATIVE NUTRIENT DYNAMICS IN LOWLAND RAIN FOREST AND RUBBER STANDS

Using the estimations of nutrients in different compartments *viz.* precipitation, throughfall, live vegetation (both tree crop and ground cover), litter and soil of each ecosystem, an attempt has been made to prepare a compartment model of nutrient balance for each ecosystem in the Hedigalla study sites in Sri Lanka. The nitrogen levels in the precipitation were very low but that of P and K were unusually high. Throughfall from the RF and YR sites had the highest levels of nutrients. The natural forest with an uneven-aged mixed vegetation

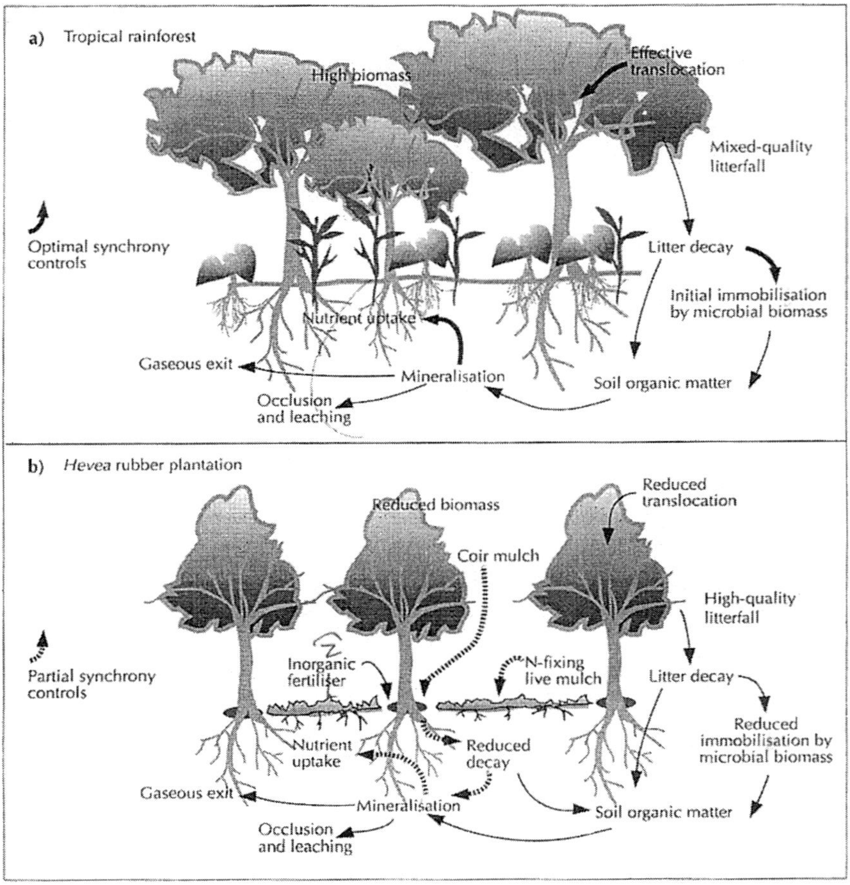

Fig. 10 Diagrammatic representation of the synchrony controls operating in a tropical rain forest and a young rubber plantation.

has a relatively high biomass contributed to by seedlings, saplings, and young and mature trees of different species occupying every available niche. This dense vegetation having a high above ground biomass may have contributed to the high level of nutrients in the throughfall of the RF site. The high concentration of nutrients in the throughfall in the YR stand could either be a clone-specific or age related phenomenon of the PB 86 clone. As the young rubber plantations are fertilized frequently, it seems likely that their leachates also contain relatively larger quantities of nutrients.

A substantial amount of nutrients are removed by way of latex during tapping in both MR and SH sites. Concentrations of N, P, K, and

Mg in latex of well managed MR site are greater than those of the SH site. This difference could be a result of a complex interaction of clonal differences, levels of management and other environmental and biological factors. Concentrations of nutrients in leaf litter in the three rubber sites were higher than those of natural forests. Natural forest species are known to possess a number of nutrient conservation mechanisms (Jordan, 1985) such as nutrient withdrawal from leaves during senescence.

The soil nutrient level in the RF site is higher than in all other sites despite regular addition of macronutrients to the well-managed *Hevea* rubber plantations. The three rubber sites have lower levels of nutrients while their biomass have higher nutrient concentrations. The selected and improved rubber clones may be more nutrient-demanding on the one hand and on the other hand, due to lower tree density, loss of fertility through erosion and leaching may also have contributed to lower soil nutrient levels in the rubber soils.

The schematic diagrams depicting nutrient stocks in different ecosystem compartments reveal that higher concentration of nutrients are held in the biomass in both natural and man-made ecosystems (Figs. 11 and 12). It is also evident that both qualitative and quantitative nutrient input from ground vegetation is significant in the rubber sites. Therefore, the selection of suitable shade tolerant ground cover and its proper management would help to arrest the leaching of inorganic fertilizer added to the rubber plantations. The ground vegetation would act as a temporary reservoirs of added nutrients which otherwise would have been washed away during rainstorms. Weeding and/or pruning of this ground vegetation and incorporating them as mulches at the time of inorganic fertilizer adding would help to synchronize the nutrient input and its slow release for uptake by rubber trees. It will also minimize soil erosion. Manipulation of spatial arrangement and density of rubber trees and intercropping with other utility shrub/herbaceous species of compatible growth characteristics need to be experimented taking the cue from the RF site which should be taken as the model for efficient nutrient conservation mechanism at the ecosystem level.

The potential for improving the regulation and synchronization of nutrient fluxes between compartments by manipulation of ground vegetation is clearly evident from the research described herein. This can be accompanied by a shift from the classical monoculture planting of rubber to multi-species cropping taking a cue from the traditional home garden systems which the smallholders are more accustomed to. It is heartening to note that the diversification of rubber plantations is already underway in most rubber growing countries and the research

RAINFALL

N	N.D
P	2910
K	8900
Ca	10416
Mg	2180

LIVE BIOMASS

	Leaves	Young twigs & petioles	Older twigs
N	15100 - 24400	5400 - 11400	3320 - 7700
P	930 - 1170	340 - 720	240 - 620
K	8120 - 17350	3320 - 14850	2250 - 10650
Ca	6570 - 14140	7870 - 14360	3880 - 8440
Mg	1550 - 3330	460 - 3070	260 - 1700

THROUGHFALL

N	N.D.
P	3343
K	32867
Ca	12261
Mg	2937

LITTER

	Forest leaves	Woody material
N	10730	7100
P	260	550
K	2900	2050
Ca	9050	6620
Mg	2280	1460

SOIL

	d1	d2	d3
NO_3	6.1	-	-
NH_4	100.0	-	-
N	2150.0	1770.0	2330.0
P	10.7	8.0	7.51
K	55.5	61.5	58.2
Ca	142.5	181.8	133.0
Mg	19.3	23.7	14.5
Soil carbon	1.8	1.9	1.9
pH	4.6	4.6	-
CEC	7.2	7.4	7.7

Fig. 11 A compartment model of nutrient balance (ppm) in a lowland rain forest (RF) in Hedigalla, Sri Lanka (d1–d3 are three soil depths from the surface up to 30 cm.; N.D. Not Determined).

RAINFALL	
N	N.D.
P	2910
K	8900
Ca	10416
Mg	2180

LIVE VEGETATION	Rubber leaves	Woody material
N	29100	12000
P	2050	1935
K	14600	14590
Ca	9530	4420
Mg	1615	870

LATEX	
N	6320
P	2450
K	11570
Ca	310
Mg	1840

THROUGHFALL	
N	N.D.
P	4948
K	11640
Ca	9665
Mg	1808

SOIL	d1	d2	d3
NO_3	12.6	-	-
NH_4	42.0	-	-
N	1550.0	1830.0	1670.0
P	13.5	12.2	19.5
K	53.7	47.6	53.1
Ca	114.0	53.9	27.7
Mg	11.0	9.3	4.5
Soil carbon	1.6	1.5	1.4
pH	4.5	4.5	-
CEC	7.8	6.7	6.9

LITTER	Rubber leaves	Woody material	Grasses
N	20270	8130	14540
P	1390	410	1010
K	3950	2860	
Ca	7980	10750	
Mg	2340	1680	

Fig. 12 A compartment model of nutrient balance (ppm) in a mature rubber plantation (RF) in Hedigalla, Sri Lanka (d1–d3 are three soil depths from the surface up to 30 cm.; N.D. Not Determined).

Plate 1 The mature rubber (MR) study site with randomly placed canopy litter traps. Note the regular pattern of rubber trees and the herbaceous ground vegetation.

Plate 2 A young rubber (YR) plantation with a well established legume cover crop cleared only around the base of the rubber trees for fertilizer application.

should also take into account the nutrient dynamic aspects of the selected suite of species. This will help the increasing numbers of small-holding rubber growers worldwide, to diversify their holdings in a more ecologically and socio-economically sustainable manner in this era of fluctuating fortunes for the farmers.

REFERENCES

Anon., 1972. Cycle of nutrients in rubber plantation, *Planters Bulletin*, **120:** 73–81.

Brown, S., Anderson, J.M., Woomer, P.L., Swift M.J., and Barrios, E., 1994. Soil biological processes in tropical ecosystems, In: *The Biological Management of Tropical Soil Fertility*, (P.L. Woomer, and M.J. Swift, eds.), John Wiley & Sons, New York, 15–46 pp.

Gunatilleke, I.A.U.N., and Perera, K.V.M., 1989. Contribution of Soil Biological Processes and Environment to Soil Fertility in a Rain Forest and *Hevea* Rubber Plantations, Final Report submitted to the Natural Resources, Energy and Science Authority of Sri Lanka, (Grant No. RG/85/Ag/05), 148 p.

Haut, T.K., and Pushparajah, E., 1985. Studies on Nitrogen in Malaysian soils, V. Mineralization of leaf litter nitrogen and its availability to rubber seedlings, *J. Rubb. Res. Inst. Malaysia*, **33**(1): 26–36.

Jordan, C.F., 1985. *Nutrient Cycling in Tropical Forest Ecosystems*, John Wiley & Sons, Chichester, 190 p.

Krishnakumar, A.K., Gupta Chandra, Singh, R.R., Sethuraj, M.R., Potty, S.N., Thomas, E., and Krishna Das, 1991. Ecological impact of rubber, (*Hevea brasiliensis*) plantations in north-east India: II. Soil properties and biomass recycling, *Indian J. of Nat. Rubber Research*, **4**(2): 134–141.

Norhayati, M., and Lau, C.H., 1990. *Soil fertility changes following land clearing and rubber cultivation*, International Soil Science Congress, Kyoto, Japan.

Pushparajah, E., 1981. Nitrogen Cycling in South-East Asian Wet Monsoonal Ecosystems, In: *Proceedings of a Regional Workshop arranged by the SCOPE/UNEP*, (R. Wetselaar, J.R. Simpson, and T. Rosswall, eds.), International Nitrogen Unit of the Royal Swedish Academy of Sciences and the Chiang Mai University, Australian Academy of Science, Canberra.

Pushparajah, E., Mahmud, bin, H., and Abdul W., 1978. Manuring in relation to covers, *Proc. Rubb. Res. Inst. Malaysia Planters Conf.*, 1977, RRI, Malaysia, 150–165 pp.

Pushparajah, E., and Tan, K.H., 1979. Legumes in the nitrogen economy in rubber cultivation, *Proc. Conf. Soil Microbiology and Plant Nutrition*, Kuala Lumpur, University of Malaya Press.

Samarappuli, L., 1996. The contribution of rubber plantations towards a better environment, *Bulletin of the Rubber Research Institute of Sri Lanka*, **33**: 45–54.

Samarappuli, I.N., Ekanayake, A., and Naranpanawa, A. 1996, Natural Rubber Industry in Transition, Proc. IRRDB Seminar on Agronomy, Farming Systems and Technology, Colombo, 117–129.

Scholes, M.C., Swift, M.J. Heal, O.W.P.A. Sanchez, Ingram, J.S.I. , and Dalal R., 1994. Soil Fertility Research in Response to the Demand for Sustainability, In: *The Biological Management of Tropical Soil Fertility*, (P.L. Woomer, and M.J. Swift, eds.), John Wiley & Sons, New York, 1–14 pp.

Shorrocks, V.M., 1965. Mineral Nutrition, Growth and Nutrient Cycle of *Hevea brasiliensis* I. Growth and Nutrient Content, *J. Rubb. Res. Inst. of Malaya.*, **19**: 32–47.

Sivanadyan, K., and Norhayati, M., 1992. Consequences of transforming tropical rain forests to Hevea plantations, *Planter.*, **68:** 547–567.

Tan, K.H., Pushparajah, E., Shepherd, R., and Teoh, C.H., 1976. *Calopogonium caeruleum,* a shade–tolerant leguminous cover for rubber, *Proc. Rubb. Res. Inst. Malaysia,* Planters Conference, Kuala Lumpur, 1976.

Watson, G.A., 1960. Cover Plants and the Soil Nutrient Cycle in *Hevea* Cultivation, *Proc. Nat. Rubb. Conf. Kuala Lumpur,* 352–361 pp.

Watson, G.A., Wang, P.W., and Narayan, R., 1964. Effect of cover plants on soil nutrient status and growth of *Hevea, J. Rubb. Res. Inst. Malaya.,* **18:** 80–161.

Woomer, P.L., and Swift, M.J., 1994. *The Biological Management of Tropical Soil Fertility,* John Wiley & Sons, New York, 243 p.

PART—II: BIOTIC DYNAMICS

— Litter layer dynamics and soil microbial biomass in reforested areas after degradation by bauxite mining in eastern Amazon. *Enir S. Costa* (UCI, USA), *Flavio J. Luizao,* *Regina C. Luizao* (INPA, Brazil) and *Andrew McMillan,* (UCI, USA).

— Litter and microbial dynamics in relation to nutrient cycling in Oil-Palm (*Elaies guineenis* jacq) ecosystem. *N.O. Aisueni* and *C.E. Airede* (NIOR, Nigeria).

— Decomposition dynamics and nutrient flow in pine forest plantation in central Java. *Herman A. Verhoef* (VU, the Netherlands) and *Bintoro Gunadi* (OSU, USA).

— Tree plantation and community structure and function of beneficial macro- and micro-arthropods of soil-litter system. *M. Vikram Reddy* (KU, India).

— Tree performance, soil plant-litter characteristics and faunal activity in some tropical plantations. *F.K. Akinnifesi* (UEMA, Brazil), *G. Tian* (IITA, Nigeria) and *B.T. Kang* (USA).

— Earthworms in tropical tree-plantations: effects of management and relations with soil carbon and nutrient use efficiency. *Xiaoming Zou* and *Grizelle Gonzalez* (UPR, USA).

CHAPTER 6

LITTER LAYER DYNAMICS AND SOIL MICROBIAL BIOMASS IN REFORESTED AREAS AFTER DEGRADATION BY BAUXITE MINING IN EASTERN AMAZON

Enir S. Costa[a], Flávio J. Luizão[b], Regina C. Luizão[b], Andrew McMillan[a]

[a]UCI, Department of Earth System Science, Rowland Hall 220, 92697-3100, Irvine CA, USA

[b]INPA, Departamento de Ecologia, Caixa postal 478 CEP 69011-970 Manaus, AM, Brazil.

ABSTRACT

The replacement of primary forest with managed ecosystems of smaller biomass involves major changes to the litter layer dynamics, nutrient cycling and soil microbial biomass. In the Brazilian Amazon, there has been an increased effort to rehabilitate previously degraded areas by reforestation programmes. Under these schemes, some soil parameters, such as soil microbial biomass, reach levels similar to that of the original vegetation within less than two years. However, others, such as soil organic carbon, can take much longer, not reaching former levels even after 10 years of reforestation. In this chapter, we review the dynamics of the litter layer. In particular, we discuss the role of organic matter decomposition which regulates litter layer mass and the remineralization of nutrients. These processes have a distinct seasonality and are therefore highly dependent on climatic variables. We then discuss how the influence of these climatic variables can be altered by the removal of the forest canopy. The subsequent replacement of the forest by an ecosystem of initially lower biomass will to some extent restore the former microclimatological conditions. However, different times are required for the various soil and litter parameters to recover to the same levels of the original forest. We discuss how different land use changes affect these recovery times. Finally, we discuss the results of a study at a site degraded by bauxite mining. Litter mass, soil microbial biomass, and other soil and litter variables were measured under various reforestation treatments. Bauxite mining involves the removal of the entire topsoil layer as well as the vegetation. We describe the effects of topsoil replacement and compare natural regeneration with plantation of native tree species as strategies for forest rehabilitation.

Key Words: soil microbial biomass; soil organic carbon; reforestation; bauxite mining.

In the last 30 years, the Brazilian Amazon has experienced a considerable increase in the prospecting of minerals, especially bauxite and in the establishment of industrial mining projects. Brazil has about 11.5% of the total bauxite reserve of the world, and is the third biggest producer, after Guinea and Australia. In Pará State alone there are 2.5 billion tons of bauxite yet to be exploited (Fernandes and Portela, 1990). Bauxite mining involves the removal of the entire vegetation and topsoil layer. Excavation of this magnitude causes soil impoverishment, erosion and soil toxicity, thus affecting the flora, fauna, water quality, and people of the region in many ways (Griffith, 1980). To mitigate such effects, programmes involving reforestation with native tree species have increasingly been implemented.

The main purpose of this paper is to review the effects of reforestation in degraded lands. We review the undisturbed tropical forest ecosystem with respect to litter production, accumulation/ deposition and decomposition, soil microbial biomass and organic carbon in Amazon as well as their implications on soil nutrient dynamics. Next we discuss disturbance associated with land use changes (such as bauxite mining) with respect to these parameters. Finally, we explore the results from a case study where the litter layer dynamics, soil microbial biomass and soil organic matter content were measured at a site degraded by bauxite mining. The response of these parameters was compared across various reforestation treatments (topsoil replacement versus no topsoil replacement; natural regeneration versus reforestation).

UNDISTURBED TROPICAL FORESTS: LITTER PRODUCTION, MASS AND DECOMPOSITION, NUTRIENT CYCLING AND SOIL MICROBIAL BIOMASS

Litter Production and Litter Mass

Litter may be defined as the set of dead organic materials derived predominantly from the aboveground vegetation (leaves, twigs and small branches, reproductive material, and coarse wood) (Madge, 1965).

The fine (or small) litter includes the wood fraction up to 2 cm in diameter (Proctor, 1983). However, the definition can also be expanded to include root detritus, which represents an input of organic matter in quantities similar to, or greater than, that of materials derived from aboveground litter-fall (Vitousek 1984; Luizão *et al.*, 1992a).

Measurements of total annual litter production in Amazon rain forests generally fall within the range of those measured in other tropical forests, between 5.8 and 12.4 t ha^{-1} yr^{-1} (Table 1). However, a distinct seasonal pattern of litterfall exists in Amazon rain forests whereby the largest portion falls during the drier months (Klinge and Rodrigues, 1968; Franken, 1979; Dantas and Phillipson, 1989; Luizão, 1989; and Scott *et al.*, 1992). The timing of the peak litter fall has been related to rainfall events after dry periods (Luizão, 1989).

The mass of the litter layer in the Amazon rain forest also exhibits strong seasonal patterns. In the dry season, the combined effects of increased litterfall and slower decomposition rates result in a transient accumulation of litter. In the rainy season, these effects are generally reversed, leading to reduced litter masses (Luizão and Schubart, 1987). The litter mass represents the net balance between detritus production (aboveground litterfall plus addition of dead roots) and removal by biological decomposition. Evidence from a tropical rain forest in Panama suggests that this effect is more closely related to the rate of decomposition than to the rate of litterfall (Wieder and Wright, 1995).

Litter Decomposition and Litter Quality

During the wet season, the rate of litter decomposition is enhanced due to the increase in microbial activity, the penetration of fine roots, and the increase in the biological activity of macro-arthropods, especially termites (Luizão and Schubart, 1987). Microorganisms, especially fungi and bacteria, play an important role in litter decomposition throughout the year. However, their role is of greatest importance during the dry season, when macro-arthropod activity is less intense. During the wet season, the initial decay of litter is greatly enhanced by the increased activity of macro-arthropods. In particular, a large portion of litter removal and organic matter breakdown has been attributed to termite (Isoptera) activity. For instance, Luizão and Schubart (1987) found that in the wet season, termites were responsible for over 40 per cent of the removal of decomposing leaves in a terra-firme forest in central Amazon. This initial processing of organic matter, along with the more favorable moisture conditions, is probably responsible for the higher population and activity of microorganisms hat are present during the rainy season.

Table 1 Fine litter production (t ha^{-1} yr^{-1}) in various tropical rain forests

Localization	*Total*	*Leaves*	*Wood*	*Reproductive*	*Fine*	*Authors*
Australia	9.7	4.9	–	–	–	Spain (1984)
	7.3	4.1	2.3	1.3	–	Lowman (1988)
Brazil-Belém	8.0	–	–	–	–	Dantas and Phillipson (1989)
	9.9	8.0	1.30	0.60	–	Klinge (1977)
	7.3	6.1	0.88	0.31	–	Silva and Lobo (1982)
Brazil-Manaus	7.9	6.4	1.03	0.47	–	Franken *et al.*, (1979)
	7.3	5.6	1.10	0.40	–	Klinge and Rodrigues (1968)
	8.3	5.4	1.56	0.42	0.79	Luizão (1989)
	7.8	5.4	1.62	0.55	0.14	Luizão (1995)
Brazil-Maracá	9.3	6.3	1.34	1.21	0.42	Scott *et al.*, (1992)
French Guyana	8.7	5.8	–	–	–	Puig (1979)
Congo-Yapo	9.6	7.1	–	–	–	Bernhard (1970)
Panama	11.4	–	–	–	–	Golley *et al.*, (1975)
Sarawak	8.8	5.4	–	–	–	Proctor *et al.*, (1983)
Venezuela	10.3	7.6	2.3	0.40	–	Cuevas and Medina (1986)
	5.8	–	–	–	–	Jordan and Murphy (1982)
Zaire	12.4	–	–	–	–	Laudelout and Meyer (1954)

Smith *et al.*, (1998a) have discussed the relationship between litter quality and decomposition rates. Various indices such as the litter C:N ratio, lignin: lignin+holocellulose ratio and C:lignin ratio have been related to rates of decomposition. Generalizations for the Amazon rain forest are difficult to arrive at (Smith *et al.*, 1998b), probably due to the wide variety of chemical compositions of plants from different forest ecosystems. A potentially confounding factor in studies that attempt to relate decomposition to litter quality is that many of these studies have employed litterbags with mesh sizes that do not permit the entry of the macrofauna. Hence, the true rates of decomposition may be underestimated to some extent (Smith, 1998c).

Litter Nutrient Dynamics

Contrary to earlier popular ideas that the majority of nutrients in the Amazonian rain forests were stored in the aboveground biomass, more recent studies suggest that the litter layer also represent an important reservoir of plant nutrients (Brinkmann, 1989; Anderson and Spencer, 1991). On old, heavily-weathered soils, regular inputs of new nutrients from soil parent material are small or negligible (Schubart *et al.*, 1984; Brinkmann, 1989). Under such conditions, the atmospheric deposition (both dry and wet) becomes an important source of nutrients, especially for base cations, compensating for the small losses from the ecosystem caused by leaching. However, most of the nutrient demand of the plants is met through the biologically-mediated remineralization of nutrients from the organic matter. Generally, the nutrients are rapidly and efficiently cycled in the rain forest ecosystem and most nutrients have a very short residence time in the soil-litter system, being quickly remineralized and made available to the plants. The concentration of mineral nutrients in the primary forest litter in Amazon is generally high for nitrogen, but lower for phosphorus and base cations, when compared to other tropical rain forests (Proctor, 1984). Thus, plants of this region are limited by nutrient availability (Vitousek, 1984). Often, nutrients are strongly translocated and retained in the plants prior to foliar abscission. Some exceptions to this general pattern exist: for instance, in Maracá, northern Amazon, Scott *et al.*, (1992), although registering high translocation for some nutrients, found that nutrient concentrations in the litter, especially for phosphorus and calcium, were much higher than the average for the Amazon. However, in general, the rapid cycling of nutrients in tropical rain forests is achieved through high decomposition rates, which are made possible by high temperatures and high annual rainfall.

Soil Microbial Biomass

In central Amazon, soil microbial biomass (biomass-C, measured by the fumigation-incubation and fumigation-extraction techniques) has been found to be strongly related to soil moisture, showing marked seasonal differences, with the highest values found in the wet season. Litter quality interacts strongly with soil moisture to influence the size of the microbial biomass, and is partially determined by the extent of previous microbial activity. Luizão (1989b) following a year-long seasonal pattern of soil microbial biomass in central Amazonia, found a negative correlation between microbial biomass and litter production after adjusting for a lag period of two months. Thus, when there was higher litter production and deposition, during the dry season, a decrease in the microbial biomass was observed. The cause of this correlation was thought to be a positive feedback loop whereby a lowered decomposition rate resulted in reduced input of new organic compounds to the soil and, consequently further reductions in soil microbial biomass. After a few months of accumulation on the soil surface during the dry season, the litter layer is quickly decomposed in the wet season, when higher soil moisture triggers soil biota activity including those of the decomposers. The release of carbon and nutrients from litter, together with the release of root exudates, favor microbial biomass growth (Alexander 1977; Luizão *et al.*, 1992).

To summarize, the major controls on litter decomposition (and thus nutrient dynamics) can be viewed as interactions between litter input and litter quality, the soil microclimate (moisture and temperature), and the resulting activity of macrofauna, microfauna and microflora. We now address how forest disturbances and subsequent land management practices affect these interactions with reference to studies conducted in the Amazon and other tropical forests. It should be noted that there is a marked paucity of published data pertaining to litter dynamics in reforested regions of the Amazon.

REPLACEMENT OF PRIMARY FOREST BY LOW-BIOMASS ECOSYSTEMS

In the case of bauxite mining, not only is the forest vegetation removed but also the upper layers of the soil. This has major implications on the rate of litter production, the seasonal pattern of the litter layer mass, and the dynamics of soil nutrients and carbon. The extent to which these parameters return to their previous states under reforestation will depend largely on the choice of land use following disturbance. Before discussing the specific case of degradation by bauxite mining, we consider the effect of canopy removal, and replacement of the forest with a lower biomass ecosystem, such as a pasture or agroforestry system.

Disturbances Associated with the Removal of the Forest Canopy

Conversion of primary forest to other land use generally involves the removal of the forest canopy. Any large-scale reduction in the canopy of a forest leads to a decrease in the rate of evapotranspiration, photosynthesis, latent heat, biological diversity, and changes in the hydrological cycle (Vieira *et al.*, 1993; Alho, 1991; May *et al.*, 1994). It also affects the soil microclimate by increasing temperature and by decreasing soil moisture. These should have opposing effects on the rate of decomposition. Further, the increased exposure of the soil and litter to rainfall events can accelerate leaching, leading to removal of nutrients at rates which may be more rapid than nutrient remineralization via microbial decomposition. In the long term, canopy removal leads to an overall decrease in the organic matter of the soil (Woomer *et al.*, 1994).

Changes in Litter Layer Dynamics Following Canopy Removal

In the Brazilian Amazon there are only a few studies of litter production and accumulation in areas that had their original vegetation changed from a forest to pastures or other managed ecosystems (Table 2). Although large areas of primary forest have been converted to pastures in Amazon region there is little available information regarding the litter dynamics of pastures, annual crops or even forest regrowth.

The most obvious effect of canopy loss is the immediate reduction of the litter production rate, with a sharp decrease in the annual input of organic matter and nutrients to the soil system. During subsequent reforestation, or natural regeneration, the rate of litter production will slowly increase but the changes in species composition and climate-induced plant phenologies will result in greatly altered seasonal patterns of litter quantity and quality (Ewel, 1976). Smith *et al.*, (1998c) found that despite the similarity in leaf litterfall and forest floor litter mass between a primary forest and a 35–37 year old plantation forest, the mean residence times for both fine litter and total leaf litter were shorter in the undisturbed forest. In a four year old agroforestry system in Manacapuru, 60 km northwest of Manaus, the stock of fine litter on the soil surface was only half that of native forests but twice as large as that of neighboring cassava plantations (Luizão *et al.*, 1999a).

Litter Quality and Decomposition Rates

In primary forest and also in reforested areas, the tree species composition not only affects the magnitude of the litter input but also the attractiveness of the litter to decomposing organisms. Exotic leguminous species, such as *Leucena leucocephala* (Lam) de Wit, are often utilized in plantation forests for their nitrogen-fixing capability (Didham, 1998), and are easily decomposed on the soil surface. However, in some studies (e.g. Smith *et al.*, 1998c), the relationship between litter chemistry and decomposition rates can be masked by the exclusion of macrofauna from the litterbags used to estimate decay rates.

CHANGES TO QUANTITIES OF SOIL ORGANIC MATTER

In tropical soils, organic matter is the main source of nutrients and, due to its high cation exchange capacity, it is also important for the retention

Table 2 Litter production (or litter-layer on soil surface-LL) and decomposition time in managed ecosystems in the Brazilian Amazon.

Localization	*Ecosystem*	*Litter production* (t ha^{-1})	*Litter decomposition* (half life)	*Authors*
Manaus	5-years abandoned pasture (LL)	5.0	280 days	EEC (1994)
	4-year-old pasture (LL)	6.8	55 days	
	Forest (LL)	6.4	135 days	
Manaus	10-year-old *Bertholletia excelsa* plantations	1.3	518 days	Kato (1995)
	5-year-old *Bertholletia excelsa* plantations	0.67	518 days	
	Forest (control)	–	83 days	
	Agroforestry 1	2.0	25–116 days	Gallardo-Ordinola (1999)
	Agroforestry 2	2.3	25–116 days	
Capitão Poço	8-year-old Cocoa in understory	5.8	–	Muller (1987)
Capitão Poço	3-year-old fallow field	5.0	–	Dantas and Phillipson (1989)

of nutrients in highly weathered soils (Uehara and Guilman, 1981). In the Manacapuru study (Luizão *et al.*, 1999b), the amount of easily decomposable material was higher in the agroforestry system than in the cassava plantation. In situations where a new soil cover is rapidly developed, such as a well-managed pasture or leguminous plantation, the soil carbon contents may have recovered after a few years (Cerri *et al.*, 1991). However, this new organic matter may have a smaller quantity of readily mineralizable nutrients than the soil under the original land cover (EEC, 1994; Luizão *et al.*, 1999b).

Effect of Disturbance on Soil Organisms

Little information is available on how forest disturbance affects the activity of soil macrofauna, but it is evident that substantial changes occur both in the species composition and relative abundance of the different functional groups of the new communities following disturbance (Dantas, 1979; Oliveira and Franklin, 1993). The new vegetation may favor some groups of soil fauna, but the diversity of soil organisms is generally lower in disturbed areas. In a comparison of agroforestry, agroforestry/pastoral systems and secondary forest in

Manaus, Brazil, the highest macrofauna densities and biomasses were associated with the most diverse agroforestry systems (Tapia-Coral *et al.*, 1999). For all treatments, the predominant order was Isopoda, followed by Diplopoda, but their abundance decreased sharply during the dry season.

Soil microbial biomass is also affected by disturbances such as slash and burn agriculture and other practices involving extensive canopy removal, but recovers quickly once a new soil cover is established (Luizão *et al.* 1992b). However, that seems not to be the case of short-term crops, such as cassava (*Manihot esculenta* Crantz) plantations, which is one of the most common agricultural crops in central Amazonia. Referring again to the study of small farms in Manacapuru (Luizão *et al.*, 1999a) the soils under cassava plantations showed lower values of microbial biomass than both agroforestry systems and native forests. Changes were related to the quality of litter produced in each system. For instance, the litter layer of the agroforestry systems (over 60% as fine woody material) produced more easily decomposable litter than the cassava plantations. Consequently, carbon and nutrients were of greater availability to the soil microbial biomass.

Agroforestry systems can, in fact, have higher soil microbial biomass than secondary forest. Levels of soil microbial biomass in four different, 5 year old, agroforestry systems were studied by Tapia-Coral *et al.*, (1997). All of the agroforestry systems had higher amounts of microbial biomass than the 10 year old secondary growth system used as a control. The highest values of microbial biomass were found in a multistrata system, consisting of perennial woody species, grown for wood and fruit, including mahogany and brazil nut, and annual crops, including papaya, cassava and rice.

In rain forests, part of the nutrient demand from plants is met by the litter layer by means of the aboveground fine root mass and associated mycorrhizae. This could be especially important for N and P (Herrera *et al.*, 1979). Disturbance is known to limit the rate of mycorrhizae-mediated nutrient transfer, both through the removal of the plants on which the fungi depend for their carbon source and through the mechanical damage to the hyphal mycelium (Janos, 1980; Allen, 1991). Allen *et al.*, (1998) found that the rate of arbuscular mychorrhizal infection was related to plant phenology and that the highest infection rates occurred during the rainy season. This suggests that soil moisture changes associated with disturbance may delay nutrient uptake by plants through reduced mycorrhizal activity.

Tian (1988) states "the loss of soil biota in the degraded soil is caused mainly by the difference in microclimate, the depletion of soil organic matter, and deterioration of soil physical properties". Yet, the soil microbial biomass shows a surprising capacity for recovery, and, depending on the type and diversity of the new ecosystem, a return to previous levels can be expected within a few years.

A CASE STUDY: LITTER LAYER, SOIL MICROBIAL BIOMASS, AND SOIL NUTRIENTS IN REFORESTED AREAS DEGRADED BY BAUXITE MINING IN EASTERN AMAZON

A project entitled "Systems of Production and Studies of Degraded Land, Reconstitution and Rehabilitation of Forest Ecosystems" was initiated in 1987 with support of United Nations Education, Scientific and Cultural Organization (UNESCO), Economic European Community (EEC), the Deutsche Gesellschaft für Technische Zusammenarbeit (GTZ), and Mineração Rio do Norte S.A. (MRN). The main purpose of this project was to study the process of degradation and rehabilitation in ecosystems disturbed by anthropogenic action (plantation agriculture, pasture, and mining) in the Amazon region. To reach this objective, the project monitored different techniques used for the rehabilitation of degraded areas. Mining companies were encouraged to participate in the project by adopting a policy of rehabilitation following their bauxite exploitation activity in the Eastern Amazon region.

The following section addresses how different approaches to reforestation facilitate forest recovery. We refer to the study of Costa *et al.*, (1998) where changes in microbial biomass, soil nutrients, carbon, and litter were monitored at a bauxite mine where several rehabilitation treatments had been implemented under the reforestation project described above. The study took place in Saracá Mine, in Eastern Pará State, Brazil. The treatment plots included: a 1-year-old reforested plot where no topsoil replacement occurred; a 1-year-old reforested plot with topsoil replacement; an 11-year-old reforested plot that was allowed to regenerate naturally; an 11-year-old reforested plot; and a primary forest that was used as a control. Both the 11-year-old treatments were on sites where the topsoil had been replaced. Samples were collected in the wet season (April, 1995), the dry season (August, 1995) and in the transitional period (November, 1995).

We first consider the changes that take place in the litter mass, litter nutrients, soil microbial biomass and soil organic carbon following the first year of reforestation. We then describe the effects of topsoil replacement in the initial stages of soil recovery. The effects of reforesting degraded sites with a diverse selection of native tree species will then be compared (with respect to these same parameters) to those sites that are allowed to regenerate naturally. The relative merits of litter layer biomass and nutrient, soil microbial biomass and soil organic carbon and nutrients will be discussed in terms of their ability to serve as indicators of soil rehabilitation in reforested mining sites in Eastern Amazon.

Effects of Bauxite Mining and Reforestation on Litter Layer and Soil Microbial Biomass

Bauxite mining in the Amazon region is conducted only after the complete removal of the primary forest and subsequent excavation to reach the bauxite layer. This process involves drastic changes to the soil system. Strips of primary forest were harvested and the soil extracted and transported to an adjacent strip awaiting topsoil replacement prior to reforestation.

Visually, there appeared to be greater vegetation cover, biomass and litter mass in the plots with topsoil replacement. Indeed the litter mass was greater in the topsoil replaced site by a factor of 1.8, 8.8 and 41 for the dry, transitional and wet season, respectively. However, the absolute magnitude of these differences was small compared to the much larger masses observed in the older sites and in the primary forest (Table 3).

Because the 11-year-old, naturally regenerated plots had fewer species and a lower basal area, we expected the litter mass to be greater than in the 11-year-old reforested plots. However, during both wet and dry season no significant differences between these treatments were found. One possible cause of this similarity is the predominance of the pioneer species *Cecropia* spp (Cecropiaceae) in the naturally regenerated plot. The leaves of *Cecropia* spp are large, coarse and more difficult to decompose than most of the species planted in the 11-year-old reforested plots. They therefore remained on the soil surface for longer periods during the wet and dry seasons. The litter mass of the primary forest exceeds that of the 11-year-old treatments forests only during the dry season. This is possibly due to the very dry conditions resulting in a reduced decomposition rate in the months prior to sampling. Other seasonal differences were either not present or difficult to detect given the high variability of the measurements.

Soil microorganisms are extremely sensitive to environmental change. Yet, the soil microbial biomass as a whole can exhibit a considerable capacity for recovery following disturbance. In plots degraded by bauxite mining, after one year of reforestation, microbial biomass was always smaller than in the primary forest (Table 3), but the differences were significant only in the transitional period. There were no significant differences among treatments with and without topsoil replacement following mining. This may be attributed to the high variability of the soil sample as well as to the plasticity of the

Table 3 Litter layer (g m^{-2}) and soil microbial biomass by fumigation-extraction method (μg C g^{-1} soil) in: 1-year-old reforested plots without (RWTH1) and with (RWTS1) topsoil replacement; 11-year-old natural regeneration (NR11); 11-year-old reforested plots (REF11); primary forest (PF) in Saracá Mine, Porto Trombetas-PA, in the wet season, dry season and transitional period. Values are represented by means ±SE (n = 3).

Plots	*Litter layer* (g m^{-2})			*Microbial biomass* (μgC g^{-1} soil)		
	Wet	*Dry*	*Transitional*	*Wet*	*Dry*	*Transitional*
RWTH1	3.6 (±0.90)c	12.9 (±4.6)b	9.2 (±1.1)c	327 (±191)a	420 (±110)a	781 (±55.4)c
RWTS1	150.1 (±27.8)bc	36.0 (±6.0)b	91.1 (±14.2)bc	536 (±130)a	338 (±49.7)a	486 (±121)abc
NR11	352.5 (±32.8)a	262.3 (±44.8)a	564.5 (±16.7)ab	665 (±110)a	443 (±74.3)a	1056 (±65.9)a
REF11	354.5 (±46.9)a	205.0 (±88.4)a	659.0 (±278.6)a	650 (±18.1)a	478 (±87.8)a	910 (±51.3)ab
PF	268.9 (±57.7)ab	319.7 (±19.4)c	418.8 (±4.4)abc	896 (±148)a	555 (±27.5)a	898 (±62.8)ab

Means followed by similar letters are not significantly different ($p < 0.05$) according to the Tukey test.

microbial population, which allows rapid adaptation to changes in environmental conditions. Though the magnitude of the microbial populations was not significantly different among treatments, the function and composition of the microbial population in the 1-year-old reforestation sites with and without top soil replacement are probably very different as a consequence of the differences in the amounts and quality of soil organic matter in both sites. The lack of significant differences among the 11-year-old treatments and between these and the primary forest plots seemed to indicate that after one decade microbial biomass approached the original levels of the primary forest, independent of treatment. This study suggested that, in terms of soil microbial biomass, the soil is fully recovered within two years (Costa *et al.*, 1998).

The Effect of Reforestation on Carbon and Nutrient Concentrations in Litter and Soil

Soil organic carbon was significantly lower in the treatment without topsoil replacement than in the treatment with topsoil replacement and higher in the forest control than in all treatments (Table 4). This indicated that more than a decade is needed for the soil carbon content to reach its original levels under either the natural regeneration or reforestation treatments.

Table 4 Organic carbon from upper layer (0-10 cm) in: 1-year-old reforested plots without (RWTH1) and with (RWTS1) topsoil replacement; 11-year-old natural regeneration (NR11); 11-year-old reforested plots (REF11); and primary forest (PF) in Saracá Mine, Porto Trombetas-PA, in the wet season, dry season and transitional period. Values are represented by means ±SE (n = 3).

Plots	*Organic carbon %*		
	Wet	*Dry*	*Transitional*
RWTH1	0.05(±0.05)d	0.18(0.06)c	0.16(±0.00)c
RWTS1	2.29(±0.23)c	2.95(±0.40)b	2.35(±0.16)b
NR11	2.98(±0.10)abc	3.74(±0.17)b	3.48(±0.13)b
REF11	3.51(±0.14)ab	3.95(±0.18)b	3.23(±0.21)b
PF	3.92(±0.39)a	6.51(±0.16)a	5.07(±0.53)a

Means followed by similar letters are not significantly different ($p<0.05$) according to the Tukey test.

In the dry season, top soil replacement had no effect on litter mass nitrogen concentration (Table 5). In the wet season, lack of material in the litter layer precluded chemical analysis in the treatment without

Table 5. Litter nutrient concentrations (N, P, K, Ca and Mg), in 1-year-old reforested plots without (RWTH1) and with (RWTS1) topsoil replacement; 11-year-old natural regeneration (NR11); 11-year-old reforested plots (REF11); and primary forest (PF) in Saracá Mine, Porto Trombetas-PA in the wet season and dry season. Values are represented by means ±SE (n = 3).

Plots	Litter Nutrients %				
	Wet season				
	N	*P*	*K*	*Ca*	*Mg*
RWTHS1	*	*	*	*	*
RWTS1	1.37 (±0.11)bc	0.01 (±0.00)a	0.05 (±0.01)bc	0.52 (±0.07)b	0.08 (±0.02)a
NR11	1.40 (±0.15)bc	0.02 (±0.00)a	0.03 (±0.80)c	1.06 (±0.20)a	0.11 (±0.02)a
REF11	1.15 (±0.05)c	0.01 (±0.00)a	0.08 (±0.00)ab	1.05 (±0.01)a	0.11 (±0.00)a
PF	1.92 (±0.00)a	0.02 (±0.00)a	0.09 (±0.00)a	0.47 (±0.07)b	0.14 (±0.00)a
Plots	*Dry season*				
	N	*P*	*K*	*Ca*	*Mg*
RWTHS1	2.14 (±0.72)a	0.02 (±0.00)a	0.17 (±0.05)a	0.89 (±0.06)ab	0.14 (±0.02)a
RWTS1	1.64 (±0.18)a	0.02 (±0.00)a	0.17 (±0.06)a	0.75 (±0.06)bc	0.11 (±0.04)a
NR11	1.24 (±0.04)a	0.02 (±0.00)a	0.08 (±0.01)a	0.77 (±0.03)bc	0.12 (±0.00)a
REF11	1.17 (±0.05)a	0.01 (±0.00)a	0.08 (±0.00)a	1.08 (±0.10)a	0.14 (±0.01)a
PF	1.97 (±0.08)a	0.02 (±0.00)a	0.15 (±0.02)a	0.41 (±0.06)d	0.14 (±0.01)a

Means followed by similar letters (within each season) are not significantly different ($p<0.05$) according to the Tukey test.
*insufficient litter for chemical analyses.

Table 6 Table 6. Soil nutrients (0–10 cm) in: 1-year-old reforested plots without (RWTH1) and with (RWTS1) topsoil replacement; 11-year-old natural regeneration (NR11); 11-year-old reforested plots (REF11); and primary forest (PF) in Saracá Mining, Porto Trombetas-PA in wet season and dry season. Values are represented by means ±SE (n = 3).

Plots	*Soil Nutrient Concentration*									
	Wet season					*Dry season*				
	N	*P*	*K*	*Ca*	*Mg*	*N*	*P*	*K*	*Ca*	*Mg*
	(%)	(*mg kg^{-1}*)	(*meq 100g^{-1} soil*)			(%)	(*mg kg^{-1}*)	(*meq 100g^{-1} soil*)		
RWTH1	0.02b	88b	0.01	0.03c	0.01d	0.01b	84b	0.01d	0.07c	0.03d
	(±0.01)	(±0.10)	(±0.01)	(±0.01)	(±0.01)	(±0.01)	(±4.50)	(±0.01)	(±0.02)	(±0.17)
RWTS1	0.08b	138a	0.09abc	0.46c	0.16cd	0.12ab	139a	0.06cd	0.39bc	0.12cd
	(±0.02)	(±7.94)	(±0.02)	(±0.07)	(±0.03)	(±0.02)	(±5.50)	(±0.01)	(±0.12)	(±0.02)
NR11	0.10b	150a	0.11ab	1.53ab	0.38ab	0.17ab	160a	0.10abc	1.79ab	0.39ab
	(±0.03)	(±3.26)	(±0.02)	(±0.35)	(±0.08)	(±0.01)	(±5.00)	(±0.01)	(±0.58)	(±0.06)
REF11	0.23a	168a	0.16a	2.34a	0.49a	0.12ab	166a	0.15a	2.47a	0.59a
	(±0.02)	(±6.74)	(±0.01)	(±0.24)	(±0.04)	(±0.04)	(±6.74)	(±0.03)	(±0.45)	(±0.10)
PF	0.21a	146a	0.06bc	0.07c	0.11cd	0.27a	153a	0.13ab	0.14c	0.18bcd
	(±0.01)	(±8.67)	(±0.03)	(±0.23)	(±0.00)	(±0.02)	(±6.75)	(±0.01)	(±0.02)	(±0.05)

Means followed by similar letters are not significantly different ($p < 0.05$) according to the Tukey test.

topsoil. Additionally, no difference was found between the nitrogen content of the two 11-year-old treatments (reforested versus naturally regenerated). However, nitrogen concentrations were always higher in primary forest litter compared to other treatments during the wet season. This may reflect the relatively higher demand for soil nitrogen in an immature forest compared to a mature tropical forest on Oxisol, where nitrogen is not considered limiting (Vitousek, 1984). Phosphorus concentrations were always very low in the litter-layer in all treatments and no significant differences were observed between any treatments in the dry and rainy season.

The soil of the 1-year-old plantations without topsoil replacement always contained lower concentrations of nutrients, particularly for phosphorus and potassium (Table 6). In the two 11-year-old treatments, soil exchangeable calcium (Ca) and magnesium (Mg) were more than double that of the 1-year-old treatments in both seasons. This may reflect the differing leaf litter chemistry of species that are prevalent in more recently disturbed areas. Previous studies of reforestation in the central Amazon have shown that pioneer species such as *Cecropia* spp. and some agroforestry planted species (Gallardo-Ordinola, 1999; Lucas *et al.*, 1993; Tapia-Coral, 1997) produce leaf litter rich in base cations, in particular, Ca. These species have a high capacity to extract Ca and Mg from soil compounds, incorporate these nutrients as biomass, and re-deliver them to the soil surface in a more available form as plant litter.

To conclude, the results of this study indicated that reforestation with native tree species (and to a lesser extent, the naturally regenerated forest) allowed the recovery of the soil litter cover and microclimate within a few years. This in turn led to the recovery of the soil microbial activity and the decomposition process. Litter layer dynamics are highly variable and unlikely to serve as good indicator of forest recovery. Topsoil nutrient availability is improved in one to two decades due to the mobilization of nutrients from the soil, particularly Ca and Mg, promoted by planted or pioneer species. Topsoil replacement prior to the plantation of native tree species can speed up the soil rehabilitation process. The organic carbon content of the soil is the slowest to recover and may therefore act as the ultimate indicator of progress in the rehabilitation process.

REFERENCES

Alexander, M., 1977. *Introduction to soil microbiology*, 2nd Ed., John Wiley and Sons, New York, 472 pp.

Alho, C.J.R., 1991. Maneje com cuidado: frágil, *Ciência Hoje*, Volume Especial Amazônia, 100–107 pp.

Allen, E.B., Rincon E., Allen, M.F., Perezjimenez., A., and Huante, P., 1998. Disturbance and seasonal dynamics of mycorrhizae in a tropical deciduous forest in Mexico, *Biotropica*, **30**(2): 261–274.

Allen, M.F., 1991. *The ecology of Mycorrhizae*, Cambridge University Press, Cambridge, UK. 184 pp.

Anderson, J.M., and Spencer, T., 1991. *Carbon, nutrient and water balances of tropical rainforest ecosystems subject to disturbance: management, implications and research proposals*, Paris: UNESCO, MAB Digest Nº7, 97 pp.

Bernhard, R.F., 1970. Étude de la litière et de as contribution au cycle des éléments minéraux en forêt ombrophille de Côte-d'Ivoire, *Oecologia Plantarum*, **5:** 247–266.

Brinkmann, W.L.F., 1989. Hydrology, In: *Tropical Rain Forest Ecosystems*, (H. Lieth, and M.J.A. Werger, eds), Elsevier, Amsterdam, **5:** 89–98.

Cerri, C.C., Volkoff, B., and Andreux, F., 1991. Nature and behaviour of organic matter in soils under natural forest, and after deforestation, burning and cultivation, near Manaus, *Forest Ecology and Management*, **38**: 247–257.

Costa, E.S., Luizão, R.C., and Luizão, F.J., 1998. Soil microbial biomass and organic carbon in reforested sites degraded by bauxite mining in the Amazon, *Advances in GeoEcology*, **31:** 443–450.

Cuevas, E., and Medina, E., 1986. Nutrients dynamics within Amazonian forest, I. Nutrient flux in fine litter fall and efficiency of nutrient utilization, *Ecologia*, **68**: 466–472.

Dantas, M., 1979. Pastagens da Amazônia Central: Ecologia e fauna do solo, *Acta Amazonica*, **9** (supl. 2): 1–54.

Dantas, M., and Phillipson, J., 1989. Litterfall and litter nutrient content in primary and secondary Amazonian 'terra firme' rain forest, *Journal of Tropical Ecology*, **5:** 27–36.

Didham, R.K., 1998. Altered leaf-litter decomposition rates in tropical forest fragments, *Oecologia*, **116**(3): 397–406.

EEC, 1994. *Dynamics of soil organic matter in the Amazon ecosystem and after deforestation: basis for efficient agricultural management*, Contract CEE Nº TS 22-0301-F(SMA). ORSTON/ CNRS-CPB/CSIC/CENA/INPA. Final Report, Brussels, 130 pp.

Ewel, J., 1976. Litterfall and leaf decomposition in a tropical succession in eastern Guatemala, *Journal of Ecology*, **64:** 293–308.

Fernandes, F.R.C., and Portela, I.C.M.H.M., 1990. *Recursos Minerais da Amazônia (alguns dados sobre a situação e perspectiva)*, Projeto Especial-DTA (Desenvolvimento da Tecnologia Ambiental), CNPq-CETEM, 23 pp.

Franken, M., 1979. Major nutrients and energy contents of the litterfall of a riverine forest in central Amazonia, *Tropical Ecology*, **20**(2): 211–224.

Gallardo-Ordinolla, J.L.E., 1999. *Produção e qualidade da liteira em sistemas agroflorestais e seus efeitos sobre as propriedades químicas do solo.* Dissertação de Mestrado. INPA/FUA. Manaus, AM. 97 pp.

Golley, F.B., McGinnis., J.T., Clements, R.G., Child., G.I., and Duever, M.J., 1975. *Mineral Cycling in a Tropical Moist Forest Ecosystem*, University of Georgia Press, Athens.

Griffith, J.J., 1980. Recuperação conservacionista de superfícies mineradas. Uma Revisão de Literatura. Sociedade de Investigações Florestais, Viçosa. Boletim Técnico 2, 51 pp.

Herrrera, R., 1979. *Nutrient Distribution and Cycling in a Amazonian Caatinga Forest on Spodosol in Southern Venezuela*, Ph.D. Thesis, University of Reading, U.K.

Janos, D.P., 1980. Mycorrhizae influence tropical succession, *Biotropica*, **12:** 56–64.

Jordan, C.F., and Murphy, P.G., 1982. Nutrient Dynamics of a Tropical Forest Ecosystem and Changes in the Nutrient Cycle due to Cutting and Burning, Annual Report to US National Science Fundation, Institute of Ecology, University of Georgia, Atlanta, 122–166 pp.

Kato, A.K., 1995. *Dinâmica da entrada de nutrientes via liteira em plantios de castanheira-do-brasil (Bertholletia excelsa H.B.K) em ecossistemas de pastagem degradadas e de floresta primária*, Tese de Doutorado. INPA-FUA Manaus-AM, 180 pp.

Klinge, H., 1977a. Fine litter production and nutrient return to the soil in three natural forest stands in eastern Amazon, *Geo. Ecol. Trop.*, **1:** 159–267.

Klinge, H., 1977b. Preliminary data on nutrients release from decomposition leaf litter in a neotropical rainforest, *Amazoniana* VI, 193–202.

Klinge, H., and Rodrigues, W.A., 1968. Litter production in an area of Amazonian terrra firme forest, Parts I and II, *Amazoniana*, **1:** 287–303.

Laudelout, O.H., and Meyer, J., 1954. Les cycles d'elements mineraux et de matiere organique Congolaise, *International Congress of Soil Science Transactions*, **5:** 267–272.

Lowman, M.D., 1988. Litterfall and leaf decay in three Australian rainforest formations, *Journal of Ecology*, **76:** 451–465.

Lucas, Y., Luizão, F.J., Chauvel, A., Rouiller, J., and Nahon, D., 1993. The relation between biological activity of the rain forest and mineral composition of soils, *Science*, **260:** 521–523.

Luizão, F.J., 1995. *Ecological studies in contrasting forest types in Central Amazon*, Ph.D. Thesis, University of Stirling, U.K., 288 pp.

Luizão, F.J., 1989. Litter production and mineral element input to the forest floor in a Central Amazonia forest, *GeoJournal*, **19**(4): 407–417.

Luizão, F.J., and Schubart, H.O.R., 1987. Litter production and decomposition in a terra firme forest of Central Amazonia, *Experientia*, **43:** 259–265.

Luizão, F.J., Luizão, R.C., and Chauvel, A., 1992. Premiers resultats sur la dynamic desbiomasses racinaires et microbiennes dans un latosol d'Amazonie centrale (Bresil) sous foret et sos paturage, *Cahiers ORSTON, Ser. Pedologie*, **27:** 69–79.

Luizão, F.J., Desjardins, T., Luizão, R.C.C., and Palheta, C., 1999a. *Biomassa microbiana e propriedades físico-químicas do solo em sistemas agroflorestais da Amazônia central.* XX Congresso Brasileiro de Microbiologia. Livro de Resumos, MS-072, 300 pp.

Luizão, R.C.C., 1989b. *Variações temporais da biomassa microbiana e aspectos da ciclagem do nitrogênio em solos de floresta natural e de sistemas manejados na Amazônia central*, Dissertação de Mestrado INPA/FUA, Manaus, 67 pp.

Luizão, R.C.C., Bonde, T.A., and Roswall, T., 1992b. Seasonal variation of soil microbial biomass—the effects of clear—felling a tropical rainforesft and establishment of pasture in the central Amazon, *Soil Biology and Biochemistry*, **24:** 805–813.

Luizão, R.C.C., Costa, E.S., and Luizão, F.J., 1999b. Mudanças na biomassa microbiana e nas transformações de nitrogênio do solo em uma sequência de idades de pastagens apos derruba e queima da floresta na Amazônia Central, *Acta Amazonica*, **29**(1): 43–56.

Madge, D.S., 1965. Leaf fall and litter disappearance in a tropical forest, *Pedobiologia*, **5:** 273–288.

May, P.H., Pooles, A.D., Ferraz, J.B., Rocha, R.B., and, Motta, R.S., 1994. Alternativas energéticas para a polo siderúrgico de Carajás, In: May, P.H. Estudos sobre Opções Energéticas para a polo Siderúrgico de Carajás. Ministério da Integração Regional e Banco Mundial. Rio de Janeiro.

Müller, A.A., 1987. *Produção de liteira e retorno de P, K, Ca e Mg ao solo em agroecosssistemas de cacau e em regeneração natural.* Dissertação de Mestrado INPA/FUA. Manaus, AM. 72 pp.

Oliveira, E.P., and Franklin, E., 1993. O efeito do fogo sobre a mesofauna do solo: Recolonização em áreas queimadas. *Pesquisa Agropecuária Brasileira*, **28**(3): 327–369.

Proctor, J., 1983a. Tropical forest litterfall. I Problems of data comparison, In: *Tropical Rain Forest: Ecology and Management*, (Sutton, T.C. Whitmore, and A.C. Chadwick, eds.), Blackwell Scientific Publications, Oxford, 267–273 pp.

Proctor, J., Anderson, J.M., Fogden, S.C.L., and Vallack, H.M., 1983b. Ecological studies in four contrasting lowland rain forests in Gunung Mulu National P, Sarawak. II Litterfall, Litter standing crop, and preliminary observations on herbivory, *Journal of Ecology*, **71:** 261–283

Proctor, J., 1984. Tropical forest litterfall. II. The data set, In: *Tropical rainforest: the Leeds Symposium*, (Sutton, T.C. Whitmore, and A.C. Chadwick, eds.), Leeds: Leeds Philosophical and Natural History Society, 83–113 pp.

Puig, H., 1979. Production of litière en forèt Guyanaise: resultats preliminaires, *Bulletin Societè Histoire Naturelle Toulouse*, **115:** 338–346.

Schubart, H.O.R., Franken, W., and Luizão, F.J., 1984. Uma floresta sobre solos pobres, *Ciência Hoje*, **2**(10): 26–32.

Scott, D.A., Proctor, J., and Thompson, J., 1992. Ecological studies on a lowland evergreen rain forest on Maracá Island, Roraima, Brazil. II. Litter and nutrient cycling, *Journal of Ecology*, **80:** 705–717.

Silva, M.F.F., and Lobo, A.M.G., 1982. Nota sobre a deposição de matéria orgânica em floresta de terra firme e igapó, *Boletim do Museu Paraense Emilio Goeldi, Botanica*, **56:** 1–13.

Smith, C.K., Gholz, H.L., and Oliveira, F.D., 1998a. Fine litter chemistry, early stage decay, and nitrogen dynamics under plantations and primary forest in lowland Amazonia, *Soil Biology and Biochemistry*, **30**(14): 2159–2169.

Smith, C.K., Gholz, H.L., and Oliveira, F.D., 1998b. Soil nitrogen dynamics and plant-induced soil changes under plantations and primary forest in lowland Amazonia, Brazil, *Plant and Soil*, **200**(2): 193–204.

Smith, C.K., Gholz, H.L., and Oliveira, F.D., 1998c. Litterfall and nitrogen-use efficency of plantations and primary forest in the eastern Brazilian Amazon. *Forest Ecology and Management*, **109**(1–3): 209–220.

Spain, A.V., 1984. Litterfall and the standing crops of litter in three tropical Australian rainforests, *Journal of Ecology*, **72:** 947–961.

Tapia-Coral, S.,C., Luizão, R.C.C, Perin, R., and Luizão, F.J., 1997. *Relação entre tratamentos agrossilviculturais implantados sobre solos degradados e as propriedades bioquímicas do solo*, XIX Congresso Brasileiro de Microbiologia. Livro de Resumos, MS-060, 223 pp.

Tapia-Coral, S.C., Luizão, F. J., and Wandelli, E.V., 1999. Macrofauna da liteira em sistemas agroflorestais sobre pastagens abandonadas na Amazônia central, *Acta Amazonica*, **29**(3): 447–495.

Tian, G., 1998. Effect of soil degradation on leaf decomposition and nutrient release under humid tropical conditions, *Soil Science*, **163**(11): 897–906.

Uehara, G., and Guilman, G., 1981. *The mineralogy chemistry and physics of tropical soils with variable charge clays*, Wetview, Boulder, Colorado. 170 pp.

Vieira, I.C.G., Nepstad, D.D., Junior, S.B., and Pereira, C., 1993. A importância de áreas degradadas no contexto agrícola e ecológico da Amazônia, In: *Bases Científicas para Estratégias de Preservação e Desenvolvimento da Amazônia, V2*, (E.G. Ferreira, C.M., Santos, G.M., Leão, E.L.M, and Oliveira, L.A., eds.), Instituto Nacional de Pesquisas da Amazônia, Manaus, 43–53 pp.

Vitousek, P.M., 1984. Litterfall nutrient cycling and nutrient limitation in tropical forests, *Ecology*, **65**(1): 285–298.

Wieder, R.K., and Wright, S.J., 1995. Tropical forest litter dynamics and dry season irrigation on Barro Colorado Island, Panama, *Ecology*, **76**(6): 1971–1979.

Woomer, P.L., Martin, A., Albrecht, A., Resck, D.V.S., and Scharpenseed, H.W., 1994. The importance and management of soil organic matter in the tropics, In: *The Biological Management of Tropical Soil Fertility*, (Woomer, P.L., and Swift, M.J., eds.), John Wiley and Sibs., Chichester, U.K., 47–80 pp.

CHAPTER 7

LITTER AND MICROBIAL DYNAMICS IN RELATION TO NUTRIENT CYCLING IN OILPALM (*Elaeis guineensis* Jacq) ECOSYSTEM

N.O. Aisueni and C.E. Airede

Nigerian Institute for Oilpalm Research,
Benign-Akure Road
P.M.B. 1030, Benin City, Nigeria

ABSTRACT

Ground litter is often generated in oilpalm cultivation through land openings from primary and secondary forests for field establishment. Other sources are the routine field maintenance through the slashing of ground cover, pruning of senescent leaves, sloughed off trunk cells and roots, microbial proliferations, fruit harvesting and processing as well as litter returns in the palm itself when felled and replanted after its economic life span. All these result in huge tons of debris.

Through the effects of soil micro flora, micro- and macro-fauna, the litter is transformed in soil to influence it physically, chemically and biologically. Biologically, the debris when decomposed gives soil organic matter which not only improve soil texture, water holding capacity, soil temperature, soil porosity, but when mineralized, enhance the nutrient status of soils. Similarly, nutrients earlier immobilized in the microbial biomass are also mineralized, while some are synthesized, such as biologically fixed N. All these influence the oilpalm ecosystem in no small measure.

Key Words: Oilpalm, ground litter, litter decomposition, nutrient cycling and soil properties.

The oilpalm (*Elaeis guineensis* Jacq) is a lowland crop of the humid tropics whose West African origin is now firmly established (Zeven, 1965). Its natural habitats are the sources and banks of water courses, moist valleys including those of the rain forest, savanna transition zones, banks of lakes, swamps, alluvial plains and low lying islands (Chevalier, 1910; Zeven, 1965). It is from these habitats that the oilpalm was taken to man-made clearings in the primary rainforest where the palm can be seen today in the wild, semi-wild and cultivated states. Soils supporting these palms in the tropics are mainly the alluvial soils of marine and river origin, the acid sands or Ultisols and the basement complex soils or Alfisols.

Their geographical distribution as influenced by tropical climate and summarized for areas of highest bunch production (Hartley, 1988) are as follows:

1. Rainfall of 2000 mm (80 in) or more distributed evenly throughout the year with no marked dry season.
2. A mean maximum temperature of about 29–33°C (85–90°F) and a minimum temperature of about 20–24° C (72–75°F).
3. Sunshine amounting to about 5 hours per day in all months of the year and rising to 7 hours per day in some months or solar radiation of about 350 Cal cm^{-2} day^{-1}.

It is in such clearly defined climate that the oilpalm is cultivated in Asia—Malaysia and Indonesia; Africa—Nigeria, Ghana, Sierra Leone, Ivory Coast, Benin, Cameroon, and Zaire and America—Brazil, Costa Rica, Ecuador, Honduras and Columbia (Table 1). With landclearing for oilpalm development, large biomass and diverse vegetation of the tropical forest are transferred to the soil surface as litter (i.e. above and belowground litter and exudates). These are subsequently decomposed by the soil microflora, macro- and micro-fauna and converted to available nutrients and soil organic matter to boost soil fertility. Through nutrient uptake, immobilization and turnover, the palms and ground-covers temporarily take up these nutrients 50 per cent of which are returned as nutrient turnover in fronds, male inflorescence etc. (Vine, 1956; Tinker and Smilde, 1963). By the time the palm is 25–30 years old, it becomes uneconomical to be operated by the farmer (Ferwarda, 1955), hence it is felled to return all the immobilized nutrients in its large biomass back to the soil and is then replanted.

The objectives of this paper therefore, are to assess the litter inputs from the cleared forest, litter generation in the oilpalm ecosystem and their physico-chemical and microbiological implications on the soil.

Table 1 Rainfall, Temperatures and Sunshine at some oilpalm cultivating centres in the tropics.

Centre	Latitude and Longitude		Temperature			Sunshine hrs		Rainfall
			mean	mean max.	mean min.	per day	per annum	annum (mm)
Asia-Malaysia Telok Anson	4° 2'N	101°1'E	28.1	32.9	23.3			2348
Layang-Layang, Johore	1° 15'N	103°30'E	26.6	31.8	21.3	5.3	1940	2180
Kuala Lumpur	3° 7'N	101°42'E				6.1	2230	
Sumatra- Medan	3° 35'N	98°41'E	26.0	30.9	22.5	6.9	2508	2487
Africa-Sierra Leone, Njala	8° 6'N	12°6'W	26.4	31.5	21.2	5.4	1971	2822
Cote D'ivore, La Me'	5° 3'N	3°5'W	26.0	30.0	21.9	5.0	1819	1790
Nigeria-NIFOR, Benin City	6° 30'N	40'E	26.0	30.2	21.8	4.6	1692	1916
Cameroon-Lobe	4° 30'N	9°10'E	26.2	30.1	22.3	3.6	1306	3508
Zaire-Yangamaba	0° 49'N	24°29'E	23.7	29.6	19.9	5.6	2054	1822
America-Brazil-Belem'	1° 16'S	48°8'W	26.3	31.3	21.3	6.1	2220	3306
-Iguape,Bahia	12° 30'S	39°W	24.5	29.1	19.9	6.4	2323	1899
Columbia Aracataca	10° 35'N	74°9W	27.8	33.5	22.3	7.7	2792	1661
Rio Mira,Tumaco	1° 33'N	78°41'W	27.8	30.1	25.5	3.5	1285	3354
Ecuador La Concordia	0° 05'N	79°20'W	24.4	28.4	20.4	2.2	808	3188
Honduras-San Alejo, Tela	15° 43'N	87°29'W	25.0	30.3	19.7	6.9	2533	2666
Costa Rica-Quepos	9° 26'N	84°9'W	25.3	30.2	20.5			3307

(*Source*: Hartley, 1988).

GROUND LITTER GENERATION IN THE OILPALM ECOSYSTEM

Forest Inputs Through Land Clearing

Land clearing is the first step in establishing oilpalm plantations in tropical rainforests using either the slash and burn method or the bulldozer clearing technique in opening up primary or secondary forests. In either of the methods, huge tonnes of organic materials are generated as ground litter which are often left for about 3 weeks to dry before burning. Estimates of these aboveground inputs from cleared matured secondary forests of 15 years or older in the humid and subhumid tropics range from 300 to 450 t/ha of aboveground biomass and store approximately 1–2 t/ha of N, 0.7–0.25 t/ha of K, 2–3 t/ha of Ca, 300–800 kg/ha of Mg and 100–250 kg/ha of P (Bartholomew *et al.*, 1953; Ogawa *et al.*, 1965; Klinge *et al.*, 1975; Adedeji, 1984; Kyuma *et al.*, 1985; Saldrarriaga *et al.*, 1988). Available data on nutrient storage in soil and aboveground biomass are shown in Table 2. Added to these are the inputs from the dead roots and exudates from the cut vegetation with the necromass of fine root, estimated as 15 tonnes per hectare (Vitousek and Sanford, 1986, Table 3). They decompose rapidly with immediate effects on nutrient releases and labile fractions of soil organic matter. Earlier oilpalm fertilizer trials on cleared matured secondary forests in Nigeria showed higher leaf N, P, K, Mg, and Ca in control plots than the critical levels for oilpalm (Onwubuya and Iremiren, 1985). Hence they concluded that application of these nutrients, particularly N, were not necessary in first cycle oilpalms planted in soils previously under secondary forest.

Ground-Cover Inputs Through Slashing

The usual procedure in oilpalm estate management is to plant leguminous cover plants soon after land clearing and burning. Usually a mixture of *Pueraria phaseoloides, Calopogonium mucunoides* and *Centrosema pubescens* in the ratio of 2:2:1 are planted as seeds in the interlines of palms after palm lines have been delineated to ensure good ground-cover. Germination occurs within 5 days after sowing while full cover is established within 3 months. This is the practice in most corporate farms that use leguminous covers in their plantations whereas many other farms establish oilpalm plantations without planting interrow cover crops by allowing the natural vegetation or volunteer weeds to grow. These weeds are often kept under control by slashing 4 to 6 times annually (Ojuederie *et al.*, 1983). The legumes are not shade tolerant, hence they decrease with increases in the oilpalm canopy (Gray and Hew, 1968; Hew and Tam, 1970; Highlands, 1971). Correspondingly

Table 2 Nutrients stored in aboveground biomass of secondary forests (15 yr) and in surface soils (0–30 cm) at surface tropical locations (Nye and Greenland, 1960; Kyuma *et al.*, 1985; Stromgaard, 1984; Szott and Palm, 1986; Uhi and Jordan, 1981).

	Nutrients		(kg/ha)	
	N	*P*	*K*	*(Ca + Mg)*
1. Kade Ghana, Alfisol, mean annual rainfall 1650mm				
Biomass	1837	126	822	2880
Soil	4608	12	652	2953
Total	6445	138	1474	5833
2. Yangambi Zaire, Oxisol, mean annual rainfall 1854mm				
Biomass	561	73	406	563
Soil	2248	19	360	153
Total	2809	92	766	716
3. Kasama, Zambia, Oxisol, mean annual rainfall 1200mm				
Biomas	1653	322	300	nd
Soil	4283	10	133	nd
Total	5936	332	433	nd
4. Nam Phrom, Thailand, Inceptisol, mean annual rainfall 1500mm				
Biomass	1567	195	755	3784
Soil	4450	86	281	9728
Total	6017	281	1036	13512
5. San Carlos de Rio Negro, Venezuela, Oxisol.				
Biomass	1722	51	300	332
Soil	1474	210	20	10
Total	3196	261	320	342

nd = not determined.

alternative bush cover increases such that by the 14th year of field establishment, volunteer weeds become the predominant ground cover in the field. At this time the covers reach a stable state and become self-sustaining with little or no effect on the palm, and nutrients extracted from the soil are now balanced by litter returns. Soil nutrients and fertilizer inputs which without the ground cover would have been lost

Table 3 Aboveground inputs in humid forest and derived agricultural systems.

System				*Aboveground Inputs (dry matter tons/ha/yr)*	*Timing of inputs P = Pulsed, C = Continual*	*Quality of Input (% lignin)*
Natural Forest	(i)	Fertile soils		10.5 a	C	20–30b,f
	(ii)	Infertile soils		8.8 a	C	
Shifting cultivation	(i)	Cropping phase		3.2	P	<8c
	(ii)	Fallow phase (years of *Fallow)*				
			1	1.2	C	>30f
			5	4.8		
			10	7.1		
			20	9.7		
	(iii)	Fallow phase (years of *Fallow)*				
			1	4.6		
			4	6.0		
			9	7.8		
			14	9.5		
Continuous cultivation	(i)	Low input (rice-cowpea)		6.0	P	<8
	(ii)	High input (maize-bean)		9.3	P	<8
Agroforestry Systems.	(i)	Alley cropping (prunnings)		5.22	P	<12
	(ii)	Multistrata systems		6.20		
	(iii)	Managed pastures		6.39	C	

(a) Vitousek and Sanford 1986
(b) Anderson and Swift 1983
(c) Palm and Sanchez 1991
(d) Toky and Ramakrishnan 1983
(e) Sanchez *et al.*, 1989

to the oilpalm through leaching are therefore temporarily retained by the ground cover and made slowly available to the palms through litter decomposition and mineralization of soil organic matter. During the dry season however it is possible for the covers to rapidly deplete soil moisture and this might adversely affect the uptake of nutrients by the palm, particularly N and P.

Oilpalm Inputs Through Leaf Prunings, Sloughed off Cellular Detritus and Plant Exudates

The oilpalm is the main crop or plant in the ecosystem. When transplanted into the field as nursery seedlings, the palms have little or no stem but palm fronds which constitute the crown, the youngest of the fronds being an unopened leaf called the spear. Each leaf or frond consists of the rachis and leaflets numbering as much as 40 fronds in the 2 year-old palms with the production rate decreasing to between 18 and 24 leaves per year in mature fruiting palms depending on seasonal changes and soil fertility. As the palm becomes bigger and taller, older palm fronds become senescent and are pruned off the palm. As pruning continues, the presence of the oilpalm trunk covered with pruned leaf bases become evident. This trunk consists entirely of living tissues (vascular system), which wears the crown and also acts as the storage organ with fairly large quantities of carbohydrates, reducing sugars and polysaccharide and mineral nutrients particularly N and K (Table 4). By the time palms are over 15 years in the field, the trunk begins to lose its cut leaf bases which now rot and fall off with other cellular detritus of the trunk surface and trapped fallen organic materials and exudates from the crown. So also are the green leaves that subtend ripe fresh fruit bunches which must be pruned off to enable the harvester gain access to the ripe bunches. They all lead to the gradual build up of the huge tonnes of organic debris that characterize oilpalm plantations. As a result, large amounts of nutrients earlier transferred from the soil and stored or immobilized in the leaves are returned to the soil by the time the biomass decomposes into soil organic matter and is mineralized by the action of the soil microflora, macro- and micro-fauna.

Oilpalm Inputs Through its Diverse Root System

Roots of the oilpalm are adventitious and lie very close to the soil surface except the descending primaries which permeate the topsoil especially when the soil is non-concretionary. Of these roots only the non-lignified quarternaries and tips of other roots are important in the absorption of nutrients (Ruer, 1967; Tailliez, 1971). These increase in age extending for only 2.5 m in 2 year-old palms to 4.5 m in 4 year-old

Table 4 Nutrient content for palms of various ages (Gray, 1969).

Nutrient elements	Palm age (yrs)								
	1½	2½	4½	6½	8½	10½	14½	17½	27½
N content (g/palm)	76	186	160	410	353	419	765	1182	134
K content (g/palm)	155	631	1328	3285	2153	3268	3537	3307	2047
Carbohydrates									
Water soluble reducing sugars (kg/palm)	0.7	1.8	3.6	3.2	13.8	26.8	19.6	35.2	
Hydrolysable polysaccharride (kg/palm)	1.0	3.4	6.5	25.7	34.1	50.0	100.0	8.5	
Total (kg/palm)	1.7	5.2	10.1	28.9	47.9	76.8	119.8	103.7	

Each figure is from one palm on coastal alluvial soil.

Table 5 Nutrient removal, immobilization and turnover in adult palms (in kg/palm yr).

	N	*P*	*K*	*Mg*
(a) Bunch removal, Nigeria: Tinker and Smilde (1963) (Yield 9.6 tons/ha).	0.20	0.04	0.23	0.03
Malaysia: Ng (1970) (Yield 24 tons/ha).	0.49	0.08	0.63	0.14
(b) Immobilization in palms, Nigeria: Tinker and Smilde (1963)	0.18	0.024	0.11	0.104
Malaysia: (Ng, 1970)	0.27	0.022	0.47	0.072
(c) Nutrient turnover in fronds, male inflorescence etc. Nigeria: Tinker (1964)	0.63	0.073	0.38	0.25
Malaysia: (Ng, 1970)	0.53	0.076	0.69	0.19
Mean total nutrient uptake (in kg/palm/yr) Nigeria.	1.01	0.137	0.72	0.384
Malaysia:	1.29	0.18	1.79	0.31

*From Corley *et al.*, (1976).

palms and to virtually the entire field by the 6th year with decreasing root densities as the distance from the oilpalm base increases (Purris, 1956; Ruer, 1967; Tan, 1973; Gray, 1969). Hence the spreading of feeding roots during the first 5 years in the field is approximately the same as the area covered by the canopy with roots of adjacent palms older than 5 years interpenetrating in non-concretionary soils. These feeding roots and associated microflora make up the rhizosphere with rhizosphere soils recording higher microbial activities (i.e. fungal and algal numbers) than in adjacent bush fallow. Enhancement of these activities were mainly by the root exudates from the actively growing feeding roots, the old, dead and decaying roots and root cell debris as well as other plant debris in soil.

Direct and Indirect Inputs Through the Microbial Community

Diseases of the Oilpalm

Economically important diseases of seeds, leaves, root, trunk, inflorescence and fruit, caused by plant pathogenic micro-organisms have been reported in oilpalm plantations (Robertson *et al.*, 1968), resulting in the death and decay of palms/palm parts, if not controlled. Important seed diseases which affect the developing radicle and cause death of the young sprouts are the brown germ (*Aspergillus* sp. and *Penicillium* sp), and black seed rot (*Ceratocyctis paradoxa*). Other species of *Penicillium* (*P. janthinellum, P. chrysogenum*) and *Aspergillus* (*A. flavus, A. niger*) have also been shown to cause seed deterioration (Airede and Esuruoso, 1987). Deteriorated seed, rotten radicles and plumules are thus returned to the soil as organic debris. Leaf diseases of nursery seedlings and field palms result in reduction in photosynthetic surface, premature senescence, necrosis of leaves, petioles, rachis and defoliation, thereby increasing ground litter. Such leaf diseases include Anthracnose (*Botryodiplodia* sp., *Melanconium* sp., *Glomerella cingulata*), *Cercospora* leaf spot (*C. elaeidis*), seedling blight (*Curvularia clavata* and *Drechslera hawaiiensis*), spear rot (*Erwinia* Sp., *Pestalotiopsis* sp.; algal spots (*Cephaleuros virescens*) and leaf base wilt (*Ceratocystis paradoxa*). Root and stem diseases of the oilpalm like blast (*Pythium splendens* and *Rhizoctonia lamellifera*); *Fusarium* wilt (F. *oxysporum* f. sp. *elaeidis*) and *Ganoderma* trunk rot (*Ganoderma* sp.) are characterized by rotting of roots and trunk tissues, necrosis, drying out, fracture and defoliation of leaves. Dry basal rot also results in extensive bunch and inflorescence rot. Immature fruits drop from the bunches before they are ripe. In acute forms of *Fusarium* wilt, decarpitation of the crown and death follows infection by the fungus. Death of oilpalms due to *Ganoderma* trunk rot, results in the addition of fructifications of the *Ganoderma* fungus to the ground litter. Thus a large volume of palm biomass resulting from moribund and defoliated leaves, petioles, rachis, stem bulb, rotted bunches, rotted trunk tissues and fructifications of *Ganoderma* spp. is contributed to the ground litter by the diseased palms. The occasional bunch and fruit rot caused by *Marasmus palmivora*, contribute a large mass of rotting materials, rotted male and female infloresences, rotten fruits and the fructifications of the *Marasmus* fungus to the organic debris in the plantation.

Pests of the Oilpalm

The major pest of the oilpalm include nematodes, slugs, snails, mites, insects, birds and mammals (Hartley, 1988). Pests outbreak due to nematode (*Radinaphelenchus cocophilus*) mites (*Oligonyohus* sp.), caterpillars (*Homophylotis catori, Caspholia extranea, Labia viridisaima, Zonoceros cerynica*); Beetles (*Coelaenomenodera elaeidis, Oryctes monoceros, Temnoscoita* sp. *Rhyncophoros phoenincia*); grasshoppers (*Zonocero variegatus; Cantantops spisus*) Mealybugs and scale insects (*Aspidiotus destructor*), bagworms (*Metisa plana*) and rodents (*Rattus rattus, Thryonomys swinderanus*) have caused economic losses in oilpalm plantations. The pests caused defoliation of leaves, root damage and in severe cases, death of the palm. Defoliated leaves and rotted tissues contribute to the mass of organic debris in the plantations.

Precipitation Inputs

During thunderstorms, atmospheric N is fixed by electrical discharge and by photochemical reactions to release nitrogen oxides which dissolve in rain water. This is an additional source of N-nutrient in the oilpalm ecosystem. Ataga and Nuabuchi (1978) considered N in precipitation at three locations in the oilpalm belt of Nigeria which they estimated at 7.21, 6.68 and 5.65 kg N/ha/annum. The locations were NIFOR main station in Benin City, Abak and Acharu experimental stations. They showed that this amount of nitrogen was only about 25% of nitrogen removed annually through bunch harvest.

GROUND LITTER DECOMPOSITION AND NUTRIENT MINERALIZATION

This is the process of converting the ground litter generated above into available nutrients and soil organic matter. Decomposition is usually best carried out by mixed microflora (Bacteria + actinomycetes + fungi) rather than individual microbial isolates (Herman *et al.*, 1977; Kirk *et al.*, 1977). It is controlled mainly by the physico-chemical environment, including climate and soils, the resource quality, i.e. (the chemical composition of the organic material), and the nature of the decomposer community (Swift *et al.*, 1979).

Decomposition of soil organic matter (SOM) releases the organic nutrients into plant available form but as the level of organic matter declines, the rate of nutrient release is reduced and soil structure deteriorates. The maintenance of soil organic matter, therefore, is dependent on continual input of organic materials as exhibited in the oilpalm ecosystem with the oilpalm and ground cover litter (above and below ground) continuously transformed in the ecosystem. Earlier studies of nutrient turnover in palm fronds, male inflorescence etc. in Nigeria by Tinker (1964) gave 0.63% N, 0.073% P, 0.38% K and 0.25% Mg while in Malaysia, Ng (1970) gave 0.53% N, 0.76% P, 0.69% K and 0.19% Mg (See Table 5). Also in another study, felled/pruned palm parts of the oilpalm were found to be fertility boosters when incubated for nutrient mineralization and greenhouse studies (Aisueni, 1991).

Nutrient levels as high as 0.01% N, 0.01% P, 0.81% K, 0.38% Ca and 0.10% Mg were mineralized from palm parts which significantly enhanced the growth of oilpalm seedlings when used as soil amendment.

NUTRIENT REMOVAL, IMMOBILIZATION AND TURNOVER IN THE OILPALM

In the oilpalm, total plant analysis was used amongst other methods to determine the essential nutrients, amount immobilized and rate of uptake of mature oilpalm, in the field. Recent data are summarized by Corley *et al.* (1976) where three demands may be distinguished as:

a. Nutrients removed from the area entirely.

b. Nutients taken up and not returned by the palm since they are immobilized in the palm itself.

c. Nutrients recycled to the soil in leaf litter, dead leaves and male inflorescence.

Data shown in Table 5 reveal that soils supporting oilpalm must be able to supply some 1.3 kg N, 0.2 kg P, 1.8 kg K and 0.3 kg Mg to each palm per year in Malaysia and 1.0 kg N, 0.14 kg P, 0.7 kg K and 0.4 kg Mg to each palm per year in Nigeria. Of these nutrients about 50% are jointly removed in bunch export and immobilization in palms while the remaining are returned as turnover in ground litter i.e. fronds, male inflorescence etc. (Table 5). It is this nutrient removed through bunch harvest and export that must be replaced by the organic nutrient pool, precipitation N, biological N_2 fixation, inorganic fertilizer application in addition to other nutrient transformations in soil to avoid specific nutrient deficiency symptoms in large fruiting palms.

NUTRIENT RETURNS IN CLEARED OILPALM BIOMASS AT REPLANTING TIME

Replanting of oilpalm plantations is usually carried out after about 25–30 years due to harvesting problems and other economic considerations. For this to be done, the old palms must be clear-felled and burnt or left to rot so as to return all nutrients earlier immobilized in its above and below-ground parts. When this is done large quantities of oilpalm biomass (particularly the trunks and fronds) become available which amounts to 85 t/ha dry matter with additional 16 t/ha below-ground roots. These contain a pool of nutrients (Table 6) equivalent to 642 kg N, 58 kg P, 1384 kg K and 156 kg Mg. In terms of inorganic fertilizers, this mass of nutrients is equivalent to approximately 3.06 tons of sulphate of ammonia (A.S), 0.37 tons of Christmas Island Rock Phosphate (CIRP), 2.77 tons of muriate of potash (MOP) and 1.0 ton of kieserite ((Haron *et al.*, 1996). According to them the nutrient capital in the old palms at replanting could theoretically provide standard fertilizer applications for immature palms of N, K, and Mg for 6 to 7 years and about 2 years for P. Teoh (1993) and Mohd Hashim *et al.* (1993) also suggested that by using the zero burning practices the requirements of inorganic fertilizers could be reduced by 20–30% for up to 4 to 5 years during the immaturity

Table 6 Nutrient content, and fertilizer equivalent of oilpalm biomass at replanting (From Haron *et al.*, 1996).

Palm residue	*Dry matter* (*Tons/ha*)	*Nutrients* (kg/ha)			
		N	*P*	*K*	*Mg*
Aboveground biomass	85.226	577	50	1255	141
Belowground biomass	15.89	65	8	129	15
Total	101.15	642	58	1384	156
Fertilizer equivalent		AS	CIRP	MOP	KIES
Aboveground biomass		2748	318	2510	908
Belowground biomass		310	51	258	96
Total kg/ha		3058	369	2768	1004

*AS = Sulphate of Ammonia, CIRP = Chrismas Island rock Phosphate; MOP = Muriate of potash; KIES = Kieserite.

or early mature phase of the palms. They based these estimates on availability of nutrients only in the above-ground biomass which will be further enhanced if contributions from the root biomass (i.e. 20% of total above-ground biomass) is included.

Earlier published data showed that total biomass accumulation within the first 10 years of fallow varied from 4 to 160 t/ha and could reach a steady state after 15 years in the Congo basin (Bartholomew *et al.*, 1953) and 7 years in the Rocky Mountain regions of the United States. Nye and Greenland (1960) and Scott (1977) predicted that litter fall in fallows would quickly attain that of forests in 5 years. While Young (1989) estimated that 8.4 t/ha/yr of above-ground and 5.8 t/ha/yr of below-ground inputs would be required to maintain soil organic matter levels in humid forests. These forest biomass figures are far exceeded by the oilpalm biomass (85 t/ha aboveground dry matter + 16 t/ha belowground dry matter) by the time it is due for replanting.

Ameliorative Effects of Oilpalm Cultivation

Soil Microbial Activity

The decomposition and mineralization of the ground litter generated from the various sources above are microbiologically driven resulting in increased microbial activity. The leguminous covers if successfully innoculated with *Rhizobia* fix atmospheric nitrogen symbiotically for the use of the oilpalm in soil. Other microbial associates of the oilpalm were investigated in surveys carried out by Aisueni (1987a) and found to consist of mostly the facultative bacteria namely *Bacillus*, *Enterobacter* and *Pseudomonas species* as well as the cyanobacteria namely *Gloeocapsa*, *Nostoc*, *Anabaena* and *Calothrix* species. These organisms showed nitrogenase activities when tested in the laboratory by acetylene reduction assay (ARA) and isotopic ^{15}N fixation technique to confirm them as atmospheric nitrogen fixers which showed that the nitrogen balance in the oilpalm ecosystems could be partially maintained by the free-living microorganisms living in association with the oilpalm.

Even when oilpalm parts were incubated for heterotrophic and photosynthetic N_2 fixation using the acetylene reduction assay technique (Aisueni, 1987a), nitrogenase activity ranges of 0.67 to 476.06 n-moles ethelene produced per day per gram sample, was recorded for oilpalm which was an amount much greater than the nitrogenase activities in grasses known for the restoration of fertility in tropical fallowlands (Henzell and Norris, 1962; Nye and Greenland, 1960). Similarly, nitrogenase activities were recorded for soils cultivated to oilpalm (ARA technique) and their algal and fungal populations were higher than in adjacent fallow lands (Aisueni, 1987b). These activities portrayed the oilpalm as a good fallow crop that can maintain its soil fertility in line with those of fallow lands.

Soil Chemical Activity

Through the cultivation of oilpalms, huge tonnes of above and below ground litter are generated which when decayed or are burnt increase soil available nutrients. Similarly the slashed ground cover and pruned palm parts return immobilized nutrients to the soil. They lead to increased cation exchange capacity of soil, buffering capacity, chelating capacity and soil pH. Added to the increased nutrients so generated are the atmospheric nitrogen fixed into ammonia for palm use symbiotically by *Rhizobia* of the leguminous cover crops and non-symbiotically by the associated free-living microorganisms (Aisueni, 1987a).

Soil Physical Activity

The foliage and litter of both the oilpalm trees and ground cover reduce rain drop impact on the soil, slow down movement of run off water, reduce erosion and increase water infiltration and percolation capacities. Soil aggregation increases with decreased soil crusting and bulk density.

DELETERIOUS EFFECTS OF PLANTATION PRACTICES ON THE BIOTA AND SOILS

The various activities and processes which can lead to pest disease outbreaks and to deleterious changes in the biota and soils are, felling, stumping, beating down, heaping, burning, replanting, development of weeds/natural covers, establishment of leguminous cover crops, harvesting, nutrient demand and employment and pesticide application.

Effect of Felling and Stumping

Felling of trees is usually followed by cutting of the felled vegetation and heaping. The stumps of the trees remain in the ground. Felled trunks and leaves contribute to the ground litter and organic debris and hence source of nutrients that are mineralized.

However, before the total decomposition of this biomass large trunks which are too heavy to be cut up and removed, make maintenance difficult and involve the use of heavy tractors and very expensive equipment for maintenance. This entails gross disturbances of the soil.

The heaped debris inevitably gets covered with weeds and creepers which compete with the plant for moisture and nutrients and pests find in these heaps good hiding places.

Stumps of trees previously affected by *Ganoderma* disease, serve as points for infection of young palms. The stumps also serve as breeding sites for *oryctes* and *Rhinocerous* beetles which cause defoliation of the palms.

Effect of Burning

Improved access into the fields is advocated by burning of felled trees. This practice however has its deleterious effects. Burning leads to the destruction of beneficial plants, beneficial soil microorganisms, natural enemies, cover crops, destruction of valuable timber, loss of organic matter and the provision of conditions leading to erosion. Destruction of the cover crops prevents the accumulation of litter, loss of organic matter and hence loss of nutrients to be mineralized. Burning causes a slight depression of organic matter and nitrogen in the top soil. A rapid nitrate flush therefore occurs immediately after burning and this leads to an earlier loss of nitrogen under certain conditions. An increased need for nitrogen fertilizer will therefore be expected in the early years. Cases of losses of organic carbon at a steady rate for 8–12 years after planting from primary and secondary forests with 2.5–3.5% carbon in the top soil have been reported (Bocquet and Michaux, 1961).

Most of the microorganisms in the top soil responsible for the disintegration of organic debris and release of nutrients for mineralization, and beneficial nitrogen fixing bacteria in the nodules of leguminous cover plants are destroyed by burning.

Quantities of very valuable timber destroyed during bush burning could have been exploited very profitably and contribute substantially to the cost of establishments, fencing posts and bridges. On the other hand in forest reserves and bush fallow, there is enhanced fertility in a number of palm nutrients as a result of high biological activity.

Effect of Natural Covers and Weeds

Forest regrowths and species which follow the felling of forests, grow with tremendous rapidity, particularly at the beginning and to a lesser extent at the end of the rains (April/June and September/October). In areas which were burnt these undesirable plants like lalang (*Imperata cylindrica*), Siam weed (*Eupatorium odoratum*) and other strong growing grasses, establish themselves naturally and vigorously, increase the volume of vegetation and compete with the palm. Most grass species are upright and can easily submerge young palms.

If ring weeding is neglected, the weeds soon invade the circles around the palm, grow up through the leaves of young palms and compete with them for water, nutrient and even light. If a weed like *Eupatorium* is left unchecked, it rapidly becomes dominant, and grows to a height of 3–6 m. Once an invasion has taken place, ring weeding becomes very difficult as the root system of the palm and grasses, are intertwined. Apart from its severe competition with the palm for light, nutrients and water, *Eupatorium* can be dangerous because it easily catches fire in dry weather and destroys young palms.

With increasing labour costs, regular manual and mechanical maintenance of these weeds becomes very expensive. The use of herbicides have their deleterious effects. Careless and excessive use of herbicides can damage the young palm or its root system and cause disturbance to the soil. Adverse effects on the growth of young palms can result from negligent spraying, or intense spray drift. Symptoms associated with herbicide damage such as, immediate leaf scorch, blotching, leaf necrosis, leaflet (chlorosis) and epinastic response (Sheldrik, 1968).

Effect of Establishing Cover Crops

The primary purpose of sowing an artificial cover mixture after burning, is to provide a cover for the soil and prevent erosion and soil nutrient depletion. A secondary purpose is to check the growth of undesired plants. Leguminous covers like *Calopogonium mucunoides* and *Centrosema pubescens* are almost always chosen because they form nodules which contain colonies of nitrogen fixing bacteria which live symbiotically on the host plants. However, the cultivation needed to establish these cover crops has the deleterious effects of reducing soil fertility which normally follow tillage. Spanaarj (1957) showed levels of total exchangeable cations in cropped soils below that of control plots with a natural cover. Moreover, where these covers are neglected and not properly maintained, they start competing with the palm for the limited moisture and nutrients, especially in already impoverished lands. On such lands competition reduced soil fertility to a point where later yields were affected, while it was observed that removal of competition had early adventageous effects on yields, (Table 7) (Spanaarj, 1957). The situation was different in forest areas where, once the palm had fully established themselves, their shade effected an ascendency over the covers which resulted in yields no longer being reduced by lack of maintenance.

In later years when the palm cover closes, some cover crops begin to die and in dry weather catch fire and destroy the palms.

Failure of these covers resulting from pests and diseases can increase the population of harmful microorganisms and pests in the ecosystem.

Table 7 Yield of annually slashed plots as a percentage of the normal maintenance yield.

	Forest area	*Previously cultivated area*
First 8 years of bearing.		
Normal maintenance	100.0	100.0
Annual slashing only	79.4	69.1
Later bearing years.	(11 years)	(2 years)
Normal maintenance	100.0	100.0
Annual slashing	102.0	85.4

Diseases due to *R. solani, Fermes lignosins, G. psedudofem* the nematode *Meloidogyne javanica* and incidences of pests like the aphid, *Aphis laburni,* the ladybird beetle (*Epilachna indica*), the beetle, *Pagria signata,* the bug, (*Chauliops bisontula),* the caterpilar, *Lamprosena diemenalis,* have been reported (Edgar, 1958). The covers may also serve as alternate hosts for other pathogens and pests of the oilpalm.

Dead tissues resulting from diseased covers, increase organic debris on the soil surface and increase nutrients to be mineralized.

Nutrient Demands and Uptake

Nutrient uptake and usage is an important physiological process in the growth and development of plants. Nutrients taken up by plants are used for production and accumulation of dry matter which is distributed and immobilized in all organs (Tinker and Smilde, 1963a). Therefore by the time one oilpalm cropping cycle of 25–30 years is completed, considerable quantities of nutrients have been removed and can be found in the bunches, stored in the trunks, leaves and roots or returned to the soil via dead leaves and roots. Where there was no burning, as in fallow systems, the earliest changes are a rapid release of potassium from the vegetation followed by substantial increases in magnesium and calcium in the top soil (Kowal and Tinker, 1959). In burnt areas however, there is an early build up of K followed by a gradual reduction of cations and phosphorus and the three cations, K, Mg and Ca in later years. The most serious steady losses during the life of the palm, were found to be in K. This steady depletion of K in later years due to the proportion diverted to bunches, lead to the onset of deficiency symptoms and low level of bunch yield, which in turn leads to need for the application of fertilizers. The lower cation contents of oilpalm fields compared to bush fallow is attributed to an increased cation mineralization, while the lower level of available P and base nutrient elements could be attributed to the drain on soil nutrient residues by oilpalm cultivation (removal of fresh fruits bunches, trunks and leaves) (Tinker, 1976; Tinker and Smilde, 1963).

Deleterious Effect of Fertilizer Application

With continuous cropping, several soils which were originally adequately or marginally deficient in trace elements are more likely to be greatly depleted and lead to increased incidence of metabolic disorders associated with trace elements imbalance, suboptimal performance of agricultural crops, and under extreme conditions, total crop failure. Therefore, for maintenance of steady growth of plants after planting, adequate water, and a balanced supply of fertilizer is required. In most oilpalm growing areas, growth and yield is improved by the application of N, P, K and Mg.

However, incorrect use of fertilizers can cause serious damage to crops. Insufficient application leads to deficiency symptom. Excessive use of certain fertilizers will also lead to deficiency of other elements. The application of N and K with Mg, tend to induce Mg deficiencies, typified by the deep orange chlorosis of the leaflets.

Excessive application or application in toxic quantities can lead to root and leaf damage. There have been reported cases of death following application of fertilizer, particularly sulphate of ammonia, in the planting hole before or at planting. In other cases, leaves paled and dried off prematurely (Spanaarj and Gunn, 1959). Toxic effects of sulphate of ammonia is usually seen on the youngest fully opened leaf which blackens off, while other leaves become dried up and eventually die. Symptoms of toxic effects following application of boron are terminal chlorosis progressing along the leaflet tips and margins, necrotic sporting, inhibition of flowering and growth retardation. Debris resulting from damaged tissues increase ground litter.

Effect of Harvesting

The oilpalm produces oily fruit bunches which are routinely harvested and transported from fields to oil processing mills. As these fruits are high in nutrients, their continual removal from the fields drains the soil nutrient reserves. Thus nutrient deficiency symptoms in terms of K, Mg and P commonly occur when fields are not properly managed. Lower levels of C, P and base nutrient elements have often been reported in oilpalm soils as opposed to adjacent bush fallow (Aisueni, 1987b; Tinker, 1976; Tinker and Smilde, 1963b; Ng, 1970). These lower levels were attributed to oilpalm cultivation (removal of fresh fruit bunches, trunks and leaves.)

Effect of Replanting

A field of oilpalm should be replanted when either a large number of palms have become so tall that they cannot be conveniently or economically harvested or the yield is so low that replacement with high yielding young palms would give a much higher return. The major disease of old field palms is *Ganoderma* trunk rot. In replanting programmes, stumps and trunks of felled field palms, especially when poisoned and left to rot *in situ*, produce many fructifications of *Ganoderma* spp. and are infested by *oryctes* and *Rhycophoros* beetles. These pests and diseases cause serious damage to the young replant. The fungus and pests are however very important in the disintegration of the stumps.

Pest and Disease Outbreaks

One of the changes in the ecosystem accompanying the opening of tropical forests for plantations, is the incidence of pests and diseases. Growth of pests and diseases has to be checked or balanced by natural enemies and else the oilpalm plantation has to suffer economic losses. This is not the case in fallow systems in forests, where the level of pest and disease incidence is observed to be relatively lower as a result of the balance obtained through natural selection and survival of resistant varieties in the grooves. Hence, in selecting for high yield, cultivated plantations have gained an advantage at the expense of loss in disease resistance (Stainer, 1929).

Effect of Pest and Disease Control

Oilpalm pests and diseases have been effectively combatted with fungicides and insecticides. However, indescriminate use of most of these chemicals (particularly the contact insecticides), have their deleterious effects on the environment. Many are not specific and so the non-target insects and natural enemies are destroyed. Others are highly toxic to the oilpalm and to humans and must be handled with great care. The oilpalm is sensitive to formulations containing copper, mercury and gamma BHC (Lindane) like Fernasan-D. Therefore, such phytotoxic seed dressings should not be used for the oilpalm. Most are persistent in the soil and create the risk of accumulation in the environment.

Effect of Cultivation on Microbiological and Physico-chemical Properties of the Soil

Changes in the microbiological properties of soil accompany oilpalm cultivation: Aisueni (1987b), observed changes in the microbiological properties of soils supporting oilpalm cultivation that count of total fungi and most probable number (MPN) of algae were higher in oilpalm fields than in bush fallow while total bacteria decreased when compared to adjacent bush fallow (Table 8). Increased count of fungi and algae could result from ground litter generated from diseased plant parts like leaves rachis, rotted root tissues, seeds, fruits, bunches and trunk tissues. These parts will contribute to the increase in the population of such fungal and algal pathogens like *C. virescence, R. lameliferna, P. splendens, C. paradoxa, F. oxysporum* and *Ganoderma* spp. in the soil. Plant residues and organic debris generated in the ecosystem by pests and diseases undergo breakdown in soil through the action of these soil micro-organisms, ultimately this residue form humus. The humic material imparts to soil most of the desirable features affecting nutrition, water holding capacity, structure, aeration, pH and provides nutrients to be mineralized by beneficial soil micro-organisms and source of energy for the soil biota.

Table 8 Microbial counts of some oilpalm fields in NIFOR main station.

Microbial counts (CFU)[a]	*Bush fallow (control)*	*Oil palm*	*Fields (years)*	
		10	20	30
Total Bacteria X 10^5	10	10	6	5
Total Fungi X 10^3	8	8	14	28
MPN of algae X 10^3	45	220	204	780

a = Bacterial and fungal populations epressed in colony forming units (CFU).

However, apart from the role beneficial micro-organisms play in the mineralization of nutrients in the soil, they could have deleterious effects. They affect the availability and deficiency of certain nutrients. They easily immobilize the mineralized nutrients through the build up of their body cells when not immediately utilized for plants growth. The application of raw organic matter to the soil around higher plants for example, is often known to bring about temporary deficiency of nitrogen as a result of microbial utilization of N incident to breakdown of organic matter. Aisueni (1991) observed that N and P content of liquid extracts of degrading residues, decreased with incubation time, while the organic N and P of the solid fraction increased simultaneously. Microbes compete successfully with higher plants for phosphorus when the ratio of P to C in soil organic matter is low.

Moreover, deleterious organic substances, toxic to plants may also be formed in the breakdown of plant residues. These include certain acids, amino acids, glycosides and other organic toxins which may be exuded from the root or be lost from the foliage. Changes in physico-chemical properties of the soil have also been shown to follow cultivation of the oilpalm (Khalid *et al.*, 1993; Aisueni, 1987). Oilpalm fields showed increases in total nitrogen but decreases in P, total C and base nutrient elements (Na, K, Mg and Ca) compared to adjacent bush fallow. The most serious steady losses were found to be in K, Mg and Ca. The lower levels of C were attributed to increased carbon mineralization while the lower levels of available P and base nutrient reserves were attributed to the drain on soil nutrient reserves by removing of fresh fruit bunches, trunks and leaves (Tinker, 1976; Tinker and Smilde, 1963; Ng, 1970).

REFERENCES

Adedeji, F.O., 1984, Nutrient cycles and successional changes following shifting cultivation practice in moist semideciduous forests, *Nigerian Forest Ecology Management*, **9:** 87–99.

Airede, C.E., and Esuruoso F., 1987. Deterioration of shelled oilpalm kernels caused by seed borne fungi, *J. of Science of Food and Agriculture*, **40:** 293–304.

Aisueni, N.O., 1987a. Potential for dinitrogen fixation in oilpalm, (*Elaeis guineensis* Jacq), *Niger Agric. J.*, **22:** 25–29.

Aisueni, N.O., 1987b. Impact of oilpalm (*Elaeis guineensis* Jacq) cultivation on some chemical and microbiological properties of an ultisol, *Bio. Fertil,.* **4:** 167–169.

Aisueni, N.O., 1991. Felled/Pruned oilpalm (*Elaeis guineensis* Jacq) parts as fertility boosters in the oilpalm ecosystem, *Nigerian J. of Palms and oil seeds*, **12:** 87–91.

Anderson, J.M., and Swift M.J., 1983. Decomposition of tropical forests, In: *Tropical rainforests: ecology and management*, (Sutton S.L., T.W. Whitmore, and A.C. Chadwick, eds.), Special Publication 2; British Ecological Society, Oxford Blackwell Scientific, 289–309 pp.

Ataga, D.O., and Nuabuchi, E.E., 1978. Nitrogen in precipitation at three locations in the oilpalm belt of Nigeria, *J. Niger. Inst. Oil Palm Res.*, **5**(220): 21–24.

Bartholomew, W., Meyers, V.J., and Laudelout, H., 1953. *Mineral nutrient immobilization under forest and grass fallow in the Yagambi region*, Serie Scientifique No. 57, Brussels: Institute National Pour L'Etude Agronomique du Congo.

Bouquet, M., and Michaux, P., 1961. Choir d' une methode d' ouvertune en foet pour une plantation de palmier a huile, *Oleagineux*, **16:** 149.

Chevalier, A., 1910. Les Vegetauz utiles de K'Afrique tropicale francais. Fasc VIII. Documents surle palmier a'huile Paaris, In: *J.W.Afr. Inst. Oil palm Res.*, J.M. Waterston (ed.), **1**(1): 24–59.

Corley, R.H., Hardon, V.J.J., and Wood, B.J., 1976. *Development in crop Science (1): Oil palm Research*, Elsevier Scientific Publishing Company.

Edgar, A.T., 1958. *Manual of rubber planting (Malaya)*, Inc. Soc. of Planters; Kuala Lumpur, FAO (1966).

Ferwarda, J.D., 1955. *Questions relevant to replanting in oilpalm cultivation*, Thesis presented to the Agricultural University Wageningen, Holland, F.A.O., 1966. Agricultural Development in Nigeria, 1965–1980 (Rome), 1966.

Gray, B.S., 1969. *A study of the influence of genetic, agronomic and environmental factor, the growth, flowering and bunch production of the oilpalm on the West coast of West Malaysia*, Ph.D. Thesis, University of Aberdeen, 747 pp.

Gray, B.S., and Hew, O.K., 1968. Cover crop experiment in oilpalms in the West coast of Malaysia, In: *Oil palm Developments in Malaysia*, (P.D. Turner, ed.), Incorporated Soc. of planters, Kuala Lumpur.

Haron, K.Z., Kakaria, Z., and Anderson, J.M., 1996. Management of palm residues using various replanting techniques in oilpalm plantations. PORIM International Palm Oil Congress 23–280 Sept., Kuala Lumpur, Malaysia, 241–253 pp.

Hartley, C.O.S., 1988. *Oil Palm*, 3rd edn. Tropical Agricultural Series, Longman Scientific and Technical. Copublished in the United States with John Wiley and Sons Inc., New York.

Henzell, C.F., and Norris, D.O., 1962. Processes by which nitrogen is added to the Soil/ Plant system, (In a review of nitrogen in the tropics with particular reference to pastures), *Bull. Commonwealth Bureau Pastures and Field Crops*, **46:** 1–18.

Herman, W.A., mcGill, W.B., and Dorman, J.O., 1977. Effects of initial chemical composition on decomposition of roots and three grass species, **57:** 205–215.

Hew, C.K., and Tam, T.K., 1970. *Agronomy annual report 1969*, Harrisons and Corsfield Oil Palm Research Station, Banting, Malaysia, 120 pp.

Highlands, 1971. Annual Report (1970), Highlands Research Unit. Kiang, Malaysia, 152 pp.

Khalid, H., Zakari Zin. Z., and Anderson, J.M., 1993. Management of palm residues using various replanting techniques in oilpalm plantations, In: *Proceedings 1993 Porim Int. Palm oil Congress*, 247–253 pp.

Kirk, T.K., Connors, W.J., and Zeikus, J.G., 1977. Advances in understanding the microbiological degradation of lignin, In: *Recent Advances in Phytochemistry*, (F.a. Loewus, and V.C. Runeckles, ed.), Plenum Press. N.Y., **11:** 369–394.

Klinge, H., Rodrigquez, W.A. Bruning, Z., and Fittkau, E.J., 1975. Biomass and structure in a central Amazon rain forest, In: *Tropical Ecological systems*, (F.B. Golley, and E. Medina eds.), New York, Springer Verlag, 115–122 pp.

Kowal, J.M.L., and Tinker, P.B.H., 1959. Soil changes under a plantation established from high secondary forest, *J.W. Afr. Inst. Oil. Palm Res.*, **2:** 376.

Kyuma, K.T., Tulaphitak, and Pairintra, C., 1985. Changes in soil fertility and tilth under shifting cultivation 1: General description of soils and effect of burning on soil characteristics, *Soil Science and Plant Nutrition*, (Japan), **31:** 227–238.

Mohd Hashim T., Teoh, C.H., Camaruzaman, A., and Mohd, A., 1993. Zero burning—an environmentally friendly replanting technique, In PORIM Congress Proceedings of 1993, Kuala Lumur, Malaysia, **10:** 185–194.

Ng, S.K., 1970. Greater productivity of the oilpalm (*Elaeis guineensis* Jacq) with efficient fertilizer practices, In: *Role of fertilization in the intensification of agricultural production*, Proc. 9th Congress Int. Inst., 357–376 pp.

Nye, P.H., and Greenland, D.J., 1960. *The soils under shifting cultivation*, Commonwealth Bureau of pasture Fld crops Tech. Comm. No. 51, Harpedenden, England, 1–156 pp.

Ogawa, H., Yoda, K., Ogino, K., and Kira, T., 1965. Comparative ecological studies on three main types of forest vegetation in Thailand, 2: Plant Biomass, *Nature and Life Southeast Asia*, **4:** 49–80.

Ojuederie B.M., Iremiren G.O., and Utulu. S.N., 1983. Effects of various interrow slashing regimes and size of weeded rings on the early, growth flowering and bunch yield of the oilpalm, *J. Niger. Inst. Oil palm Research*, **24:** 322–334.

Onwubuya I.I., and Iremiren, G.O., 1985. Response of the Oil palm to nitrogen fertilizers in a secondary forest, *Niger. Agric. J.*, **19**/20: 1–6.

Palm C.A., and Sanchez P.A., 1991. Nitrogen release from some tropical legumes as affected by lignin and polyphenol contents, *Soil Biology and Biochemistry*, **23:** 83–88.

Purvis, C., 1956. The root system of the oilpalm, its distribution, morphology and anatomy, *J. of WAIFOR*, 1. (4) 60–82.

Reur. P., 1967. Repartition on surface du systems radiculaire du palmier a huile, *Oleagineux*, **22:** 5345–537.

Robertson, J.S., Prendergasts, A.G., and Sly, J.M.A., 1968. Diseases and disorders of the oilpalm (*Elaeis guineensis* Jacq) in West Africa, *J. Nig. Institute Oilpalm Research*, **4:** 381–409.

Saldrarriaga J.G., West, A.G., Tharp, M.L., and Hhl. C., 1988. Longterm, consequence of forest succession in the upper Rio Negro of Culumbia and Venezuela, *J. of Ecology*, **76:** 938–956.

Sanchez, P.A., Palm, C.A., Szott, L.T., Cuevas, E., and Lal., R., 1989. Organic input management in tropical agroecosytem, In: *Dynamics of soil organic matter in tropical ecosystems.* (D.C. Coleman, J.M. Oades, and G. Ucheara, eds.), Hawaii, U.S.A., Niftal Project, 125–152 pp.

Scott G., 1977. The importance of old field succession biomass increments to shifting cultivation, Great plains-Rocky mountain, *Geographical Journal*, **6:** 318–327.

Sheldrick, R.D., 1968. The control of groundcover in oil plam plantations with herbicides. **2**: Screening trials for herbicides suitable for ring weeding, *J. Nig. Inst. Oil Palm Res.*, **4** (16): 417–424.

Spanaarj, L.D., 1957. Mixed cropping in oilpalm cultivation, *J.W. Afri. Inst. Oil Palm Res.*, **2:** 244.

Spanaarj, L.D., and Gunn, J.S., 1959. The development of transplanting techniques for the oilpalm in West Africa, Inst. Oil Palm Res., **2:** 281.

Stainer, P., 1929. Les. ennemies de L' Elaies. Agric et Elevage an Congo Belge, **3** (26): 405–407.

Stromgaard, P., 1984. The immediate effect of burning and ash fertilization, *Plant and Soil*, **80:** 307–320.

Szott, L.T., and Palm, C.A., 1986. Soil and vegetation dynamics in shifting cultivation fallows, In: *Simposio do Tropico Umido*, Bekem, Brazil EMBFAPA, **1:** 360–379.

Swift, M.J., Heal, O.W., and Anderson, J.M., 1979. *Decomposition in terrestrial ecosystem*, Oxford, England: Blackwell Scientific.

Teoh C.H., 1993. Waste or by-products? The recycling aspects of oilpalm cultivation and oilpalm processing, In: *1993 PORIM Palm Oil Congress*, Kuala Lumpur, Malaysia.

Tailliez, B., 1971. The root system of the oilpalm on the San. Alberts Plantation in Columbia, *Oleagineaux*, **26:** 435–447.

Tan, K.S., 1973. Fertilizer trials on oilpalms on inland soils on Dunlop estates, In: *Advances in oilpalm cultivation*, Incorp. Soc. of Planters, Kuala Lumpur, 243 pp.

Tinker, P.B.H., 1976. Nutrient demands by the oilpalm in relation to soil nutrient content, In: *Devpts. in Crop Science Part. I. Oil palm research*, (R.H.V. Corley, I.J. Hardon, and B.J. Wood, eds.), Elsevier Amsterdam, 169 pp.

Tinker, P.B.H., 1964. Studies on soil potassiim IV Equilibrium cation activity ratios and responses to potassium fertilizer of Nigerian oilpalms, *J. of Soil Sci.*, **15:** 35–41.

Tinker, P.B.H., and K.W. Smilde, 1963. Dry matter production and nutrient content of plantation oilpalms II. Nutrient content, *Plant and Soil*, **19:** 350–36.

Toky, O.P., and Ramakrishnan, P.S., 1983. Secondary succession following slash and burn agriculture in north east India, *J. of Ecology*, **71:** 735–745.

Uhi, C., and Jordan C.F., 1984. Succession and nutrient dynamics following forest cutting and burning in Amazon, *Ecology*, **65:** 1476–1490.

Vine, B.H., 1956. Studies of soil profiles at the WAIFOR main Station and at some other sites of oilpalm experiments J.W.AFR. Inst. Oil palm Research 1. No. 4: 2–59.

Vitousek, P.M., and Sanford, R.L., 1986. Nutrient cycling moist tropical forest, *Annual Review of Ecology and systematics*, **17:** 137–167.

Young, A., 1989. *Agroforestry for soil conservation*, C.A.B. International, Oxon, U.K., 275–276 pp.

Zeven, A.C., 1965. The origin of the oilpalm, *J. Nigeria Institute for Oil Palms Research*, **4**(15): 218–225.

CHAPTER 8

DECOMPOSITION DYNAMICS AND NUTRIENT FLOW IN PINE FOREST PLANTATION IN CENTRAL JAVA

Herman A. Verhoef[1] and Bintoro Gunadi[2]

[1]Institute of Ecological Science, Vrije Universiteit, de Boelelaan 1087, 1081 HV Amsterdam, the Netherlands

[2]Soil Ecology Lab. Ohio State University, Columbus, USA

ABSTRACT

This chapter focusses on studies on decomposition and nutrient flow in pine forest plantations in Central Java, including the slope of Mount Merapi, an active volcano and the slopes of Mount Merbabu and Mount Sindoro, both dormant volcanoes. An undisturbed pine forest at Aek Nauli, North Sumatera was studied as a comparison for a short period.

Abiotic factors, related to decomposition, and diversity and dominance of forest floor vegetation are presented. Concerning the composition of the soil organisms determination was only possible till the family level, due to the absence of adequate determination literature. Acari and Collembola are among the most dominant soil arthropods.

Litter decomposition was differentiated over the different organic layers and measured using litterbags. Organic breakdown was expressed in terms of weight loss, C and N content of the organic layer and enzymatic activity of the soil microbes. The decomposition rate of freshly fallen pine needles was influenced by moisture, but the chemical composition seemed to be even more important in determining the decay rate of this material.

Concerning the elemental concentrations in rainfall, it is clear that the relatively high input of SO_4, Cl and K is closely related to volcanic activity. There are no indications of decreased vitality of natural vegetation due to this input, which may be caused by adaptation of the vegetation to these high deposition levels.

Nutrient flow through the soil is influenced by the water flow through the soil pores. Porosity, pore size distribution and, consequently, nutrient flow is influenced by Formicidae and Lumbricidae. The presence of these soil animals had a direct effect on the flow of specific elements, such as Cl and K. This can, in its turn, influence bioavailability for plant growth.

Key Words: Pine forest plantations, decomposition, nutrient flow, abiotic factors, soil organisms, atmospheric deposition.

There are about 750,000 ha of pine forest plantations (*Pinus merkusii* Jungh. & De Vriese) in Indonesia of which 50% are in Java. From the 1960s onward, especially in the 1980s, the area of pine forest plantations showed a gradual increase, as they are the main species of reforestation suggested by the government. The increase in pine forest plantations in Indonesia is about 0.5% per year and in Java from 1977 till 1981 this is about 17% per year (Anonymous, 1983).

In Central Java pine is considered an important industrial plant and is number two after teak plantations. The area under pine forest plantation forms 50% of the total forest plantations (Gunadi, 1993).

There is no clear distinction between "tree plantations" (e.g. coffee, rubber, clove, vanilla, cocoa, etc.) and "forest plantations" (e.g. pine, teak, agathis, mahogany, eucalyptus, etc.). Sometimes the term tree plantations is used to describe tree crop cultivated for non timber products and forest plantations for conventional wood production (Evans, 1992).

Besides having commercial importance there are at least three other reasons why pine forest plantations in Indonesia need to be studied.

Firstly, *P. merkusii* is a native tree of Indonesia, originating from northern Sumatra (Backer and van den Brink Jr., 1963; Tantra, 1983). In 1983 the area covered by natural pine forest in northern Sumatra was about 180,000 ha (Anonymous, 1983). Since 1986 there is a pulp and rayon factory which consumes enormous amounts of woody plant material. The area of natural pine forest around Lake Toba decreases quickly and has been replanted with eucalyptus which grow faster and consume much more water, if their root network gets access to groundwater reserves (Calder, 1986). Moreover, in contrast to the natural pine stands which frequently have palms and rattans growing in them, the eucalypt plantations form pure stands which are almost devoid of undergrowth (Gunadi, 1989a). As biodiversity is of global concern, studies on native organisms in- and ex-situ are needed for a better understanding of our, endangered life-support systems (Odum, 1989).

Secondly, pine tree strategies are unique. Distribution of pine is very broad (Satjapradja, 1977) as the trees can grow on poor soil, too (Will, 1967). Pine is one of the oldest living trees in the world. *Pinus longaeva* in the White mountains of eastern California is about 4900 years old; the individual needles of this species function for up to 45 years (Raven *et al.*, 1992). In most species the needles can be retained on the tree for more than one year (Millar, 1974). The decomposition of pine needles is relatively slow compared to broad-leaved plants; 50 per cent breakdown of the needle takes about 7 years (Mason, 1979). This is probably due to the relatively high amounts of phenolic compounds (Kuiters, 1987) and the low pH of the organic layer. Due to the toughness of the needle, decomposition is retarded as well (Swift *et al.*, 1979). Pine litterfall generally exhibits low nutrient concentrations. For needle litter from *P. merkusii* on a high rainfall site in Central Java, average nutrient concentrations were 9.2, 1.9, 0.4, 2.5, 0.4, and 6.0 mg g^{-1} dry wt of Ca, Mg, Na, K, P, and N respectively (Bruijnzeel, 1985). In low nutrient ecosystems, plants grow slowly and use nutrients efficiently, minimizing demand for or loss of nutrients, resulting in smaller amounts of nutrients being cycled (Vitousek and Sanford, 1986; Hobbie, 1992). Slow growing trees are often associated with low seed germination and dispersal (Finegan, 1984). Species from low nutrient ecosystems shed fewer leaves and produce poorer quality litter that decomposes more slowly (Proctor, 1984; van Schaik and Mirmanto, 1985). These species invest in C-based secondary metabolites such as terpene and polyphenol which are effective defences against herbivores (Coley *et al.*, 1985). Based on this information, it is interesting for multidisciplinary scientists to understand more deeply the ecosystem processes in pine forests, especially in the tropics where data are limited and problems are more complicated and unique because of a higher diversity of organisms.

Thirdly, most forest plantations differ from rain forests and savanna and continue to increase rapidly (Evans, 1992). Deforestation of tropical forests is a matter of global concern not only because of tremendous waste of valuable natural resources in economically vulnerable regions, but also because of extinction of organisms and contribution to green house effects (Jepma and Blom, 1992). Ecological studies have traditionally focused attention on natural ecosystems. As it seems impossible to stop deforestation and as there are signs that reforestation using different tree species is successful (Wang *et al.*, 1991; Evans, 1992; Lugo, 1992b), comparative studies between natural forests and forest plantations are needed. Lugo (1992a) pointed out that pine forest plantations are an ideal subject at the ecosystem level. Firstly, age and species composition of the trees are known and controlled from the time the trees have been planted. Secondly, plantations are from a structural point of view "simple" ecosystems, and thus provide contrast to the inherent structural complexity of native forests. Thirdly, plantations can be used to match the age, edaphic and climatic characteristics, and site use histories of secondary forest for successional studies. Finally, understanding the ecological characteristics of plantations is important because of the high rate of tropical forest destruction; the need to restore degraded land to forest production have forced an increase in the rate of plantation establishment.

GENERAL ABIOTIC AND BIOTIC CHARACTERISTICS

Three pine forest plantations in Central Java have been studied by the second author (Fig. 1). The first one is situated on the slope of Mount Merapi, an active volcano. The second one is situated on the slope of Mount Merbabu, a dormant volcano. The latter site is strongly influenced by human activity, as litter is regularly removed by local people to be used for fuel, compost, and bedding for cattle. The third one is situated on the slope of Mount Sindoro, a dormant volcano. Recently, human activities have become a very serious problem in the plantation in Sindoro because it is now used as a recreation forest. As a comparison an undisturbed pine forest at Aek Nauli, North Sumatra was studied for a short period.

The description of these pine forests and some abiotic factors which are related to decomposition are presented in Table 1.

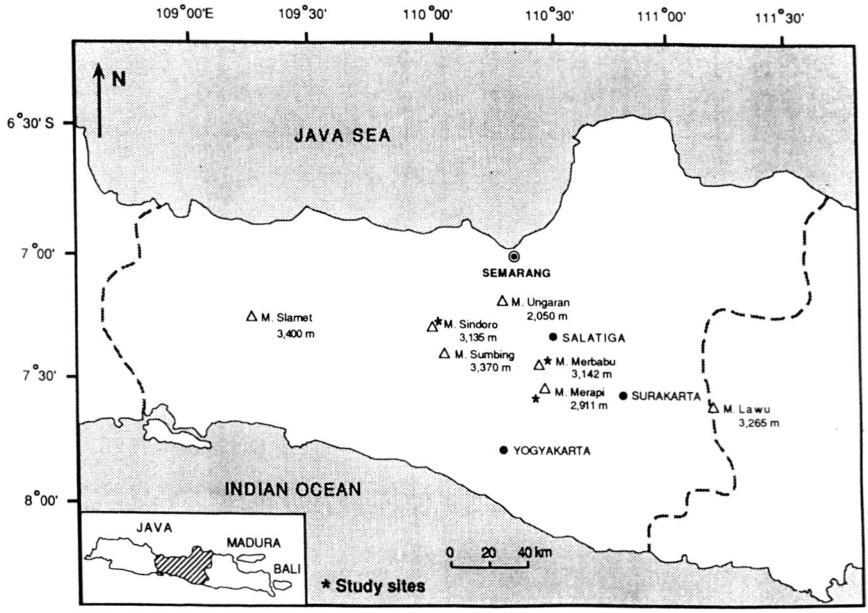

Fig. 1 The location of the study areas in Central Java, Indonesia.

Table 1 Description of four forest plantations in Central Java.

Observations	*Merapi*	*Merbabu*	*Sindoro*	*Aek Nauli*
Forest plantations				
Year of planting	1958	1963	1963	–
Number of trees ha^{-1}	230	310	–	260
Average tree height (m)	30.6	20.2	–	42.2
Average tree diameter (cm)	31.1	29.3	–	42.8
Litterfall (g dry wt m^{-2} $month^{-1}$)	74.9	33.8	–	55.2
Litter standing crop(g dry wt m^{-2})	1098.5	69.8	–	810.1
Abiotic factors				
Altitude (m a.s.l.)	800	1050	1300	1050
Degree of slope (°)	5	5	10	5
Soil type	Andisol	Andisol	Andisol	Andisol
Average air temperature (°C)	21.8	21.4	19.1	21.1
Average soil temperature (°C)	22.0	21.0	19.2	–
Average air humidity (%)	76.2	75.4	82.5	74.0
Average soil moisture (%)	27.9	9.8	21.1	–
pH H_2O				
L layer	4.5	4.4	–	4.4
F1 layer	5.3	–	–	4.7
F2 layer	5.5	–	–	5.1
H layer	5.7	5.6	–	5.4
Soil (5 cm depth)	5.8	5.7	6.7	5.4
Annual rainfall (mm)	3592	3206	±3000	2859

Since the diversity of the forest floor vegetation in the tropical pine forest plantation is very high, the composition of the undergrowth may influence the decomposition of pine needles. The composition of the undergrowth and its importance value are shown in Table 2. There were 29, 32, and 10 species of undergrowth plants in the pine forest plantations on the slope of Mount Merapi, Merbabu, and Sindoro, respectively. The most dominant undergrowth at Merapi is the grass *Centotheca lappacea,* at Merbabu the grass *Polythrias amaura,* and at Sindoro the herb *Eupatorium riparium.* The dominant forest floor vegetation in the natural pine forest at Aek Nauli were Zingiberaceae and *Chromolaena odorata.*

Information on the composition and the density of soil arthropods in pine forest plantations in Indonesia and even more so in natural pine forest is scarce. As there are no guide lines till the species level, the data shown in Table 3 presents the lowest taxonomy groups, determined according to common entomology textbooks (Bland and Jaques, 1978;

Table 2 Undergrowth and its importance value (%) in three forest plantations. The data are based on Gunadi and Notosoedarmo (1988).

Species	*Merapi*	*Merbabu*	*Sindoro*
Ageratum conyzoides L.	–	11.15	–
Axonopus compressus (Swartz.) Beauv.	–	20.89	–
Biophytum reinwardtii (Zucc.) Klotzsch	–	1.41	–
Borreria alata (Aubl.) DC.	6.88	–	–
B. distans (H.B.K.) Cham. & Schlecht.	–	2.59	–
B. laevis (Lamk) Griseb.	–	6.01	–
Breynia cernua (Poir.) M.A.	4.13	–	–
Centella asiatica (L.) Urb.	3.81	34.23	6.11
Centotheca lappacea (L.) Desv.	117.27	–	–
Chromolaena odorata (L.) R.M. King & H. Robinson	6.86	–	–
Chrysopogon aciculatus (Retz.) Trin.	–	4.39	–
Clerodendrum serratum (L.) Moon	2.14	–	–
Crassocephalum crepidioides (Benth.) S. Moore	1.92	1.56	–
Cyanotis cristata (L.) D. Don	–	1.79	–
Cyclea barbata Miers	1.92	–	–
Cyperus brevifolius (Rottb.) Hassk.	3.91	1.34	5.91
C. cyperoides (L.) O. Kuntze	6.46	–	–
C. haspan L.	–	3.14	–
C. kyllingia Endl.	1.92	–	–
Desmodium heterophyllum (Willd.) DC	–	4.77	–
Digitaria sp	–	3.59	–
Disperis javanica J.J.S.	1.92	–	–
Duranta erecta L.	–	–	5.91
Elephantopus scaber L.	2.36	–	–
Emilia sonchifolia (L.) DC. ex Wight	–	2.59	–
Erigeron sumatrensis Retz.	–	5.23	–
Eupatorium riparium Reg.	32.51	1.41	224.26
Gonostegia triandra (Bl.) Miq.	–	29.93	7.37
Hydrocotyle sibthorpioides Lmk.	–	8.77	–
Imperata cylindrica (L.) Beauv	8.80	24.72	–
Lantana camara L.	15.41	–	–
Lindernia crustacea (L.) F.v.M.	–	1.41	–
Melastoma affine D. Don	–	–	7.96
Oplismenus burmanni (Retz.) Beauv.	3.91	–	5.91
Orchidaceae	–	–	6.48
Oxalis corniculata L.	–	6.77	–
Paederia scandens (Lour.) Merr.	3.70	–	–
Panicum repens L.	–	8.80	12.03
Paspalum commersonii Lamk	–	5.79	–
P. conjugatum Berg.	35.79	3.18	19.02

Contd.

Tabel 2 Contd.

Species	*Merapi*	*Merbabu*	*Sindoro*
Phyllanthus urinaria L.	1.92	9.17	–
Piper aduncum L.	4.09	–	–
Polygala paniculata L.	–	1.34	–
Polythrias amaura (Buese) O.K.	–	38.72	–
Psidium guajava L.	3.91	–	–
Pteridophyta	3.91	–	–
Pteris ensiformis Burm. f	3.91	–	–
Richardia brasiliensis (Gomez) Hayne	–	22.19	–
Rostellularia sundana Bremek.	1.92	19.59	–
Sida acuta Burm. f.	1.92	1.34	–
Sporobolus poiretii Hitchc.	–	11.15	–
Stachytarpheta indica (L.) Vahl	1.92	–	–
Synedrella nodiflora (L.) Gaertn.	–	1.41	–
Thunbergia fragrans Roxb.	6.00	–	–
Zingiber americans Bl.	6.79	–	–

Importance value is according to Mueller-Dombois, and Ellenberg (1974). Species names are according to Backer and van den Brink (1963, 1965, 1966).

Table 3 The density of soil arthropods (m^{-2}) (collection using soil cores), based on Gunadi and Notosoedarmo (1988); Gunadi (1989b).

Taxonomic groups	*Merapi*	*Merbabu*	*Sindoro*	*Aek Nauli*
Crustacea				
Oniscoidea	72	–	135	42(*)
Arachnida				
Pseudoscorpionides	48	24	91	–
Opiliones	72	24	–	–
Acari	5608	1366	10227	684
Araneae	144	72	161	129
Pauropoda	24	–	8	–
Diplopoda				
Polydesmidae	24	–	176	3(*)
Julidae	72	–	25	29
Pentazonia	–	–	3*	–
Chilopoda				
Scolopendromorpha	48	24	11	6(*)
Symphyla	120	24	570	–
Insecta				
Diplura				
Acerentomidae	312	24	–	14
Japygidae	72	24	–	–
Campodeidae	72	24	83	72
Collembola				

Table 3 Contd.

Taxonomic groups	*Merapi*	*Merbabu*	*Sindoro*	*Aek nauli*
Onychiuridae	3379	263	3683	1726
Hypogastruridae	2732	360	2528	1481
Entomobryidae	2804	96	2142	1481
Sminthuridae	527	288	348	460
Thysanura				
Lepismatidae	24	–	–	–
Orthoptera				
Acrididae	72	24	8	–
Gryllidae	48	1(*)	82	72
Dictyoptera				
Phasmida	1(*)	–	1(*)	1(*)
Blattodea	48	48	5	29
Mantodea	1(*)	–	2	–
Dermaptera				
Labiidae	72	1(*)	3	3(*)
Isoptera				
Rhinotermitidae	216	–	–	960(*)
Liposcelidae	48	48	8	58
Thysanoptera	72	72	4	3(*)
Hemiptera				
Lygaeidae	1(*)	48	24	–
Tingitidae	24	24	–	–
Aphididae	-	48	8	14
Coccidae	72	–	–	–
Coleoptera				
Carabidae	72	48	9	72(**)
Scydmaeidae	–	–	10	–
Staphylinidae	72	72	1	86
Pselaphidae	48	–	–	–
Histeridae	24	–	–	–
Scarabaeidae	24	24	–	–
Elateridae	–	48	–	–
Cucujidae	–	24	–	–
Coccinellidae	72	24	–	–
Byturidae	24	–	–	–
Tenebrionidae	1(*)	–	–	72(**)
Chrysomelidae	48	–	–	–
Scolytidae	72	24	–	72(**)
Curculionidae	48	1(*)	28	–
Lepidoptera larvae	72	72	18	–
Diptera larvae	72	24	50	14
Formicidae	312	144	143	590

*collected using quadrants
**collected using pit falls

Borror *et al.*, 1981; Harde, 1984). There were 44, 33, 32, and 26 groups of soil arthropods found in the pine forest plantations at Merapi, Merbabu, Sindoro, and in the natural pine forest at Aek Nauli, respectively. Acari and Collembola (Onychiuridae, Hypogastruridae, Entomobryidae, and Sminthuridae) were the most dominant soil arthropods in all study areas. The comminutors, which are supposed to have an important role in decomposition, are Oniscoidea and Julidae. They were absent at the Merbabu site because of the small amount of organic matter and the low soil moisture. Important predators which were found in relatively high densities were Pseudoscorpionida, Araneae, Scolopendromorpha, Carabidae, and Staphylinidae. The less known decomposers such as Symphyla, Protura (Acerentomidae), and Diplura (Japygidae and Campodeidae) may be important in tropical pine forest plantations and in natural pine forests because of their abundance. Some soil arthropods such as Rhinotermitidae, Lepidoptera larva, Diptera larva, and Formicidae live in aggregated colonies which makes estimation of densities difficult.

It is well known that soil arthropods interact with microorganisms during the decomposition of pine needles (Anderson *et al.*, 1984; Fitter *et al.*, 1985; Teuben, 1991). In mor-soil types like in coniferous forest, fungi are more dominant than bacteria (Swift *et al.*, 1979). The dominant fungi of the litter layer in the pine forest plantation on the slope of Mount Merapi are *Penicillium citrinum, P. glabrum, Penicillium* sp., *Aspergillus ochraceus, Fusarium* cf. *oxysporum, Trichoderma harzianum, T. pseudokonigii, Mucor hiemalis, Choanephora cucurbitarum, Cladosporium* sp, *and Verticillium* sp (CBS, Baarn, 1992, private determination).

SEASONAL DYNAMICS OF DECOMPOSITION

Decomposition of litter is regulated by the interaction between environmental conditions, litter quality and decomposer organisms (Swift *et al.*, 1979). In general, the rate of decomposition in the tropics is higher than that in the temperate zone (Kimmins, 1987; Jordan, 1989). This difference may be caused by environmental conditions such as temperature and water content of the litter (Taylor and Parkinson, 1988). Usually the effect of either factor is not constant and they interact in a complicated way. Dry and wet seasons appear to have a big effect on litter turnover and mobilization of nutrients in the pine forest plantations in Central Java (L.A. Bruijnzeel, 1983, Unpublished Ph.D. Thesis, Vrije Universiteit, Amsterdam; Gunadi, 1994a). Altitude may also influence the nitrogen mineralization and cation-mobilization in tropical rain forests (Marrs *et al.*, 1988). Litter quality plays an important role in the decomposition of pine litter (Raven *et al.*, 1992). In many studies the C-to-N ratio has been used to monitor decomposition and to predict weight loss (Taylor *et al.*, 1989). However, it should be mentioned that the importance of climatic variables on decomposition rates is explained by the effects of microclimate on soil biota, as microorganisms, in interaction with soil fauna, are responsible for the decay of litter (Anderson and Ineson, 1984; Verhoef and de Goede, 1985; Verhoef and Brussaard, 1990).

The seasonal dynamics of decomposition were studied in a pine plantation at different altitudes on the slope of Mount Merapi, Central Java. The variables used in monitoring decomposition were weight loss, C content, N content of the organic soil and dehydrogenase activity of the soil microbes. As the litter in pine forests can be differentiated into an L layer and F layer (Kendrick, 1959; Millar, 1974), the decomposition processes were monitored in these different horizons to discriminate between decomposition stage. Carbon and N content of the organic soil were studied separately as the dynamics of each specific element give more information on decomposition rate than the C-to-N ratio alone (Berg and Staaf, 1981). As soil enzymes allow a reliable determination of microbial activity in the organic layer (Teuben and Roelofsma, 1990; Faber *et al.*, 1992) dehydrogenase activity was measured.

In both plots (one at 600 m and one at 800 m altitude) litterbags filled with needle litter from the L layer and the F layer were randomly distributed over the 0.2 ha area, as stratified sets of two litterbags (L bags on top of F bags), in the L and F horizon, respectively. The L and F bags were placed on top of each other as this is the natural position of the respective layers. The experimental layout is a modification from Herlitzius (1983). In the first treatment ("from-sequence treatment" *sensu* Herlitzius, 1983) all litterbags were introduced in the field at the start of the experiment and retrieved at intervals (after 2, 4, 6, 8, 10 and 12 months). These bags, which were exposed to increasing periods of time in the field, are referred to as "equal start" bags (ES bags). The total number of the ES bags was 192 (: 8 replicates × 2 layers × 6 times × 2 sites). In the second treatment (the "to-sequence treatment" *sensu* Herlitzius, 1983) removed ES bags were immediately replaced by new bags. These bags, which were exposed to a decreasing period in the field, are indicated as "equal end" bags (EE bags). These bags (also 192) were removed after 1 year and became the chronological mirror image of the equal start treatment. This provides a series of samples which had not only experienced different periods of exposure (: 2, 4, 6, 8, 10 and 12 months), but also different dates of exposure. In this way comparison of similar exposure periods during different times of the year was possible. The two treatments (ES and EE) were used to study dynamics in weight loss and changes in C and N content in the two layers of the two sites, and the importance of season versus incubation time on changes in these variables (Verhoef, 1995).

Weight loss from the L and F layers in both plots as resulted from the "equal start" treatment is shown in Fig. 2. The interaction Plot × Time × Layer (Table 4) was just significant ($P = 0.043$). The highly significant Layer × Plot interaction ($P < 0.001$) underlines the result that in the upper plot, the weight loss of the L layer was less than that in the F layer throughout the year, whereas in the lower plot, the weight loss of the L layer was mostly greater than that in the F layer.

In our study the highest weight loss during 1 year was in the midst of the wet season. In a previous study, however, on litterfall and litter turnover in the same pine forest, decomposition expressed as litter turnover (k) was relatively low (0.20) in the wet season compared to

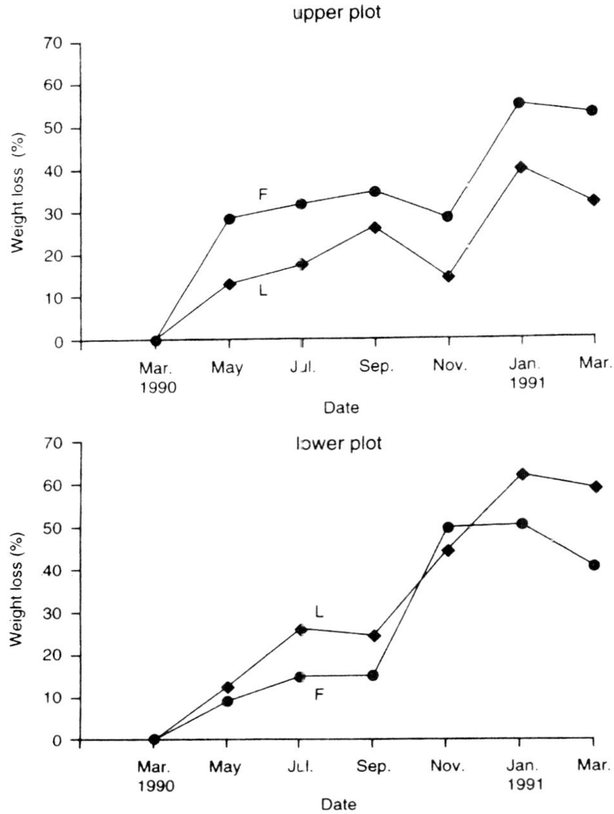

Fig. 2 Dry weight loss (% of the initial weight) in L and F layer in the upper and the lower plot. L = L layer; F = F layer. (From Gunadi *et al.* 1998: Fig 1; printed with permission of Elsevier Science Ltd.)

Table 4 F-Values from analyses of variance of the water content, dry weight loss, carbon and nitrogen content and dehydrogenase activity for plot (high and low elevation), layer (L,F) and time (2 to 12 mo.) (n = 192).

Variables	*Plot*	*Time*	*Plot × time*	*Layer*	*Layer × plot*	*Layer × time*	*Plot × tme × layer*
Water content (% wet wt)	7.1**	55.5***	10.2***	3.7 ns	0.0 ns	17.1***	2.7*
Weight loss (%)	1.6 ns	41.2***	10.4***	7.4**	104.3***	2.2 ns	2.4*
Carbon content (% dry wt)	13.3***	51.2***	3.2*	365.1***	3.9 ns	1.9 ns	0.6 ns
Nitrogen content (% dry wt)	62.3***	9.3***	2.6*	84.4***	0.8 ns	20.8***	1.5 ns
Dehydrogenase (mg TPF g^{-1} d. wt) activity	134.3***	35.0***	12.9***	15.8***	5.0*	15.5***	4.4**

*P≤0.05, **P ≤0.01, ***P ≤0.001, ns not significant. (Modified from Gunadi *et al.* 1998: Table 2; printed with permission of Elsevier Science Ltd.)

the dry season (0.80) (Gunadi, 1994a). These contrasting results should be attributed to the inaccuracy of the turnover-method in estimating the decomposition rate. It is therefore suggested that the results from the litterbag study are more reliable than those based on litter turnover.

Initial C contents in the needle material used in the litterbags were the same in both plots: 49% in the L and 38% in the F layer (Fig. 3). Carbon content decreased over the year from 49 to 35–37% in the L layer and from 38 to 28–31% in the F layer in both plots. Based on these values it is suggested that after 1 year of decomposition L (49% C) becomes F (38% C). All interactions with layer (Table 4) were not significant, while the main effect of layer was highly significant

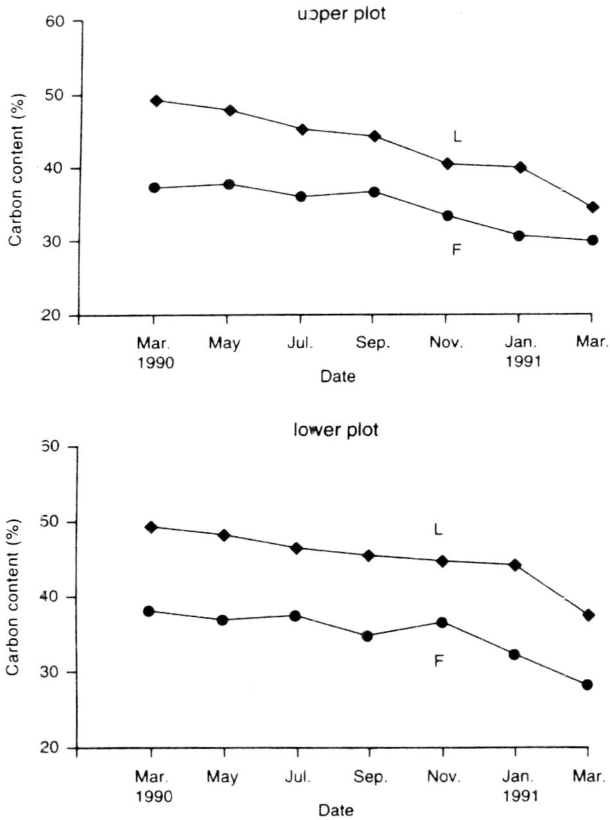

Fig. 3 Carbon content (% dry weight) in L and F layer in the upper and the lower plot. L = L layer; F = F layer. The data for March 1990 represent the value of the original material. (From Gunadi *et al.* 1998: Fig 2; printed with permission of Elsevier Science Ltd.)

(P < 0.001). So, the difference between L and F layer was very clear and more or less constant over time and plot.

Initial N contents were similar in both plots: in the L layer 0.87% and in the F layer 1.17–1.12% (Fig. 4). The interaction plot × time × layer (Table 4) was not significant, but the interaction time × layer is highly significant (P < 0.001), indicating the differing time course of N content in the two layers: the L layer showed a gradual increase and the F layer a more or less gradual decrease, which may indicate the start of net N-mineralization in the F layer. In the lower plot N content of the L layer increased to the start level of N content of the F layer, which agrees with the earlier statement that after 1 year L becomes F. However, this was less clear in the upper plot.

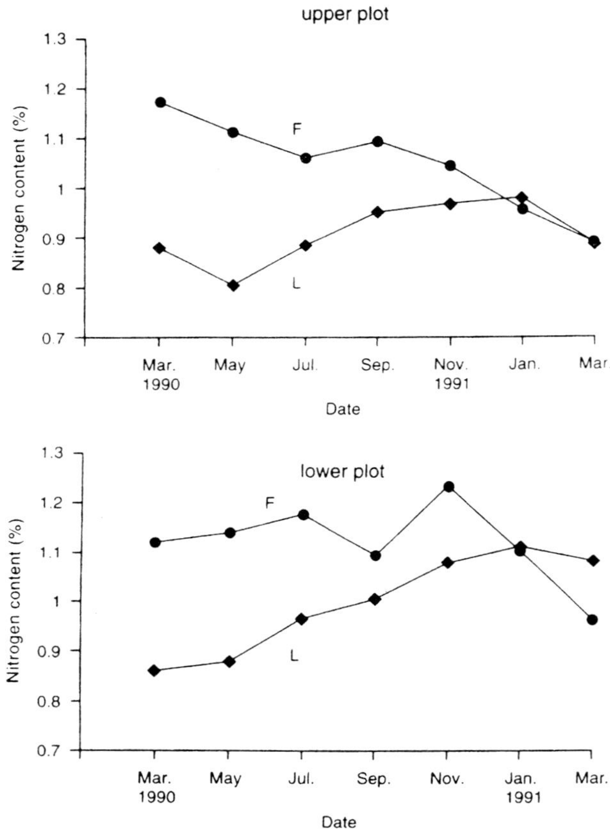

Fig. 4 Nitrogen content (% dry weight) in L and F layer in the upper and the lower plot. L = L layer; F = F layer. The data for March 1990 represent the value of the original material. (From Gunadi *et al.* 1998: Fig 3; printed with permission of Elsevier Science Ltd.)

Literature shows close relationships between C-to-N ratio and weight loss. However, absolute-C and N content, presented separately, also provide useful information. During the one year study the pattern of C content was more or less constant in both layers and plots. According to Berg and Staaf (1981) three separate phases can be distinguished for litter N over time: leaching, accumulation and release. In our study the N leaching phase was brief in the high plot in the L layer. This might be caused by less dense pine forest plantations in the upper plot, so the effect of the rainfall is greater than in the lower plot. The initial N content in the L layer was the same in both plots, about 0.87%. The accumulation phase is followed by the release phase, starting at a value of 1.0% in the upper plot and 1.1% in the lower plot. The N content, at which transition from accumulation to release occurs, is referred to as the critical N percentage. For *P. contorta* initial total N and the critical N% were found to be 0.6 and 1.0% respectively (Hodkinson, 1975), for *P. sylvestris* 0.4 and 0.7% (Staaf and Berg, 1977) and for *P. radiata* 0.6 and 1.4% (Will, 1967). We have calculated these data based on the assumption of an average of 50% C, as the original data give only the C-to-N ratio. The release phase was continued in the F layer in both plots.

Initial dehydrogenase activity in the litterbags was similar in both layers and plots (c. 100–150 mg TPF g^{-1} d.wt). As the interaction plot × time × layer is significant ($P = 0.001$), no conclusions can be drawn from the, significant, time layer interaction ($P < 0.001$). Still, there seemed to be a tendency of an increasing activity in the L layer and a more or less decreasing activity in the F layer (Fig. 5).

Dehydrogenase activity fluctuated during the year in both layers and both plots. In the lower plot maximal dehydrogenase activity in the L layer might be related to the fact that in the midst of the wet season in January the critical N% was reached. This is supported by the general positive correlation between total N content and dehydrogenase activity.

Results from the ES and EE series treatments are presented in Fig. 6. Carbon content has been chosen instead of the more commonly used total weight loss because C content showed less variation in time than weight loss. The statistical test to decide which mirror symmetry is applicable is shown in Table 5. According to test 1, the interaction treatment × time and the main effect (treatment) were not significant for C content in the L layer in both plots. This means that incubation time determined C dynamics in the L layer of both plots. According to test 2, significant effects were shown in the L and F layers of both plots. Season was not important for C dynamics in the L and F layers of both plots. Incubation time and season were not important in the F layer (tests 1 and 2).

As mentioned earlier, one of the goals of this investigation was to assess the different variables to study the decomposition of pine needle

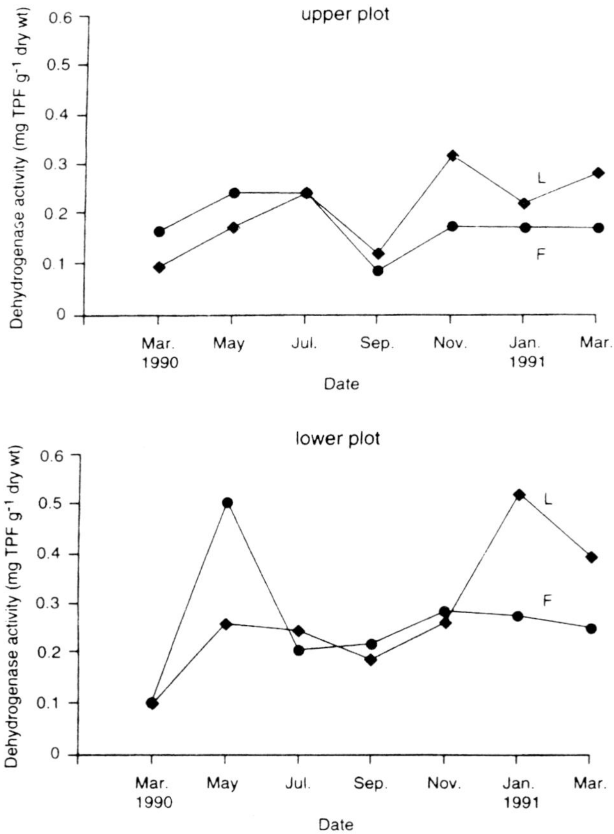

Fig. 5 Dehydrogenase activity (mg TPF g^{-1} dry wt) in L and F layer in the upper and the lower plot. L = L layer; F = F layer. The data for March 1990 represent the value of the original material. (From Gunadi *et al.* 1998: Fig 4; printed with permission of Elsevier Science Ltd.)

litter in tropical plantation forests. Herlitzius (1983) used weight loss as a variable to monitor the decomposition of hazel leaves in different temperate forests (alluvial, beech and spruce). He concluded that duration of exposure was important in the decomposition of the hazel litter in both alluvial and beech forests. For the weight loss in the spruce forest, he concluded that date of incubation was more important than duration of exposure. In our study, duration of exposure seemed more important than date of incubation in both plots. As duration of exposure is closely related to degradation stage and the date of exposure is closely related to seasonal effects, it is concluded that season (especially water content) has an important influence on the decomposition rate.

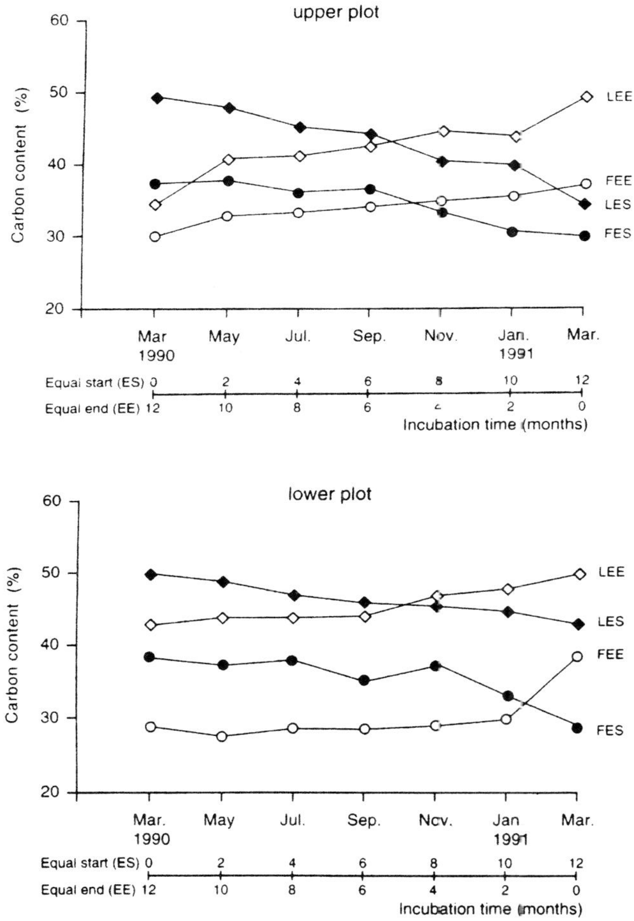

Fig. 6 Carbon content (% dry weight) in L layer (equal start: ES) L; in F layer (equal start: ES) F; in L layer (equal end: EE) L; in F layer (equal end: EE) F, in the upper and the lower plot. (From Gunadi *et al.* 1998: Fig 5; printed with permission of Elsevier Science Ltd.)

However, the chemical composition of the organic material seems to be even more important in determining the decomposition rate of freshly-fallen tropical pine litter.

THE FLOW OF NUTRIENTS

There are different ways to study nutrient flow in terrestrial ecosystems e.g. by quantifying input of nutrients into ecosystems, loss of nutrients

Table 5 F–values from analyses of variance of the carbon content from two plots and layers against treatment (equal start and equal end) and time (2 to 12 mo.). Test 1 concerns the testing of vertical axis symmetry and test 2 concerns the testing of horizontal axis symmetry. (n = 96).

Locations	*Treatment* (*test 1*)	*Treatment* × *time* (*test 1*)	*Treatment* × *time* (*test 2*)
Upper plot			
L layer	2.219 ns	1.645 ns	42.273***
F layer	3.196 *	3.267*	28.227***
Lower plot			
L layer	0.066 ns	1.925 ns	8.987***
F layer	114.287***	5.363***	7.405***

*P ≤ 0.05, ***P ≤ 0.01, ns - not significant. (From Gunadi *et al.* 1998: Table 4; printed with permission of Elsevier Science Ltd.)

from ecosystems, turnover of nutrients in ecosystems, and uptake of nutrients from soils by plants (Harrison *et al.*, 1990). Recently, input of nutrients via rain water into forests has been studied intensively both as a result of anthropogenic influences (Brasell and Sinclair, 1983; Core *et al.*, 1992; Wright *et al.*, 1992), and due to natural pollution from volcanoes in natural tropical rain forests (Johnson and Parnell, 1986; Veneklaas, 1990) and in forest plantations (Bruijnzeel, 1989; Riekerk and Korhnak, 1992).

Decomposition of pine needles in Central Java has been studied by Gunadi *et al.* (1998) and it was found that seasonal effects (e.g. rainfall) had more impact on decomposition than litter quality. High precipitation retarded decomposition. Furthermore, it is well known that rain water also has an important role in transferring water, nutrients and pollutants from tree canopies (McCune and Boyce, 1992). Although pine forest plantations in Java usually grow along the slope of volcanoes on relatively fertile soils, nutrient losses from the forest plantations via leaching are high. This makes these soils relatively poor in nutrients (Bruijnzeel, 1984). Input of nutrients via rain water are thus important, especially in areas influenced by sulphur from an active volcano.

Nutrient fluxes have been studied in two plots along the slope of the active stratovolcano Mount Merapi.

Within the earlier mentioned high (800 m) and the low (600 m altitude) plot twelve rain collectors of 13 cm diameter and 225 mm capacity were used. Two collectors were located outside the plot in an open area to collect rainfall and ten were located randomly inside the plot to collect throughfall. Throughfall was collected 1 m above ground level, below the undergrowth. Filters of fine-mesh (200 μm) made from nylon gauze were placed in the plastic funnels of the collectors in order to prevent contamination by debris. The accumulated litter was removed and the filters were cleaned weekly. The water containers were covered by black plastic bags to minimize algal growth. Every week the volume of throughfall was determined and samples were taken for chemical analysis.

Soil water percolate was collected from mesocosms by vacuum tube lysimeter (Fig. 7). The cup has a pore size of 1–2 μm, and is 2 cm in outer diameter, and 5 cm in length (P80, Staathliche Porcelän Manufaktur, Berlin, Germany). Before installation in the field, each ceramic cup was washed with 0.1 M HCl and subsequently 3 times with distilled water. The mesocosm is constructed of dark grey PVC with the same diameter as the rain collectors (: 13 cm diameter), 30 cm height, 0.5 cm thick, open at the top and covered by 200 μm nylon gauze, with a closed bottom, and connected with well closed dark glass bottle with a capacity of 225 mm. Mesocosms were filled with an undisturbed soil column, including the litter and fragmented layers, using a soil core and were put randomly in the soil in such a way that the soil surface of the column was level with the surrounding soil surface. Soil animals could enter and leave the mesocosms via 6 holes in the upper part of the tube. Each hole had a 3 cm diameter and was covered by 4 and 2 mm nylon gauze for the L and F layer, respectively. All mesocosms were installed randomly, below understorey vegetation with 8 replicates in each plot. The tension inside the glass bottle was set at –0.4 bar in order to suck the percolate. Therefore, a vacuum pump was used weekly.

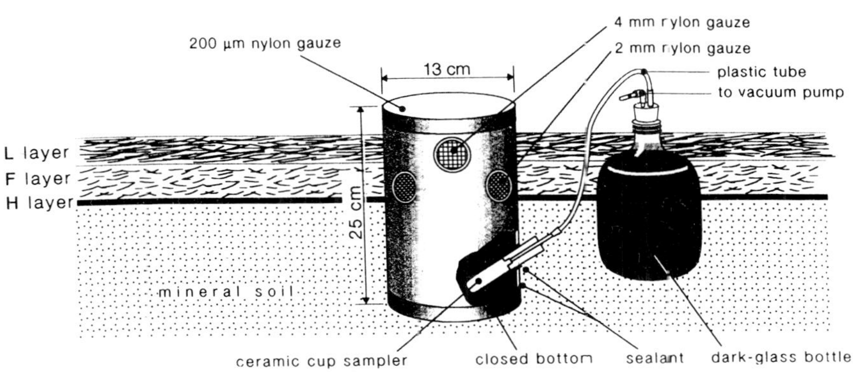

Fig. 7 Mesocosm and vacuum tube lysimeter installation in the field.

The pattern of volumes of rainfall and throughfall were the same throughout the study period (Fig. 8). Rainfall interception for the high

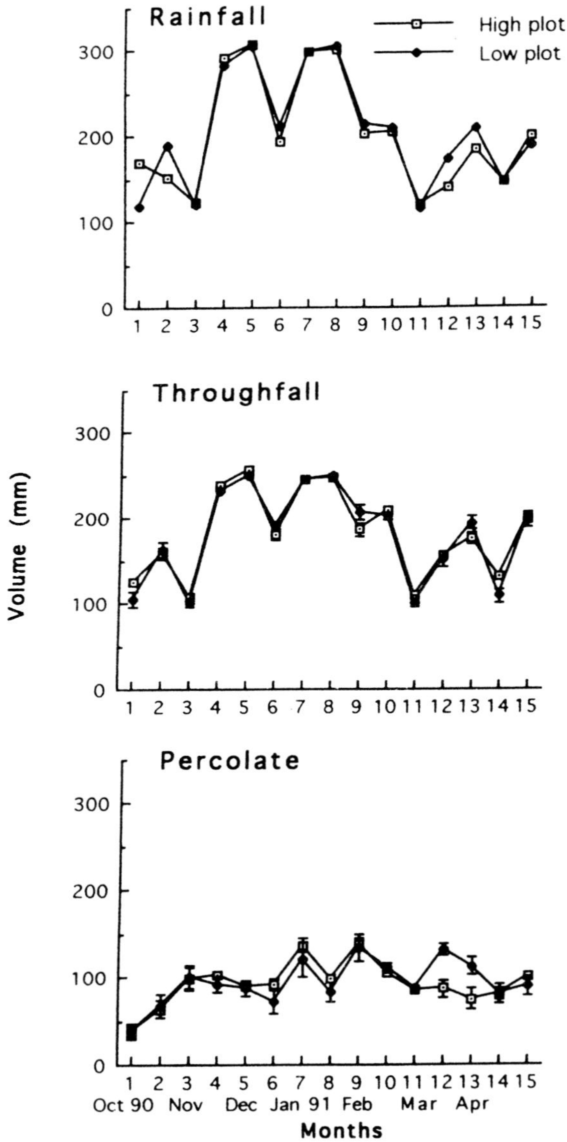

Fig. 8 Volumes of rainfall, throughfall (±S.E.), and percolate (±S.E.) in the high and low plots. The x axis is divided into 15 series collections.

and low plots were also similar i.e. 19 and 18% of incident precipitation. Maximum rainfall and throughfall occurred in the middle of the wet season, about 300 mm per two weeks.

Volumes of percolate were more constant in time compared to those of rain water inputs. The two plots also showed a similar pattern. On most occasions volumes of percolate were about half of those for rainfall or throughfall (i.e. about 100 mm per two weeks). As the effects of additional interception by understorey vegetation (i.e. below the throughfall gauges) and the gauze lids of the mesocosms will be small, the reduced volumes of percolate may reflect enhanced evaporation from the litter in the mesocosms associated with the removal of surrounding ground vegetation during instalment. In general the spatial variation in percolate volumes was higher than that of the throughfall volumes.

pH and concentrations of SO_4, NH_4, NO_3, Cl, K, Ca, and Mg in rain water fluctuated more or less in the same manner in both plots (Fig. 9). Fluctuations in concentrations of SO_4 probably bear a close relation to the volcanic activity of Mount Merapi.

The initially high concentrations of NO_3 and NH_4 were observed during a relatively dry spell at the start of the wet season.

Cl, K and Ca patterns differed between the plots. The peaks in K concentrations occurred in the same period as the peak in Cl concentrations. These high input phenomena may be of maritime or volcanic origin. In the high plot Ca concentrations were stable during the study period and lower than those in the low plot.

The pHs of throughfall and percolate in both plots were slowly increasing like in rainfall (see Fig. 10 for the high plot and Fig. 11 for the low plot). The pH of percolate was always higher than that in throughfall. This is due to the fact that mineral soil is relatively alkaline although the organic matter developed from the pine needles is relatively acid.

As in rainfall, SO_4 concentrations in throughfall and percolate at both plots were initially low and stable. Then SO_4 concentrations started to fluctuate in throughfall and percolate in both plots as in rainfall.

Patterns of NH_4 concentrations in throughfall and percolate were similar in both plots (Figs. 10 and 11). The pattern of NH_4 concentrations in throughfall and percolate in the high plot followed that in rainfall (Fig. 9).

In both plots the NO_3, Ca and Mg concentrations in throughfall were always lower and the S.E. smaller than those in percolate (Figs. 10 and 11).

As conservative elements, Cl and K have more or less the same pattern during this study. In most of the measurements, Cl and K

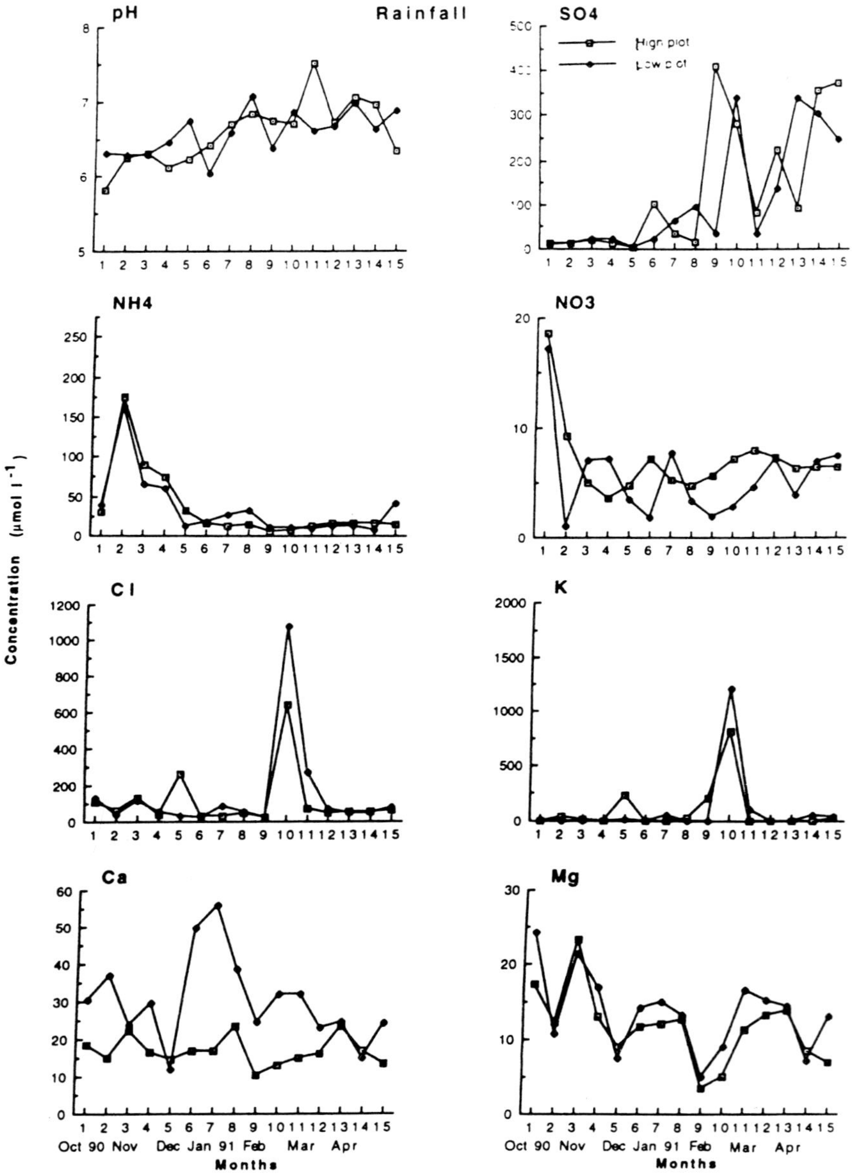

Fig. 9 Dynamics of pH and element concentration in rainfall in both plots.

concentrations in throughfall were lower than those in percolate (Figs. 10 and 11).

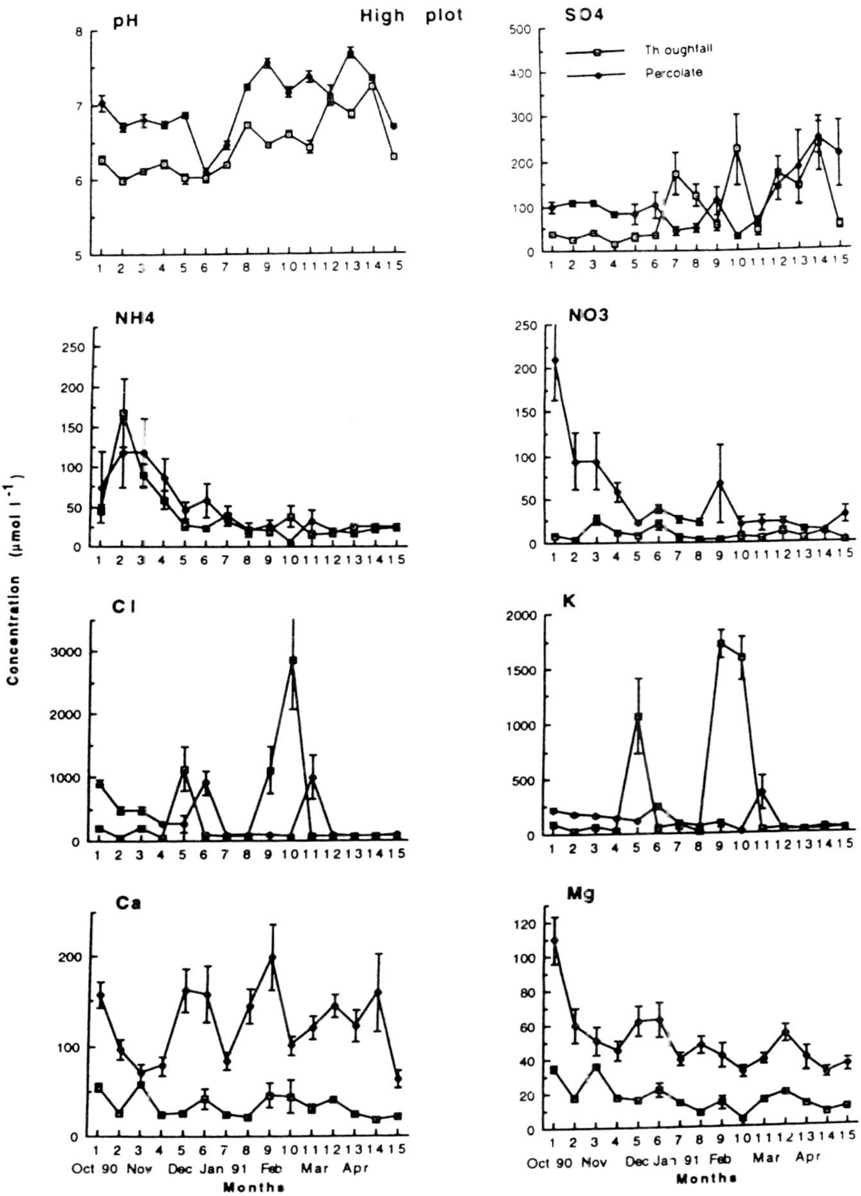

Fig. 10 Dynamics of pH and element concentration in throughfall and percolate in the high plot. The results are ± S.E.

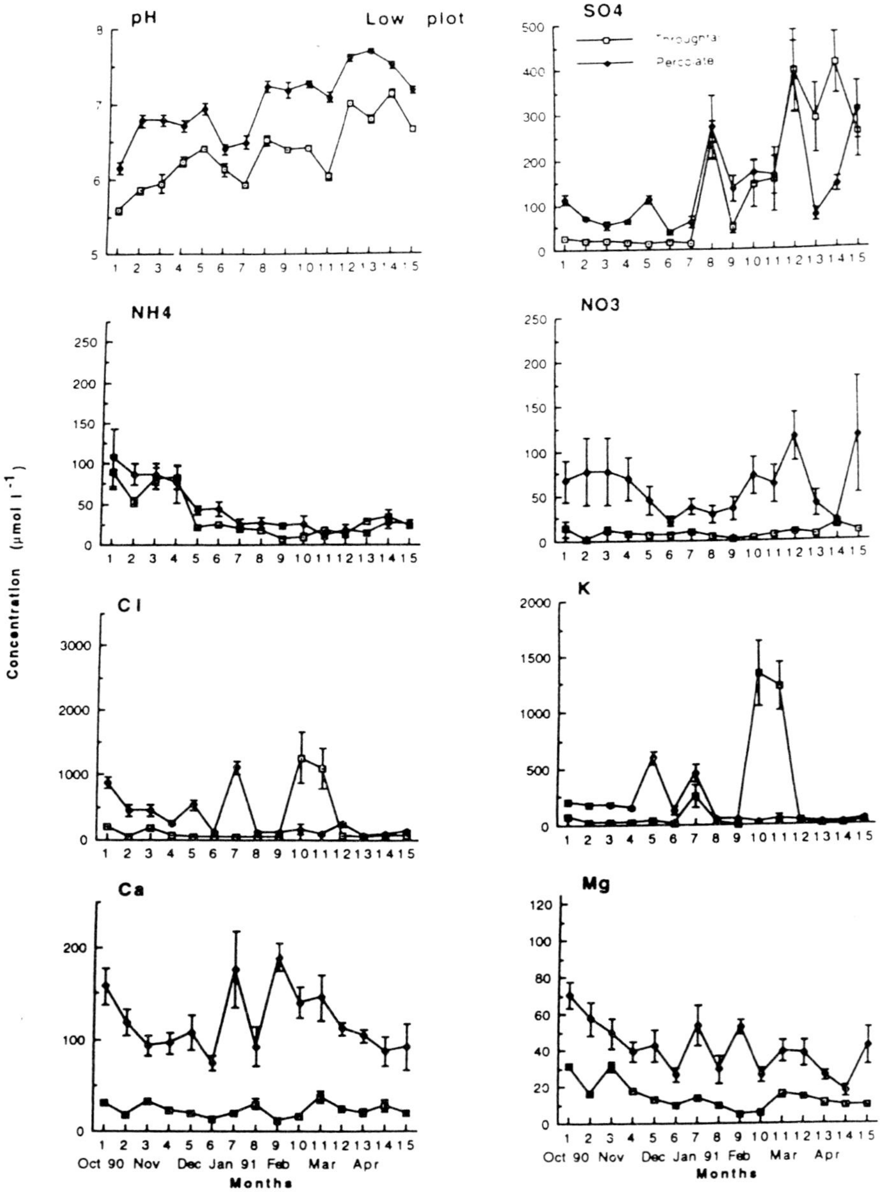

Fig. 11 Dynamics of pH and element concentration in throughfall and percolate in the low plot. The results are ± S.E.

Rainfall interception in this study was 19 and 18% of incident rainfall and was lower compared to the results calculated by Bons in Bruijnzeel (1988) for a 31-year-old *P. merkusii* forest plantation in West Java (25–28%). This may be due to the fact that although the trees at the Merapi site had about the same age, the thinning of branches by local people before the beginning of the study reduced the canopy biomass and made the interception lower. Another reason may be the fact that rainfall intensities and total rainfall for the sites in West Java were considerably lower (Bons, personal communication).

Element concentrations in rainfall, throughfall, and percolate in the wetter period (monthly rainfall of about 250 mm) were comparable with the drier period (monthly rainfall of about 150 mm). This is probably because the study took place in the wet season. It is well known that due to volcanic activity HCl and H_2SO_4 concentrations in rain water may increase (Darzi and Winchester, 1982; Smith *et al.*, 1982). The dominance of HCl in volcanic "acid rain" makes it distinctly different from anthropogenic acid rain which usually contains HNO_3, H_2SO_4 and $(NH_4)_2$ SO_4 (Johnson and Parnell, 1986; van Breemen and van Dijk, 1988). Comparison of element concentrations with other research in the tropics (Table 6) shows that the high concentrations in rainfall and throughfall of Cl in this study were similar to the results from the slope of the Masaya volcano in Nicaragua. The SO_4 concentrations in rainfall in this study were about 4 times higher than the former results. SO_4 and Cl concentrations in rainfall in this study were much higher than other researches done by Bruijnzeel (1989) in a Javanese plantation forest or by Veneklaas (1990) in a Colombian montane rain forest. Both studies were carried out in upland volcanic terrain with active vulcanism, although the volcanoes were not as close to the sampling sites as in the present study. Moreover, contrary to these two examples, the present study was done on the slope of a continuously active volcano.

There was no negative correlation between SO_4 -pH or Cl -pH as shown in other studies (Johnson and Parnell, 1986; Veneklaas, 1990). On the other hand, cation concentrations, especially K, were high in rainfall and may have neutralized the pH. Also there were significant positive correlations between Ca-pH and Mg-pH. Following the high concentrations of SO_4 and Cl in rainfall in this study, the concentrations of both elements in throughfall increased as reported by the authors mentioned in Table 6, except for SO_4 in the high plot. This might be caused by a heterogeneous input of SO_4 via rainfall; local input following the wind direction may have influenced the distribution of SO_4 on the canopy in the high plot. The very high percolate data on SO_4, NO_3, Cl, K, Ca, and Mg of Johnson and Parnell (1986) are difficult to compare

Table 6 Comparison of pH and elements input (kg ha^{-1} $year^{-1}$) by rainfall and throughfall with other research in pine forest plantations.

Location	*pH*	*SO_4–S*	*NH_4–N*	*NO_3–N*	*Cl*	*K*	*Ca*	*Mg*
Central Java[1]								
Rainfall	6.6	66.5	19.1	3.4	193.5	89.4	31.9	9.0
Throughfall	6.4	50.6	15.2	3.4	256.6	262.5	35.4	10.7
West Java[2]								
Rainfall	5.8	5.2	8.1	4.2	12.4	9.7	13.6	1.8
Throughfall	5.9	7.2	5.2	1.7	25.1	40.3	13.9	4.9
Fiji[3]								
Rainfall	5.5	8.7	5.3	5.5	13.5	1.7	0.8	0.5
Throughfall	5.7	28.7	2.3	1.7	156.4	13.5	5.7	11.5
The Netherlands[4]								
Rainfall	5.1	21.0	23.7	13.7	5.0	4.7	4.9	4.4
Throughfall	4.2	28.6	38.0	11.7	53.3	15.2	5.5	6.4

[1]Present study; [2]Bons (in prep.) 31-year-old *Pinus merkusii* forest plantation; [3]Waterloo (in prep.) 11-year-old *Pinus caribaea* forest plantation; [4]Berg *et al.*, (in prep.) 35-year-old *Pinus sylvestris* forest plantation.

with these results as they do not mention clearly the kind of lysimeter used in their research; the porous cup alundum lysimeter they use may have lead to differences with the vacuum tube lysimeter used in this study. However, the strong element accumulation in the soil surface of Masaya volcano may be a possible reason.

The relatively high inputs of SO_4, Cl and K in rainfall in both plots had a close relation with the volcanic activity. Since Mount Merapi is one of the most active volcanos in Indonesia, its continuous activity causes a high input of elements via rainfall into the forest plantation. In this case the evidence that elemental input through rainfall may decrease with increasing altitude as found in Hawaii (Eriksson, 1957) is rejected. It was reported by Kellman *et al.* (1982), that volcanic activity appeared to be the most likely explanation for the high input of cations recorded in a rainfall study, although their research was far away from the volcano; the nearest volcano Volcan Fuego is located near Guatemala City, approximately 300 km west of the research location at Siguatepeque, Honduras.

As in literature most data on elemental input are presented as kg ha^{-1} $year^{-1}$ instead of kg ha^{-1} $week^{-1}$, the comparison with other research is calculated on annual basis. For Table 6 average values of rainfall and throughfall data within 3 years in the high and low plots have been used. Two comparisons have been made: with elemental input from other research in tropical pine forest plantations and with other research focused on anthropogenic influences. As shown in Table 6, the average elemental input by rainfall and throughfall in this study is much higher compared to research in the tropical pine forest plantations in Sumedang, West Java (Bons in prep.) or in Korokula, Fiji (Waterloo in prep.). This can be explained by the fact that studies in West Java and Fiji were not done along the slope of an active volcano. The cyclone influences in Fiji caused a high Cl input from the ocean by throughfall till about 150 kg ha^{-1} $year^{-1}$. Still, this amount is half the Cl input in this study. Comparisons with anthropogenic pollution give information to evaluate the vulnerability of ecosystems. Volcanic activity is a kind of natural pollution to forests. In The Netherlands the estimated atmospheric input of NHx and NOx by industrial and agricultural practices is amongst the highest in the world. The high input of NH_4 and NO_3 is an important reason for the decreasing vitality of both coniferous and deciduous forests in Europe (van Breemen and van Dijk, 1988). Most other elements especially SO_4, Cl and K were much higher in this study compared to the anthropogenic pollution in The Netherlands. However, there is no indication that high elemental input (SO_4, NH_4, Cl, K, Ca and Mg) decreased the vitality of natural vegetation in the area as was also noticed by Sieffermann (unpublished data). A plausible explanation for this phenomenon may be that the vegetation on the slope of Mount Merapi is already adapted to this natural pollution. The first eruption of Mount Merapi was recorded in 1006 (Kartodirdjo *et al.*, 1975 *in* Triyoga, 1991).

Time scale of only about 20 years for the vegetation in Europe to adapt to the anthropogenic pollution, including the acid rain, may be too short. Although the chemical agents in this type of atmospheric deposition are the same, forest decline by means of manmade pollution is more dangerous than natural pollution because in the former case the vegetation is not adapted to that situation.

SOIL ANIMALS AND NUTRIENT FLOW

Recently many authors have reviewed the role of soil animals in nutrient cycling (Verhoef and Brussaard, 1990; Wolters, 1991; Shaw *et al.*, 1991; Lee and Pankhurst, 1992; Roth, 1992). Available results are mainly based on laboratory studies with microcosms, while in the field the contribution of soil animals may be different. Odum (1989) criticized models on ecological succession based on microcosm studies in which respiration of the organisms is usually underestimated. Consequently, as resources such as space and nutrients become depleted in the closed systems, the rate of production becomes limited by the rate of decomposition and mobilization of nutrients.

As it is difficult to study the contribution of soil animals to the flow of nutrients under natural conditions, a mesocosm experiment using systems which are closed or open to soil animals is one alternative to study the role of soil fauna in the field. Verhoef (1996) concluded that mesocosm experiments in combination with microcosm experiments are a good approach to study ecosystem processes.

In this paragraph the contribution of soil animals to the flow of nutrients (SO_4, NH_4, NO_3, Cl, K, Ca, and Mg) through soil columns is focussed upon at different altitudes along an active volcano. Additionally, the volume and pH of the percolate was collected using vacuum tube lysimeters from the closed and open mesocosms.

The mesocosm is constructed of dark grey PVC, 13 cm diameter, 25 cm height, 0.5 cm thick, open at the top and covered by 200 μm nylon gauze, with a closed bottom connected with a well closed dark glass bottle with a capacity of 225 mm. Soil water percolate was collected from the mesocosm by vacuum porous ceramic cup lysimeter. The tension inside the glass bottle was set at –0.4 bar in order to suck the percolate. Therefore, a vacuum pump was used weekly.

There were two types of mesocosms. In the open ones soil animals could enter and leave the mesocosm via 6 sideways holes, 3 cm diameter, in the upper part of the tube. Each hole was covered by 4 mm (for the 3 upper) and 2 mm nylon gauze (for the 3 lower holes); 3 holes on a level with the litter layer and 3 on a level with the fragmented layer. In the closed mesocosms there were no holes in the upper part of the tube. Principally, the macrofauna such as Araneae, Oniscoidea, Julidae, Formicidae, and Lumbricidae could not enter the closed mesocosms.

The mesocosm was filled with an undisturbed soil column, including the litter and fragmented layers. Before installation all mesocosms were treated for one week at –5°C to decimate the soil biota. All mesocosms were installed randomly below understorey vegetation with 8 replicates of open and closed mesocosms in each plot.

In the closed systems 4 taxonomic groups of microarthropods were found in both plots at the end of the experiment, i.e. Acari, Hypogastruridae, Entomobryidae and Sminthuridae. In the open systems, besides these groups, macroarthropods (Araneae, Oniscoidea, Julidae, and Formicidae), Onychiuridae, and Lumbricidae were found in both plots (Table 7).

Table 7 Mean numbers ± S.E. (n = 8) at the end of the experiment of the dominant soil animals. *= p<0.05, N.S. = not significant, – = not present; when differences are significant the highest value is given in bold..

Taxonomic groups	*High plot*		*Low plot*	
	Closed system	*Open system*	*Closed system*	*Open system*
Acarina	12 ± 2.8	19 ± 2.9 N.S.	16 ± 4.0	27 ± 4.4 N.S.
Araneae	–	3 ± 1.0	–	6 ± 1.8 N.S.
Oniscoidea	–	2 ± 0.3	–	1 ± 0.2 N.S.
Julidae	–	2 ± 0.8	–	3 ± 1.8 N.S.
Onychiuridae	–	2 ± 0.4	–	2 ± 0.5 N.S.
Hypogastruridae	82 ± 14.0	32 ± 12.8 *	15 ± 5.2	36 ± 11.1 N.S.
Entomobryidae	8 ± 1.4	10 ± 3.2 N.S.	20 ± 9.4	15 ± 2.5 N.S.
Sminthuridae	56 ± 15.2	22 ± 8.4 N.S.	13 ± 5.6	17 ± 2.8 *
Formicidae	–	37 ± 30.6	–	8 ± 5.3 N.S
Lumbricidae	–	4 ± 1.1	–	1 ± 0.3*

(From Gunadi and Varhoef, 1998: Table 1; printed with permission of Elsevier Science Ltd.)

Comparison between closed and open systems showed that in the high plot only the mean number of Hypogastruridae differed significantly: the mean number of Hypogastruridae in the closed systems was higher than that in the open systems. In the low plot, only the mean number of Sminthuridae differed significantly: the mean number of Sminthuridae in the open system was higher than that in the closed system. Comparison between high and low plot was only possible for the open system for Araneae, Oniscoidea, Julidae, Onychiuridae, Formicidae, and Lumbricidae. Higher numbers of Lumbricidae were found in the high plot, while all other taxonomic groups had more or less the same density in both plots.

Dynamics of volumes of percolates from closed and open systems in both plots are shown in Fig. 12. The mean volumes of percolates from closed and open systems differed significantly from each other in both plots (Table 8); on the average, open systems collected about 25–30% more percolate than closed systems. This may be caused by the burrowing activities of Formicidae and Lumbricidae.

It is well known that soil macrofauna, especially earthworms, can significantly modify porosity and pore size distribution which can influence soil physical properties such as aeration and water retention, infiltration, and drainage (Carter *et al.*, 1982; van de Westeringh, 1972; van Rhee, 1969). Earthworms such as *Lumbricus terrestris,* construct permanent, vertical burrows lined with faeces, which act as conduits that enhance drainage of free water. The faeces of earthworms are often preferentially enriched in colloidal materials and usually have a high water-holding capacity (Mulongoy and Bedoret, 1989; Shaw *et al.*, 1991). Similar effects by ants have been reported by Wicken *et al.* (1976). These soil animals construct permanent structures as channel systems

Table 8 Mean, S.E. and F-values for the volume of percolate (mm) from the mesocosms in both plots.

Description	*High plot*	*Low plot*
Closed system		
Mean volume (mm)	69.1	65.8
Standard error	5.9	7.9
Open system		
Mean volume (mm)	92.5	93.5
Standard error	6.5	6.6
F-statistic closed-open system	7.161**	7.291**

n = 30,** = $p < 0.01$ (t-test). (From Gunadi and Varhoef, 1993: Table 2; printed with permission of Elsevier Science Ltd.)

throughout the underground and aboveground nests which may have long-term effects on soil hydrology. It was reported by Lee and Pankhurst (1992) that one of the key functions of soil animals is increasing water infiltration and thus reducing soil loss by erosion.

We assume that colonization of soil animals of the open mesocosms, especially by Formicidae and Lumbricidae, takes some time; this we call an adaptive period. Figure 12 shows that in both plots there is an adaptive period, as during that time the volumes of percolate were the same for both closed and open systems. After that the volumes of percolate from the open systems were higher than from the closed systems. Later the volumes of percolate from both systems are the same. This might be caused by the heavy rainfall before that period, causing saturation of the mesocosms and decreasing the contribution of the soil animals in the open systems via destruction of the nests and burrows. Till the end of this study, again open systems collected more water than closed systems.

Table 9 shows ANOVA results of all variables against three factors i.e. plot, treatment and time. The results which will be focussed upon are the ones which involve treatment as main effect, in the first and second order interactions. The dynamics of pH and nutrient output in percolate from closed and open mesocosms in the high and low plot are shown in Figs. 13 and 14. The main effect of treatment (closed and open systems) differed significantly for volume, Cl and K.

Compared with the dynamics of the volume of percolate in the closed and open systems in both plots which has been explained before, the dynamics of the output of SO_4 and Ca are difficult to explain because closed systems collect sometimes more and sometimes less SO_4 and Ca than open systems. According to Gunadi (1993), SO_4 fluctuations in percolate from the open mesocosms had a close relation to the volcanic input via rainfall and throughfall. The contribution of soil animals, however, to the SO_4 availability is not clear. Ca output in the percolate was much higher than in the throughfall (Gunadi and Verhoef, 1993). The fluctuations of the output of Ca from closed and open systems are difficult to explain. E.g. in the closed systems there were microarthropods (i.e. Acari, Hypogastruridae, Entomobryidae and Sminthuridae) which can contribute to Ca mobilization. In open systems, apart from the microarthropods mentioned, we found macrofauna such as Oniscoidea, Julidae and Lumbricidae which also can contribute to Ca mobilization. Collembola (Seastedt and Crossley, 1983) as well as isopods (Reyes and Tiedje, 1976) can digest Ca oxalate by means of mineralization by bacteria living in their guts, resulting in a higher Ca availability in the faeces (Teuben and Verhoef, 1992). Ca output was sometimes lower in the open systems than in the

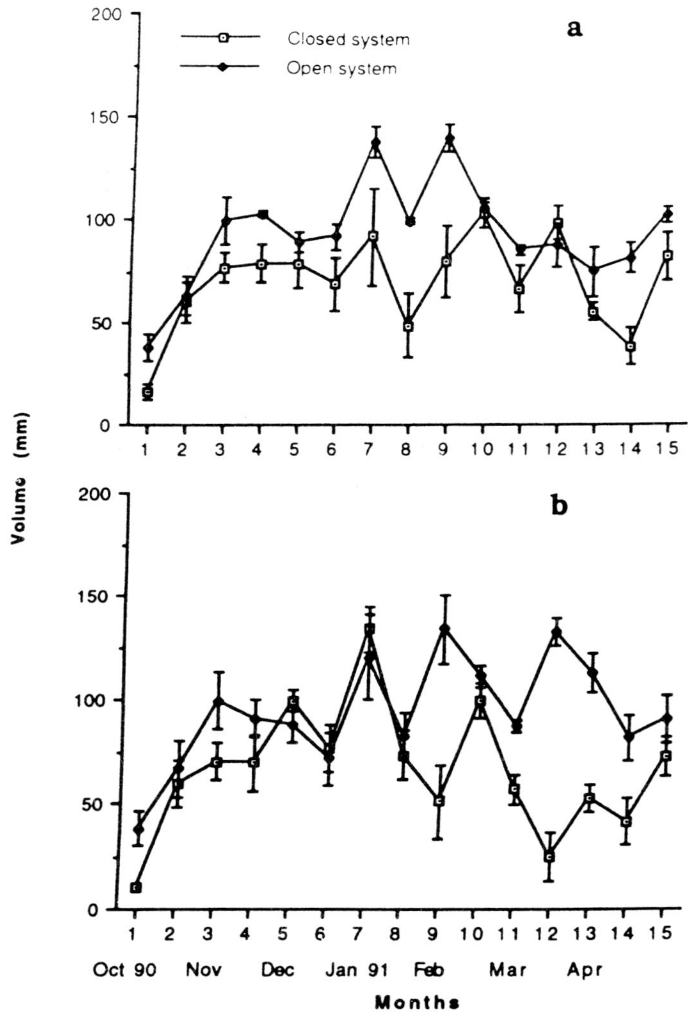

Fig. 12 Volumes of percolate ± S.E. collected from the mesocosms in the high (a) and low plots (b). The x-axis is divided into 15 collections. (From Gunadi *et al.* 1993: Fig 1; printed with permission of Elsevier Science Ltd.)

closed systems. This can partly be explained by the fact that some macroarthropods such as Oniscoidea and Julidae immobilize Ca carbonate to calcify their exoskeleton (Hopkin and Read, 1992; Sutton, 1972; Verhoef, 1993).

Table 9 ANOVA results (F-volues) for valume (mm), pH, SO_4, NH_4, NO_3, Cl, K, Ca, Mg (kg ha^{-1} 2 $weeks^{-1}$) against plot (2 levels: high and low plot), treatment (2 levels: closed and open mesocosms), and series (15 levels: 1st till 15th collections).

Sources	*Volume*	*pH*	*SO_4*	*NH_4*	*NO_3*	*Cl*	*K*	*Ca*	*Mg*
Plot	0.006N.S,	3.769N.S.	4.717N.S.	0.635N.S.	4.288N.S.	1.941N.S.	1.005N.S.	1.696N.S.	7.084**
Treatment	73.390***	0.687N.S.	10.823N.S.	0.191N.S.	0.742N.S.	44.818***	12.214***	0.681N.S.	1.914N.S
Time	15.967***	66.379***	12.444***	7.209***	1.909N.S.	28.992***	27.785***	9.182***	8.813***
Plot*treat.	0.146N.S.	7.169**	7.471**	0.012N.S.	3.556N.S.	0.389N.S.	0.976N.S.	1.104N.S.	0.286N.S.
Plot*time	0.578N.S.	10.933***	3.567***	0.527N.S.	0.854N.S.	26.403***	5.410***	1.774N.S.	1.269N.S.
Treat.*time	2.275**	1.420N.S.	3.913***	0.711N.S.	1.908N.S.	3.708***	2.350**	1.434N.S.	1.735N.S.
Plot*treat.*time	2.209**	2.017N.S.	2.918***	0.680N.S.	1.089N.S.	1.784N.S.	1.128N.S.	2.195**	1.906N.S.

n = 480,** = $p < 0.01$,*** = $p < 0.001$, N.S. = not significant. (From Gunadi and Varhoef, 1993: Table 3; printed with permission of Elsevier Science Ltd.)

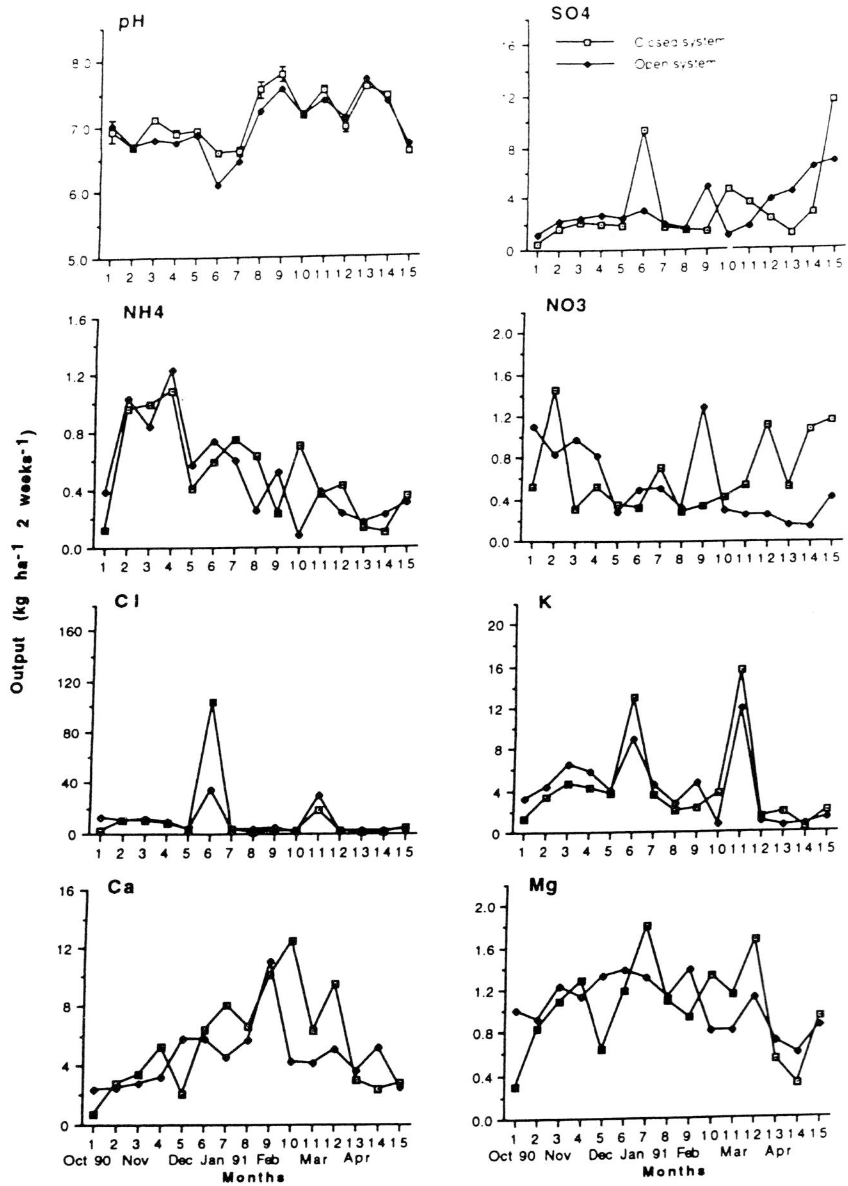

Fig. 13 Dynamics of pH and element output in percolates from the mesocosms in the high plot. (From Gunadi and Varhoef, 1993: Fig. 2; printed with permission of Elsevier Science Ltd.)

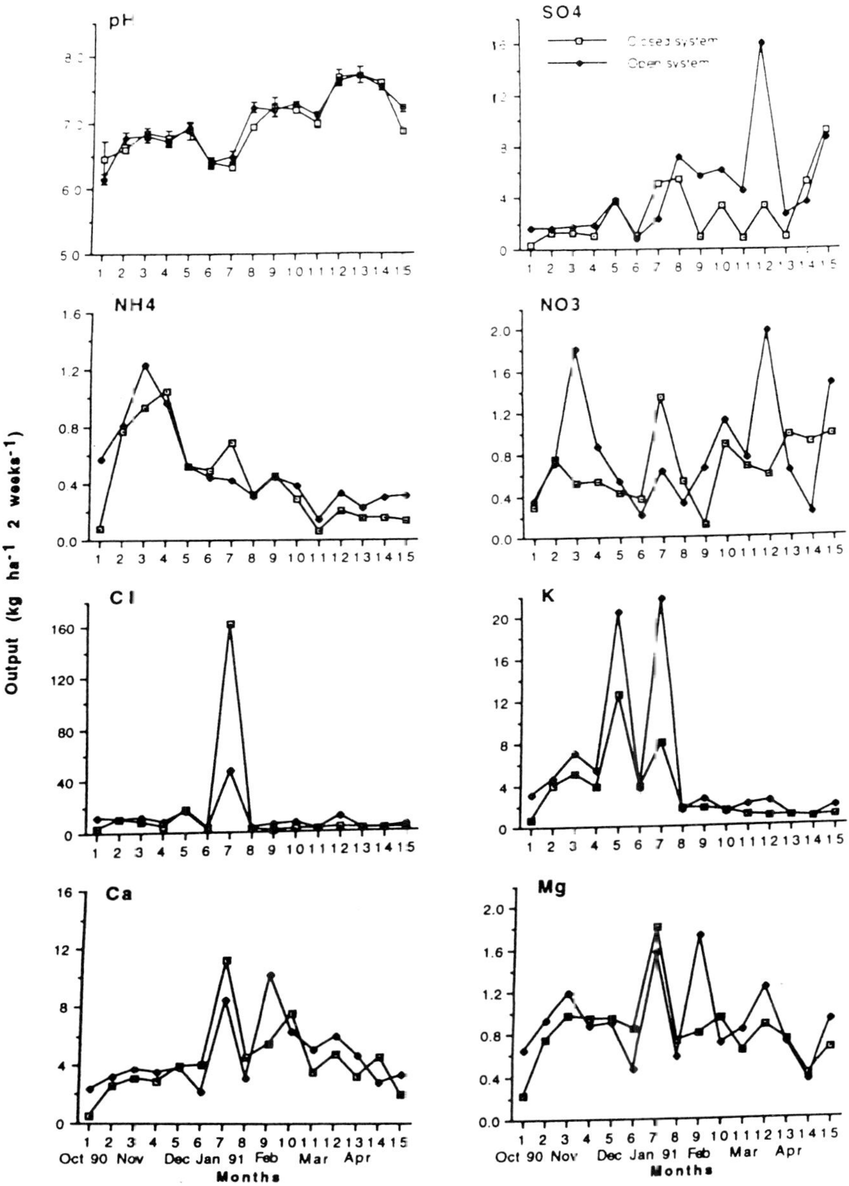

Fig. 14 Dynamics of pH and element output in percolates from the mesocosms in the low plot. (From Gunadi and Varhoef, 1993: Fig. 3; printed with permission of Elsevier Science Ltd.)

Table 10 Total elements output (kg ha^{-1} 7 $months^{-1}$) in percolates from the mesocosms.

Element	*High plot*		*Low plot*	
	Closed system	*Open system*	*Closed system*	*Open system*
SO_4–S	42.9	45.9	43.1	68.2
NH_4–N	7.9	7.6	6.3	7.4
NO_3–N	9.6	8.2	9.9	12.3
Cl	174.5	134.0	225.9	148.1
K	62.6	60.6	47.0	78.6
Ca	81.3	67.9	63.3	68.1
Mg	15.0	15.7	12.3	13.6

(From Gunadi and Varhoef, 1993: Table 4; printed with permission of Elsevier Science Ltd.).

REFERENCES

Anderson, J.M., and Ineson, P., 1984. Interaction between microorganisms and soil invertebrates in nutrient flux pathways of forest ecosystem, In: *Invertebrate-Microbial Interactions,* (J.M. Anderson, A.D.M. Rayner, and D.W.H. Walton, eds.), Cambridge University Press, Cambridge, 59–88 pp.

Anonymous (ed.), 1983. Proceeding Simpo Pinus '83. Pusat Penelitian dan Pengembangan Hasil Hutan dan Perum Perhutanik, Jakarta.

Backer, C.A., and van den Brink Jr., R.C.B, 1963. *Flora of Java,* Vol. I. N.V.P. Nordhoff, Groningen.

Berg, B., and Staaf, H., 1981. Leaching, accumulation and release of nitrogen in decomposing forest litter, In: *Terrstrial Nitrogen Cycles,* (F.E. Clark, and T. Rosswall, eds.), *Ecological Bulletins,* 33, Stockholm, 163–178 pp.

Bland, R.G., and Jaques, H.E., 1978. *How to know the insects,* Wm. C. Brown Publishers, Iowa.

Borror, D.J., de Long, D.M., and Triplehorn, C.A., 1981. *An introduction to the study of insects,* Saunders College Publishing, Chicago.

Brasell, H.M:, and Sinclair, D.F., 1983. Elements returned to forest floor in two rainforests and three plantation plots in tropical Australia, *J. Ecol.,* **71:** 367–378.

Brasell, H.M., Unwin, G.L., and Stocker, G.C., 1980. The quantity, temporal distribution and mineral-element content of litterfall in two forest types in tropical Australia, *J. Ecol.,* **68:** 123–139.

Bruijnzeel, L.A., 1984. Immobilization of nutrients of plantation forest of *Pinus merkusii* and *Agathis dammara* growing on volcanic soils in Central Java, In: *Soils and Nutrient of Perennial Crops,* (A. Thajib, and E. Pusparadjah, eds.), Malaysian Soil Science Society, Kuala Lumpur, 19–29 pp.

Bruijnzeel, L.A., 1985. Nutrient content of litterfall in coniferous forest plantation in Central Java, Indonesia, *J. of Tropical Ecology,* **1:** 353–372.

Bruijnzeel, L.A., 1988. Estimates of evaporation in plantations of *Agathis dammara* Warb in south-central Java, Indonesia, *J. Trop. For. Sci.,* **1**(2): 145–161.

Bruijnzeel, L.A., 1989. Nutrient content of bulk precipitation in south-central Java, Indonesia, *J. Trop. Ecol.,* **5:** 187–202.

Calder, I.R., 1986. Water use of eucalypts: a review with special reference to South India, *Agricultural Water Management,* **11:** 333–342.

Carter, J.A., Heinonen, J., and de Vries, J., 1982. Earthworms and water movement, *Pedobiologia,* **23:** 395–397.

Coley, P.D., Bryant, J.P., and Chapin, F.S., 1985. Resource availability and plant antiherbivore defense, *Science,* **230:** 895–899.

Core, E.S.P., Anderson, J.M., Beese F., Berg, M.P., Bolger, T., Couteaux, M.M., Henderson, M., Ineson, P., McCarthy, F., Palka, L., Raubuch, M., Splatt, P., Verhoef, H.A., and Willison, T., 1992. Mechanisms of nutrient turnover in the soil compartment of forests, In: *Responses of Forest Ecosystems to Environmental Changes,* (T. Teller, P. Mathy, and J.N.R. Jeffers, eds.), Elsevier Science Publishers Ltd, London, 342–350 pp.

Darzi, M., and Winchester, J.W., 1982. Volcanic aerosol phosphorus, chlorine, and sulphur at Kilaeua, Hawaii, *J. Geophys. Res.,* **87:** 11095–11101.

Eriksson, E., 1957. The composition of Hawaiian rainfall, *Tellus,* **9:** 509–520.

Evans, J., 1992. *Plantation forestry in the tropics,* Oxford University Press, New York.

Faber, J.H., Teuben, A., Berg, M.P., and Doelman, P., 1992. Microbial biomass and activity in pine litter in the presence of *Tomocerus minor,* (Insecta, Collembola), *Biology and Fertility of Soils,* **12:** 233–240.

Finegan, B., 1984. Forest succession, *Nature,* **312:** 109–114.

Fitter, A.H., Atkinson, D., Read, D.J., and Usher, M.B., (eds), 1985. *Ecological Interactions in Soil: Plants, Microbes and Animals,* Blackwell Scientific Publications, Oxford.

Gunadi, B., and Verhoef, H.A., 1993. The flow of nutrients in a *Pinus merkusii* forest plantation in Central Java; the contribution of soil animals, *Eur. J. Soil Biol.,* **29:** 133–139.

Gunadi, B., 1989a. Kerusakan hutan pinus: studi perbandingan di Jateng dan Sumut, Kompas, 14th May, p. III.

Gunadi, B., 1994a. Litterfall, litter turnover and soil respiration in two pine forest plantations in Central Java, Indonesia, *J. of Tropical Forest Science,* **6:** 310–322.

Gunadi. B., 1994b. Seasonal fluctuations of Collembola along the slope of a pine forest plantation in Central Java, *Acta Zoologica Fennica,* **195:** 62–66.

Gunadi, B., Verhoef, H.A., and Bedaux, J.J.M., 1998. Seasonal dynamics of decomposition in a forest plantation, (*Pinus merkusii*) in Central Java, Indonesia, *Soil Biology and Biochemistry,* **30:** 7, 845–852.

Harde, K.W., 1984. *A field guide in colour to beetles,* Octopus Books, Czechoslovakia.

Harrison, A.F., Ineson, P., and Heal, O.W., 1990. *Nutrient Cycling in Terrestrial Ecosystems,* Elsevier Science Publishers Ltd, Essex, 454 pp.

Herlitzius, H., 1983. Biological decomposition efficiency in different woodland soils, *Oecologia,* (Berlin), **57:** 78–97.

Hobbie, S.E., 1992. Effects of plant species on nutrient cycling, *Trends in Ecology and Evolution,* **10:** 336–339.

Hodkinson, I.D., 1975. Dry weight loss and chemical changes in vascular plant litter of terrestrial origin, occurring in a beaver pond ecosystem, *Journal of Ecology,* **63:** 131–142.

Hopkin, S.P., and Read, H.J., 1992. *The Biology of Millipedes,* Oxford Science Publications, Oxford.

Jepma, C.J., and Blom, M., 1992. Global trends in tropical forest degradation: the Indonesian case, In: *Issues of Environmental Policy,* (W.J.M. Heijman, and J.J. Krabbe, eds.), Wageningse Economische Studies 24, Groningen, 1–29 pp.

Johnson, N., and Parnell, R.A., 1986. Composition, distribution and neutralization of "acid rain" derived from Masaya volcano, Nicaragua, *Tellus,* **38B:** 106–117.

Jordan, C.F., 1989. Are process rates higher in tropical forest ecosystems?, In: *Mineral Nutrients in Tropical Forest and Savanna Ecosystems,*(J. Proctor, ed.), Blackwell Scientific Publications, London, 205–215 pp.

Kellman, M., Hudson, J., and Sanmugadas, K., 1982. Temporal variability in atmospheric nutrient influx to a tropical ecosystem, *Biotropica,* **14:** 1–9.

Kendrick, W.B., 1959. The time factor in the decomposition of coniferous leaf litter, *Canadian J. of Botany,* **37:** 907–912.

Kimmins, J.P., 1987. *Forest Ecology*, Macmillan Publishing Company, New York.

Kuiters, A.T., 1987. *Phenolic acids and plant growth in forest ecosystems*, Ph.D. thesis, Vrije Universiteit, Amsterdam.

Lee, K. E., and Pankhurst, C.E., 1992. Soil organisms and sustainable productivity, *Aust. J. Soil Res.*, **30:** 855–892.

Lugo, A.E., 1992a. Comparison of tropical tree plantations with secondary forests of similar age, *Ecological Monograph*, **62**(1): 1–41.

Lugo, A.E., 1992b. Tree plantations for rehabilitating damaged forest lands in the tropics, Ecosystem Rehabilitation, Vol. 2, In: *Ecosystem Analysis and Synthesis*, (M.K. Wali, ed.), Academic Publishing, The Hague, 247–255 pp.

Marrs, R.H., Proctor, J., Heaney, A., and Mountford, M.D., 1988. Changes in soil nitrogen mineralization and nitrification along an altitudinal transect in tropical rain forest in Costa Rica, *Journal of Ecology*, **76:** 466–482.

Mason, C.F., 1979. *Decomposition*, Studies Biology no. 74, Edward Arnold (Publishers) Limited, London.

McCune, D.C., and Boyce, R.L. 1992. Precipitation and the transfer of water, nutrients and pollutants in tree canopies, *Trends Ecol. Evolut.*, **7:** 4–7.

Millar, C.S., 1974. Decomposition of coniferous leaf litter, In: *Biology of Plant Litter Decomposition*, vol. 1 (C.H. Dickinson, and G.J.F. Pugh, eds.), Academic Press, London, 105–128 pp.

Mulongoy, K., and Bedoret, A., 1989. Properties of worm cast and surface soils under various plant covers in the humid tropic, *Soil Biol. Biochem.*, **21:** 197–203.

Odum, E.P., 1989. *Ecology and our endangered life-support systems*, Sinauer Associates, Inc. Publishers, Sunderland, Mass., USA.

Proctor, J., 1984. Tropical forest litterfall. II. The data set, In: *Tropical Rainforests*, (S.L. Sutton, and A.C. Chadwick, eds.), Natural History Society, London, 83–113 pp.

Raven, R.H., Evert, R.F., and Eichhorn, S.E., 1992. *Biology of Plants*, Worth Publishers, New York.

Reyes, V.G., and Tiedje, J.M., 1976. Ecology of the gut microbiota of *Tracheoniscus rathkei* (Crustacea, Isopoda), *Pedobiologia*, **16:** 67–74.

Riekerk, H., and Korhnak, L.V., 1992. Rainfall and runoff chemistry of Florida pine flatwoods, *Water Air Soil Pollut.*, **65:** 59–68.

Roth, M., 1992. Metals in invertebrate animals of a forest ecosystem, In: *Biogeochemistry of Trace Metals*, (D.C. Adriano, ed.), Lewis Publishers, Boca Raton, 299–328.

Satjapradja, O., 1977. *The genus Pinus and its importance for the tropical and subtropical world*, Ph.D. thesis, Rijks Universiteit, Gent.

Seastedt, T.R., and Crossley Jr, D.A., 1983. Nutrients in forest litter treated with naphthalene and simulated throughfall: a field microcosm study, *Soil Biol Biochem.*, **15:** 159–165.

Shaw, C.H., Lundvist, H., Moldenke, A., and Boyle J.R., 1991. The relationships of soil fauna to long-term forest productivity in temperate and boreal ecosystems: processes and research strategies, In: *Long-Term Field Trials to Assess Environmental Impacts of Harvesting*, (W.J. Dyck, and C.A. Mees, eds.), Forest Research Institute, Rotorua, New Zealand, 39–77 pp.

Smith, D.B., Zielinski, R.A., Rose Jr., W.I., and Huebert, B.J., 1982. Water-soluble material on aerosol collected within volcanic eruption clouds, *J. Geophys. Res.*, **87:** 4963–4972.

Staaf, H., and Berg, B., 1977. Mobilization of plant nutrients in a Scots pine forest mor in Central Sweden, *Silva Fennica*, **11:** 210–217.

Sutton, S.L., 1972. *Woodlice*, Ginn & Company Limited, London.

Swift, M.J., Heal, O.W., and Anderson, J.M., 1979. *Decomposition in terrestrial ecosystems*, Blackwell Scientific Publications, Oxford.

Tantra, I.G.M., 1983. Pelestarian plasma-nutfah serta taksonomi *Pinus merkusii*. Proceeding Simpo Pinus 83 (Ed. by anonymous), Pusat Penelitian dan Pengembangan Hasil Hutan dan Perum Perhutani, Jakarta, 141–143 pp.

Taylor, B.R., and Parkinson, D., 1988. Aspen and pine leaf litter decomposition in laboratory microcosms. II. Interactions of temperature and moisture level, *Canadian J. of Botany*, **61:** 1966–1973.

Taylor, B.R., Parkinson, D., and Parsons, W.F.J., 1989, Nitrogen and lignin content as predictors of litter decay rates: a microcosm test, *Ecology*, **70:** 97–104.

Teuben, A., 1991. *Interactions between animals and microorganisms during decomposition of pine litter*, Ph.D. thesis, Vrije Universiteit, Amsterdam.

Teuben, A., and Roelofsma, T.A.P.J., 1990. Dynamic interaction between functional groups of soil arthropods and microorganisms during decomposition of coniferous litter in microcosm experiments, *Biology and Fertility of Soils*, **9:** 145–151.

Teuben, A., and Verhoef, H.A., 1992. Direct contribution of soil arthropods to nutrient availability through body and faecal nutrient content, *Biol. Fert. Soils.*, **14:** 71–75.

Triyoga, L.S., 1991. *Manusia Jawa dan Gunung Merapi*, Gadjah Mada University Press, Yogyakarta, 157 pp.

Van Breemen, N., and van Dijk, H.F.G., 1988. Ecosystem effects of atmospheric deposition of nitrogen in The Netherlands, *Environ. Pollut.*, **54:** 249–274.

van de Westeringh, 1972. Deterioration of soil structure in worm free orchard soils. *Pedobiologia*, **12:** 6–15.

van Rhee, J.A., 1969. Inoculation of earthworms in a newly drained polder, *Pedobiologia*, **9:** 128–132.

Van Schaik, C.P., and Mirmanto, E., 1985. Spatial variation in the structure and litterfall of Sumatran rainforest, *Biotropica*, **17:** 196–205.

Veneklaas, E.J., 1990. Nutrient fluxes in bulk precipitation and throughfall in two montane tropical rain forest, Columbia, *J. Ecol.*, **78:** 974–992.

Verhoef, H.A., 1995. Animal ecophysiology; cornerstone for soil ecosystem studies as exemplified by studies on arthropods, *Acta Zoölogica Fennica*, **196:** 176–182.

Verhoef, H.A., 1995. Litterbag method, In: *Methods in Applied Soil Microbiology and Biochemistry*, (K. Alef, and P. Nannipieri, eds.), Academic Press, London, 479–481 pp.

Verhoef, H.A., 1996. The role of soil microcosms in the study of ecosystem processes, *Ecology*, **77**(3): 685–690.

Verhoef, H.A., and Brussaard, L. 1990. Decomposition and nitrogen mineralization in natural and agro-ecosystems: the contribution of soil animals, *Biogeochemistry*, **11:** 175–211.

Verhoef, H.A., and de Goede, R.G.M., 1985. Effects of collembolan grazing on nitrogen dynamics in a coniferous forest, In: *Ecological Interactions in Soil: Plants, Microbes and Animals*, (A.H., Fitter, D. Atkinson, Read D.J., and M.B. Usher, eds.), Blackwell Scientific Publications, Oxford, 367–376 pp.

Vitousek, P.M., and Sanford Jr, R.L., 1986. Nutrient cycling in moist tropical forest, *Annual Review of Ecology and Systematics*, **17:** 137–167.

Wang, D., Bormann, F.H., Lugo, A.E., and Bowden, R.D., 1991. Comparison of nutrient-use efficiency and biomass production in five tropical tree taxa, *Forest Ecology and Management*, **46:** 1–21.

Wicken, E.B., Broersma, K., Lavkulich, L.M., and Farstad, L., 1976. Biosynthetic alteration in a British Columbia soil by ants (*Formica fusca* Linné), *Soil Sci. Soc. Amer. J.*, **40:** 422–426.

Will, G.M., 1967. Decomposition of *Pinus radiata* litter on the forest floor. Part 1. Changes in dry matter and nutrient content, *New Zealand Journal of Science*, **10:** 1030–1044.

Wolters, V., 1991. Soil invertebrates effects on nutrient turnover and soil structure—a review, *Z Pflan. Bodenk.*, **154:** 389–402.

Wright, R.F., van Breemen, N., Emmet, B., Roelofs, J.G.M., Tietema, A., Verhoef, H.A., Hauhs, M., Rasmussen, L., Hulthberg, H., Persson, H., and Stuanes, A.O., 1992. NITREX -nitrogen saturation experiments, In: *Responses of Forest Ecosystems to Environmental Changes*, (T. Teller, P. Mathy, and J.N.R. Jeffers, eds.), Elsevier Science Publishers Ltd, London, 335–341 pp.

CHAPTER 9

TREE-PLANTATION AND COMMUNITY STRUCTURE AND FUNCTION OF MACRO- AND MICRO-ARTHROPODS OF SOIL-LITTER SYSTEM

M. Vikram Reddy

Applied Ecology Laboratory, Department of Zoology and Environmental Science Prog. Kakatiya University, Warangal–506 009, A.P., India

ABSTRACT

The effects of a mixed tree-plantation predominated by *Eucalyptus* sp. on the community structure and function of macro- and micro-arthropods of the soil-litter system in a tropical grassland in the semi-arid region of Andhra Pradesh (south India), depicted significant increase in the qualitative and quantitative composition of both the soil-surface inhabiting and the soil inhabiting macro- and micro-arthropods. Twenty-four species of soil-surface inhabiting arthropods were recorded in grassland area while 27 species in the adjacent tree-plantation area. Shannon index of general diversity, species richness and evenness indexes of different groups of the soil-surface arthropods were higher in tree-plantation area. The soil micro-arthropods were significantly higher in densities in the tree-plantation area. The micro-arthropods were more common in the surface layer compared to the sub-layer of grassland area whereas in tree-plantation area, they were recorded in larger numbers in the sub-layer. Seasonality of relative population abundance in different groups of soil-surface arthropods sampled monthly by pit-fall traps in the grassland and adjacent tree-planted areas showed that densities of most of the groups were lower during summer and higher during the rainy season, being less abundant during winter. Similarly, seasonal population abundance of soil microarthropods in the grassland and tree-planted areas showed that, these arthropods were minimum during summer and maximum during rainy season. Both the soil-surface inhabiting arthropods and the soil inhabiting microarthropods were recorded in larger densities in tree-plantation area compared to grassland.

This article also reviews litter decomposition and mineralization processes in relation to microarthropod abundance in different plantation forests in tropical and sub-tropical India. It shows wide variation in the decomposition rate and in mineral concentrations of decaying litter during decomposition. The decomposition rate of coniferous litter was low while that of deciduous leaf litter was high. The densities of the microarthropods, mostly Collembola and Acarina inhabiting the litter bags placed in plantation forests showed significant correlations with mass-loss and concentrations of some of the plant-nutrients in decaying litter, indicating their probable role in the fragmentation and comminution during litter decomposition process.

Key Words: Tree-plantation, grassland, soil macro-and micro-arthropods, litter decomposition, nutrient cycling.

In the tropics many types of lands have been used for tree plantations particularly the fast growing tree plantations, the largest proportion being grassland or savanna as much of such land is well suited for tree planting (Evans, 1982). Afforestation of these areas, i.e., conversion of natural savanna and grassland to tree-plantation adds to the ecological diversity and influences the whole foodchain (Lamotte, 1979), particularly the soil macrofauna (Lavelle *et. al.*, 1994) of the savanna and brings about major environmental changes affecting biodiversity and biotic activity in the soil litter system. The visible change on the soil surface when grassland is converted to tree-plantation is the accumulation of litter leading to the physico-chemical changes of the surface soil. Effects of practices of reforestation such as replacement of natural forests by new tree-plantation of different species, on soil arthropod communities are dependent on various factors such as, a) the nature of tree species either native or exotic used for the plantation, b) the management techniques used to install and to manage the tree plantation and c) the previous history of the reforested area. (Deharveng, 1996; Sousa *et. al.* 1997). However, most of the attempts that have examined the effects, have dealt with the first factor only, and did not consider the other two (Sousa *et al.* 1997). Besides there are a few reports on establishment of forests in formerly tree-less steppe regions and grasslands (afforestation), on faunal succession during reforestation and afforestation (Kuhnelt, 1963), on zooamelioration in afforested arable land (Szujeki *et al.,* 1980), and on impoverishment of the soil fauna after reforestation with conifers (Kuhnelt 1963). Moreover, very little information is available on the effect of tree-plantation on structure and function of soil faunal communities in tropics (Evans, 1982; Felker, 1986; Reddy, 1984a).

Seasonal changes in the population densities of soil arthropods are influenced by different types of vegetation covers, climate, physical and chemical properties of soil, nature and depth of the litter and humus and a variety of other environmental factors (Price, 1973; Reddy, 1984a; Reddy, 1995; Ananthakrishnan, 1996). Environmental factors such as temperature, moisture, humidity and availability of food affect their horizontal and vertical distribution. These arthropods play a significant role in litter decomposition and nutrient cycling in terrestrial ecosystems (Swift *et al.*, 1975; Reddy, 1984a; Reddy, 1995). Thus, it is essential to understand the seasonal population dynamics of these arthropods to assess their intrinsic ecological role in plantation forest ecosystems. The present paper attempts to review and compare the qualitative and quantitative composition of beneficial soil macro- and micro-arthropods, both soil-surface inhabiting and soil inhabiting forms, their seasonal community structure with particular reference to grassland and the adjacent tree-planted systems of 10 to 12 years of tropical semi-arid nature, and sort out the effects of tree-plantation on the seasonal patterns of population structure and distribution of these beneficial soil arthropods, and finally their role in plant litter decomposition and nutrient dynamics.

SOIL-SURFACE INHABITING ARTHROPRODS

Qualitative and Quantitative Composition

The macro-arthropods sampled by pit-fall traps in the grassland area, included Arachnida (Scorpionidae, Pseudoscropionidae, Araneae and Acarina), Myriapoda (Diplopoda, Pauropoda and Chilopoda) and insecta (Apterygota: Thysanura and Collembola, and Pterygota: Dermaptera, Orthoptera, Heteroptera, Dictyoptera, Isoptera, Hymenoptera, Coleoptera and their larvae, and Lepidopteran larvae) whereas in the adjacent tree plantation area predominated by *Eucalyptus* sp., Psocoptera and Homoptera were recorded more in addition to the arthropod taxa recorded in the grassland (Table 1). Among all soil-surface arthropods, the pterygotes were dominant, of which Formicidae was most abundant in both the areas (Mboukou-kimbatsa *et al.*, 1998). *Monomorium indicum* Forel was numerically dominant among Formicidae followed by *Crematogaster* sp. Collembola, the subdominant group was dominated by *Lepidocyrtus* sp. Among Acarina, Cryptostigmata being dominated by *Lancetoppia willmanni* Hammer was most abundant, and among Araneae, *Storena* sp. was dominant in both the areas. However, among Coleoptera, *Coleolissus* sp. was dominant in the grassland whereas *Pachycera* sp. was dominant in the tree-plantation (Reddy and Venkataiah, 1990b).

Total numbers of arthropods were 1.5 times more abundant in tree plantation area compared to grassland (t-test: $P < 0.01$). Among them, all the groups except Scorpionidae, *Lamellobates palustris* (Cryptostigmata), Thysanura, Gryllidae, and *Odontotermes* sp. (Isoptera) were more abundant in tree-plantation area (Reddy and Venkataiah, 1990b). This was probably due to the variation in vegetation cover and development of many undergrowth species, influencing both

Table 1 Mean abundance of various taxa of soil-surface inhabiting arthropods in grassland and tree-plantation areas (Reddy and Venkataiah, 1990b)

Arthropod taxa	*Grassland area*		*Tree-plantation area*	
	Number	*% of total*	*Number*	*% of total*
Myriapoda	24	0.9	32	0.7
Scorpionidae	12	0.5	4	0.1
Total Araneae	141	5.5	227	5.3
Total Acarina	462	18.1	724	17.0
Mesostigmata	82	3.2	134	3.1
Prostigmata	106	4.2	174	4.1
Cryptostigmata	274	10.8	416	9.8
Total Apterygota	658	25.9	1,054	24.8
Thysanura	13	0.6	11	0.3
Total Collembola	645	25.3	1,043	24.5
Entomobryidae	345	13.5	581	13.6
Isotomidae	150	5.8	178	4.1
Sminthuridae	117	4.6	222	5.2
Onychiuridae	28	1.1	56	1.3
Hypogastruridae	5	0.2	6	0.2
Total Pterygota (including larvae)	1,251	49.1	2,221	52.1
Dermaptera	2	0.08	5	0.1
Orthoptera	32	1.2	25	0.6
Dictyoptera	22	0.8	80	1.9
Psocoptera			4	0.01
Isoptera	93	3.6	57	1.3
Homoptera			6	0.2
Hymenoptera	697	26.6	1,090	25.8
Coleopteran larvae	367	14.4	889	20.9
Lepidopteran larvae	16	0.6	20	0.5
Total arthropods	2,548		4,262	

micro-and macro-climatic conditions that probably supported a greater abundance of arthropods (Reddy, 1984a; Ananthakrishnan, 1996; Mboukou-kimbatsa *et al*, 1998). Moreover, the trees possessed a greater biomass, and greater complexity and more microhabitats and created a mosaic of different living conditions for harbouring arthropods including insects (Young, 1982; Vohland and Schroth, 1999). Temperature in the tree-plantation area, was low and soil moisture was relatively higher compared to that of grassland (Venkataiah, 1988). Moreover, soil moisture was available for a longer period of time in the tree-plantation area, because of the presence of litter layer on the soil surface. Besides, the availability of more organic matter in the form of litter and humus produced every year being added continuously to the soil system of the tree-plantation, probably served as a resource for microorganisms and micro-and macro-fauna increasing the biodiversity. Many of the soil-surface arthropods are nocturnal and predatory in nature, and may have suitable habitat niches under the litter and logs in the tree-plantation area (Elton, 1966). However, Scorpionidae, and *Odontotermes* sp. were more abundant in grassland area compared to plantation area (Reddy and Venkataiah, 1990a) probably due to their habitat preference. Presence of Homoptera only in tree-plantation areas was most probably due to the availability of favourable host plants and their root as these are sap-sucking insects (Ostbye *et al.*, 1978) and Psocoptera, due to availability of its food requirements such as dead animal remains, etc.

The diversity of common arthropods comprised 24 species in grassland and 27 species in tree-plantation (Table 2). A homopteran, *Sisyphus hirtus* Wiedemann and a coleopteran, *Lasiodactylus chevrolati* Reitter were totally absent in the former area. Shannon index of general diversity (»H) for the total arthropods, Collembola, Acarina, Coleoptera, Araneae and Formicidae was greater in tree-plantation area (Table 3), suggesting greater habitat complexity, or heterogeneity of the tree-plantation area over grassland area. These findings may further indicate that the plantation of trees made the site favourable for the existence of more species of arthropods, and probably have contributed to the creation of longer food chains and food webs (Odum, 1971).

Similarly, the species richness index (d) and evenness (e) showing apportionment of the individuals among species, increased in the tree-plantation area for total arthropods (Niijima and Akiomi, 1991), Formicidae, Collembola and Acarina (Reddy and Venkataiah, 1990b), indicating relatively greater complexity of the area (Dindal, 1977).

The macro-arthropods became richer in species diversity by tree plantation. Similarly, Levieux (1982) reported greater evenness of the ant community in forests than in savanna, because of better equilibrated environmental conditions at the soil level in the forest. Moreover, the index of dominance (c) of total arthropods, Collembola, Acarina and Coleoptera was greater in the tree-plantation. However, (c) for Araneae and Formicidae was greater in grassland area (Reddy and Venkataiah, 1990b).

Table 2 Composition of some common species (excluding many unidentified species) of the soil inhabiting arthropods in grassland and tree-plantation area (Reddy and Venkataiah, 1990b)

Name of the Species	*Grassland area*	*Tree-plantation area*
Araneae:		
Storena sp.	+	+
Thanatus sp.	+	+
Acarina:		
Cosmolaelaps sp. (near) *vacua* (Michael)	+	+
Stigmaeus sp.	+	+
Lancetoppai willmanni Hammer	+	+
Eremulus avenifer Berlese	+	+
Scheloribates praeincisus interruptus Berl	+	+
Scheloribates fimbriatoides Hammer	+	+
Annectacarus lonoisetosus Bhattacharya *et al.*	+	+
Collembola:		
Lepidocyrtus sp.	+	+
Isotoma trispinata (MacGillivray 1896)	+	+
Sphaeridia sp. near *pumilis* (Krausbauer)	+	+
Isoptera :		
Odontotermes sp.	+	+
Homoptera:		
Sisyphus hirtus Wiedemann	–	+
Hymenoptera:		
Monomorium indicum Forel	+	+
Crematogaster sp.	+	+
Pachycondyla? tesserinoda (Emery)	+	+
Camponotus sp.	+	+
Coleoptera:		
Lasiodactylus chevrolati Reitter	–	+
Oryzaephilus acuminatus Halstead	+	+
? Megalodecne sp.	+	+
Mesomorphus sp.	+	+
Gonocephalum sp.	+	+
Pachycera sp.	+	+
Notocorax nervosus Mulsant and Roy	+	+
Total	24	27

Seasonality

Total Arthropods

The relative abundance of total soil-surface arthropods was minimum during summer, maximum during the rainy season and less abundant

Table 3 Different indices of species structure of the soil-surface-inhabiting arthropod communities of the grassland and the adjacent tree-plantation areas (S, total number of species; N, total number of individuals; »H, Shannon index of general diversity; d, species richness index; e, evenness index; C, index of dominance) (Reddy and Venkataiah, 1990b).

Arthropod taxa	*Grassland*						*Tree-plantation*					
	S	*N*	*H*	*d*	*e*	*c*	*S*	*N*	*H*	*d*	*e*	*c*
Total Arthropods	38	2548	1.30	10.86	0.88	0.07	41	4262	1.48	11.02	0.92	0.69
Araneae	2	141	0.43	0.83	0.90	0.41	2	226	0.45	0.85	0.91	0.37
Acarina	7	462	0.73	2.25	0.86	0.21	8	724	0.75	2.40	0.88	0.20
Collembola	5	645	0.50	1.42	0.70	0.38	5	1043	0.52	0.53	0.71	0.38
Coleoptera	7	367	0.75	2.73	0.83	0.21	8	889	0.76	2.80	0.86	0.21
Formicidae	4	679	0.49	0.99	0.81	0.37	4	1090	0.54	1.10	0.89	0.26

during the post-monsoon period and winter (Fig. 1) (Reddy and Venkataiah, 1990a). Most of the arthropods exhibited a more or less similar seasonal pattern of population fluctuation in both the areas.

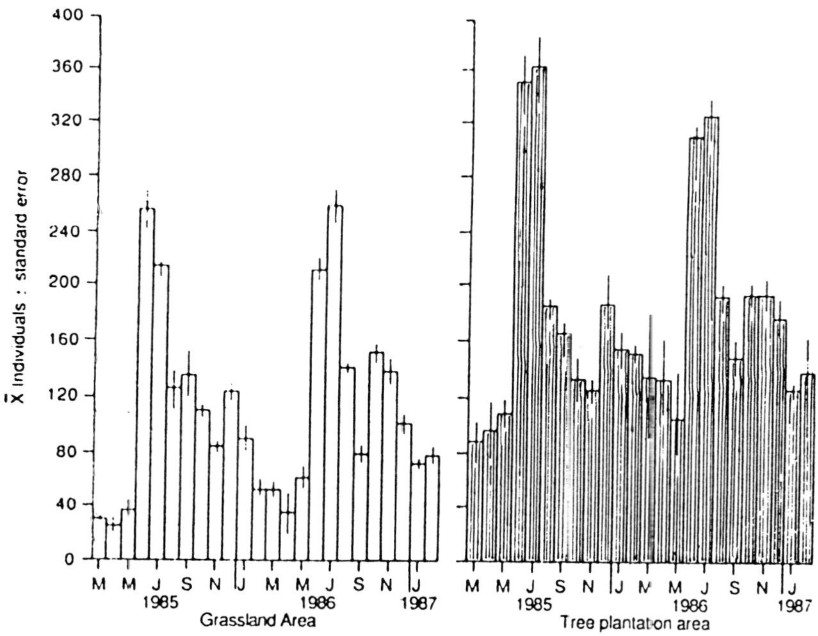

Fig. 1 Temporal variation in mean relative population abundance of total arthropods in the grassland and the adjacent tree-plantation areas (Reddy and Venkataiah, 1990b).

Moreover, the total arthropod populations showed broadly similar trends of fluctuation during the annual cycles, suggesting that their regular patterns of seasonal abundance were influenced by abiotic factors (Levings and Windsor, 1985). The differences in the pattern of abundance during the seasons were probably controlled by abiotic factors such as rainfall, temperature, and relative humidity at the soil surface (Remmiert, 1981) and soil moisture (Young, 1982; Reddy, 1984a). The seasonal variation in the population abundance of total arthropods showed significant linear correlations with various physico-chemical factors. The total arthropods occurred in higher abundance in the tree-plantation area during the annual cycles (Reddy and Venkataiah, 1990b).

FORMICIDAE

Total Formicidae were recorded in minimum numbers in summer in the grassland and in June in the tree-plantation. The abundance was maximal in early monsoon in the former area and in winter in the later, and decreased during the post-monsoon period (Fig. 2a). In the grassland *M. indicum* increased to maximum numbers during winter and decreased

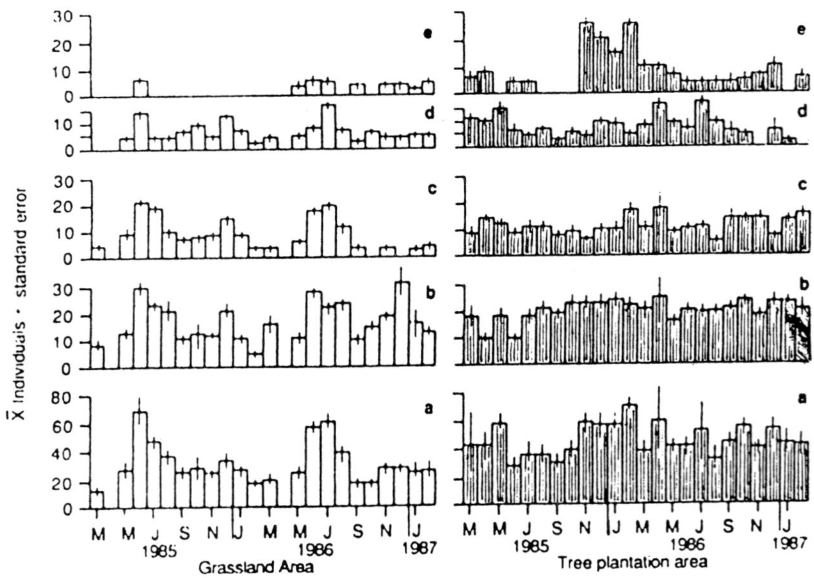

Fig. 2 Temporal variation in mean relative population abundance of total Formicidae (a), *Monomorium indicum* (b), *Crematogaster* sp. (c), *Pachycondyla*? *tesserinoda* (d) and *Camponotus* sp. (e) in the grassland and the adjacent tree-plantation (Reddy and Venkataiah, 1990b).

to a minimum during late winter. In the Tree-plantations the trend was less obvious, with maximum numbers in early monsoon and decreasing to a minimum in summer (Fig. 2b). *Crematogaster* sp. occurred in maximum numbers in the rainy season and in minimum numbers in winter and was absent for a few months in the grassland whereas in the tree-plantation, maximum numbers occurred in the rainy season (Fig. 2c). *Pachycondyla?tesserinoda* (Emery) and *Camponotus* sp. were absent for a longer period in the grassland than the tree-plantation (Fig. 2d, e). The abundance of *P.?tesserinoda* reached a peak in the rainy season and decreased to minimum density during the winter (Fig. 2d). *Camponotus* sp. was recorded in negligible numbers in both areas (Fig. 2e), although in the grassland maximum abundance occurred during the rainy season with minimum numbers or a few occurring in the summer season, followed by an increase in density during early winter (Fig. 2e) (Reddy and Venkataiah, 1990a).

While comparing tree-plantation area to the grassland area, total Formicidae, *M. indicum, Crematogaster* sp. and of *P.? tesserinoda* were recorded in lower numbers during the rainy season and in maximum numbers during the winter. Moreover, larger numbers were recorded during early summer in the former area. This was most probably due to the cooling effect of the canopy and litter cover in the tree-plantation. Changes in the diversity of ant community in coffee plantations were attributed to differences in shading and leaf litter (Perfecto and Vandermeer, 1996). The seasonality in population abundance of different species of Formicidae was different in both the areas (Reddy and Venkataiah, 1990a). None of the species investigated showed similarity of seasonality in both these areas. Such patterns of differences in seasonal abundance have been reported in different species of desert ants previously (Whitford, 1978). However, he stated that the surface activity of ants in arid conditions is controlled by a number of abiotic and biotic factors, the interaction of these factors being responsible for the seasonal community structure. Donnelly and Giliomae (1985) reported that the indigenous ant assemblage under pine plantations has fewer species and fewer ants when compared with fynbos habitats.

Collembola

Populations of total Collembola and its species viz., *Lepidocyrtus* sp., *Isotoma trispinata* (MacGillivray, 1896), *Sphaeridia* sp. nr. *pumilis* (Krausbauera) were absent during the summer in both areas, which appeared in large numbers reaching maximum density in the early rainy season, the peaks being more distinct in the tree-plantation area. Hypogastruridae occurred only during the rainy season, and decreased in numbers during post-monsoon and winter. All the Collembola except

I. trispinata showed a single peak of abundance i.e., unimodality in its seasonal abundance patterns (Figs. 3 and 4) (Choudhuri and Banerjee,

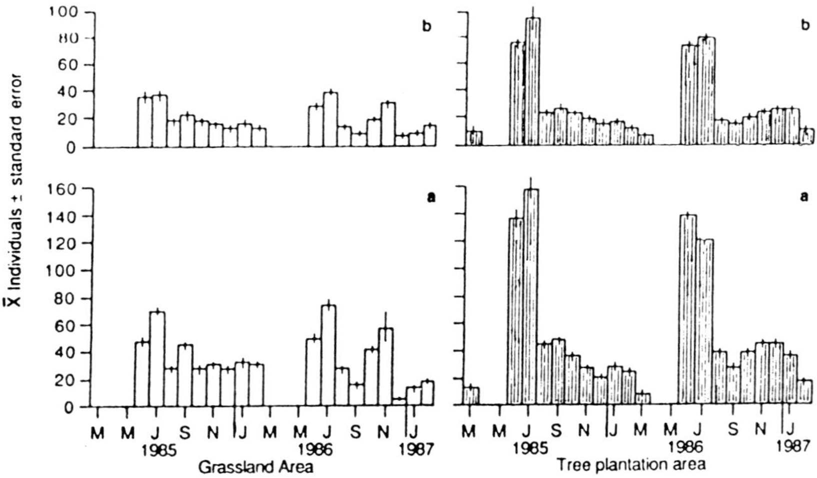

Fig. 3 Temporal variation in mean relative population abundance of total Collembola (a) and *Lepidocyrtus* sp. (b) in the grassland and the adjacent tree-plantation (Reddy and Venkataiah, 1990b).

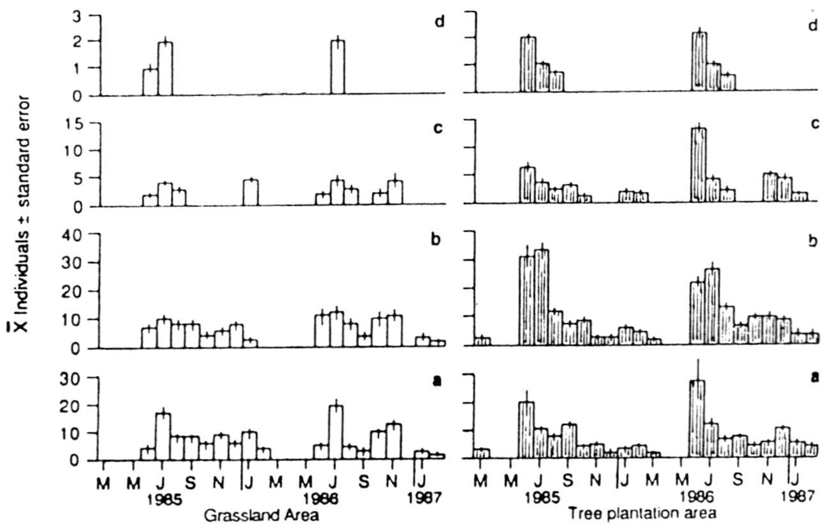

Fig. 4 Temporal variation in the mean relative population abundance of *Isotoma trispinata* (a), *Sphaeridia* sp. nr. *pumilis* (b), Onychiuridae (c) and Hypogastruridae (d) in the grassland and the adjacent tree-plantation (Reddy and Venkataiah, 1990b).

1975; Reddy and Venkataiah, 1990a) which is probably indicative of a single generation per year.

The population abundance of Thysanura was too few in numbers to reflect any seasonal fluctuation in either area.

COLEOPTERA

Overall seasonal population patterns of Coleoptera, its species, viz., *Mesomorphus* sp., *Pachycera* sp., *Gonocephalum* sp., *N. nervosus* Mulsant and Roy, *Coleolissus* sp., *O. acuminatus* Halstead, ?*Megalodecne* sp. and *L. chevrolati* Reitter and the coleopteran larvae in both areas indicated that many of them increased to maximum numbers during the rainy season and decreased during the summer (Figs. 5 and 6) (Reddy and Venkataiah, 1990a). *Gonocephalum* sp. and *N. nervosus* reached maximum numbers during the post-monsoon and *Mesomorphus* sp. and *O. acuminatus* during the winter. However, *Pachycera* sp. was in maximum density during the summer, which may indicate good adaptability to the hot summer with temperatures reaching up to 40°C. The remaining and other unidentified coleopterans were too few in number to show any seasonality (Fig. 5). Such seasonal peaks in abundance of different species of Coleoptera were probably due to diurnal activity rhythms (Dennison and Hodkinson 1984a).

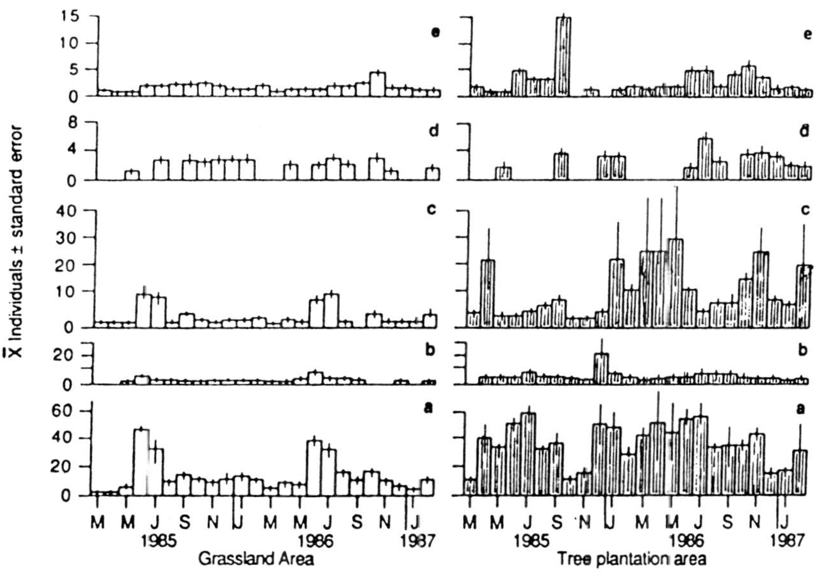

Fig. 5 Temporal variation in the mean relative population abundance of total Coleoptera (a), *Mesomorphus* sp. (b), *Pachycera* sp.(c), *Gonocephalum* sp. (d) and *Notocorax nervosus* (e) in the grassland and the adjacent tree plantation. (Reddy and Venkataiah, 1990b).

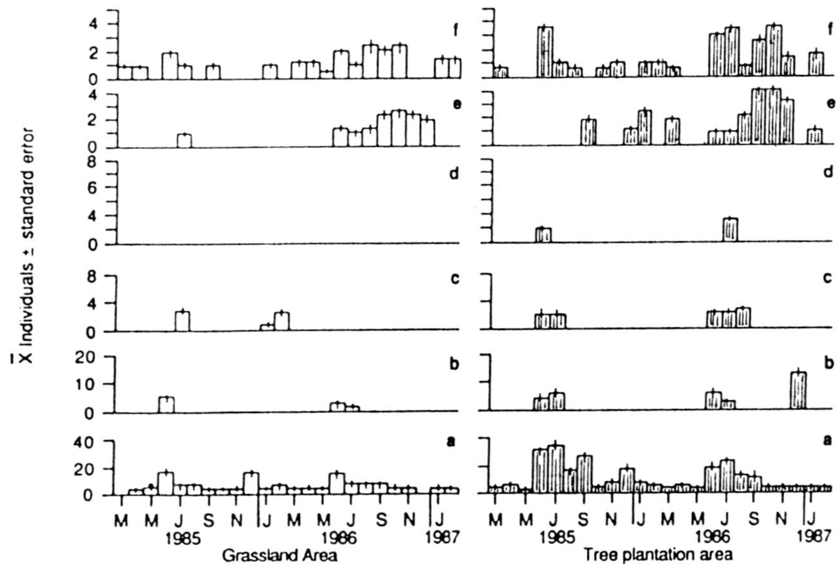

Fig. 6 Temporal variation in mean relative population abundance of *Coleolissus* sp. (a), *Oryzaephilus acuminatus* (b), ?*Megalodecne* sp. (c), *Lasiodactylus chevrolati* (d), the unidentified and less abundant *coleopterans* (e) and Coleoptera larvae (f) in the grassland and the adjacent tree-plantation (Reddy and Venkataiah, 1990b).

All the Coleoptera occurred in higher numbers in the tree-plantation area. *Mesomorphus* sp., *O. acuminatus*, ?*Megalodecne* sp. and coleopteran larvae occurred for a longer period during the annual cycle in that area compared to the grassland. *L. chevrolati* occurred only in the tree-plantation area (Fig. 6d) (Reddy and Venkataiah, 1990a), probably because of its closed canopy and microclimatic conditions compared to the open grassland area, and these differences probably have influenced the beetle community living at the soil surface (Thiele, 1977). The microclimate was probably the major factor, which influenced the microhabitat selection (Larmuth 1979). Parameter and MacMahan (1984) are of the view that it is the "food and cover" resources and not shrub architecture which affects the distribution of ground beetles. Variations in abundance and activity of coleopteran population in different habitats were previously shown to be correlated with temperature and rainfall (Grum, 1959; Grenslade, 1964; Luff, 1973). In addition to the abiotic environmental variables, availability of food is probably an important factor to influence their population increase (Dennison and Hodkinson, 1984a, b).

Araneae

Araneae and its two species, viz., *Storena* sp. *Thanatus* sp. and those of Gnaphosidae increased to maximum numbers at the beginning of the rainy season and decreased to a minimum during the post-monsoon period in both tree-plantation and grassland areas (Fig. 7). The abiotic variables probably influenced the Araneae population indirectly, may be first affecting the growth of prey populations. Araneae, being a predatory group, probably were influenced directly by the availability of food i.e., prey such as Collembola and other invertebrates.

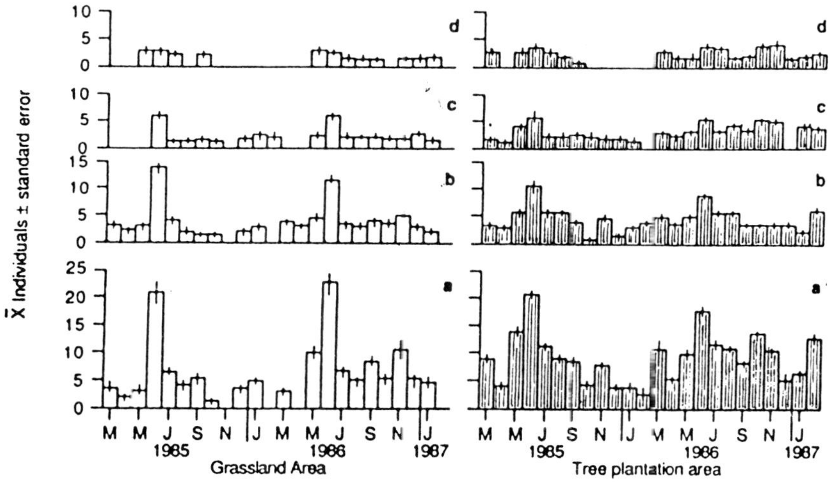

Fig. 7 Temporal variation in mean relative population abundance of total *Araneae* (a), *Storena* sp. (b), Gnaphosidae (c) and *Thanatus* sp. (d) in the grassland and the adjacent tree-plantation (Reddy and Venkataiah, 1990b).

Acarina

Interestingly, the overall abundance of Acarina and the constituent species, viz., *Cosmolaelaps* sp. nr. *vacua* (Michael), *Stigmaeus* sp., *L. willmanni*, *Eremulus avenifer* Berlese, *Scheloribates praeincisus interruptus* Berlese, *Annectacarus lonoisetosus* Bhattacharya *et al.*, *L. palustris* and *S. fimbriatoides* Hammer were absent during the summer. Their maximum abundance occurred during the rainy season and decreased during the winter in both areas (Figs. 8 and 9) (Reddy and Venkataiah, 1990a). All the species were more abundant in the tree-plantation area and interestingly, *S. fimbriatoides* was recorded only in that area.

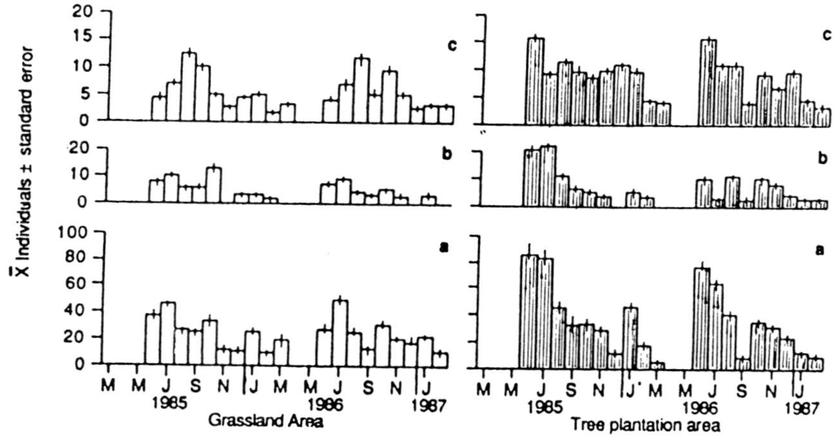

Fig. 8 Temporal variation in mean relative population abundance of total Acarina (a), *Cosmolaelaps* sp. nr. *vacua* (b) and *Stigmaeus* sp. (c) in the grassland and the adjacent tree-plantation (Reddy and Venkataiah, 1990b).

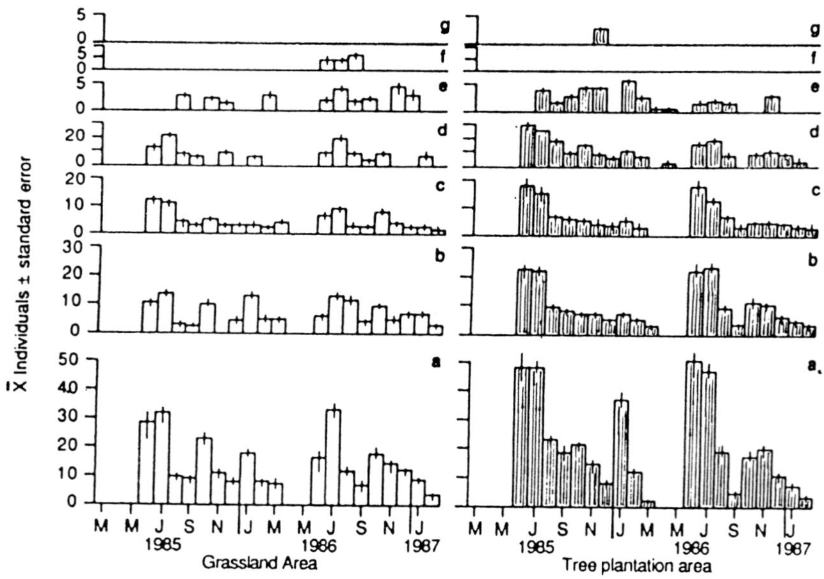

Fig. 9 Temporal variation in mean relative population abundance of total *Cryptostigmata* (a), *Lancetoppia willmanni* (b), *Eremulus avenifer* (c), *Scheloribates praeincisus interruptus* (d), *Annectacarus lonoisetosus* (e), *Lamellobates palustris* (f) and *S. fimbriatoides* (g) in the grassland and the adjacent tree-plantation (Reddy and Venkataiah, 1990b).

MISCELLANEOUS GROUPS

Most of the other less abundant group such as Scorpionidae. Pseudoscorpionidae, Diplopoda, Pauropoda, Chilopoda, Dermaptera, and Blattidae (Fig. 10), and *Odontotermes* sp. (Isoptera), Heteroptera, *S. hirtus* (Homoptera), Gryllidae, Psocoptera and Lepidoptera larvae (Fig. 11) were too few in number to show any seasonal fluctuation in both the grassland and tree-plantation areas (Moeed and Meeds, 1985; Reddy and Venkataiah, 1990a).

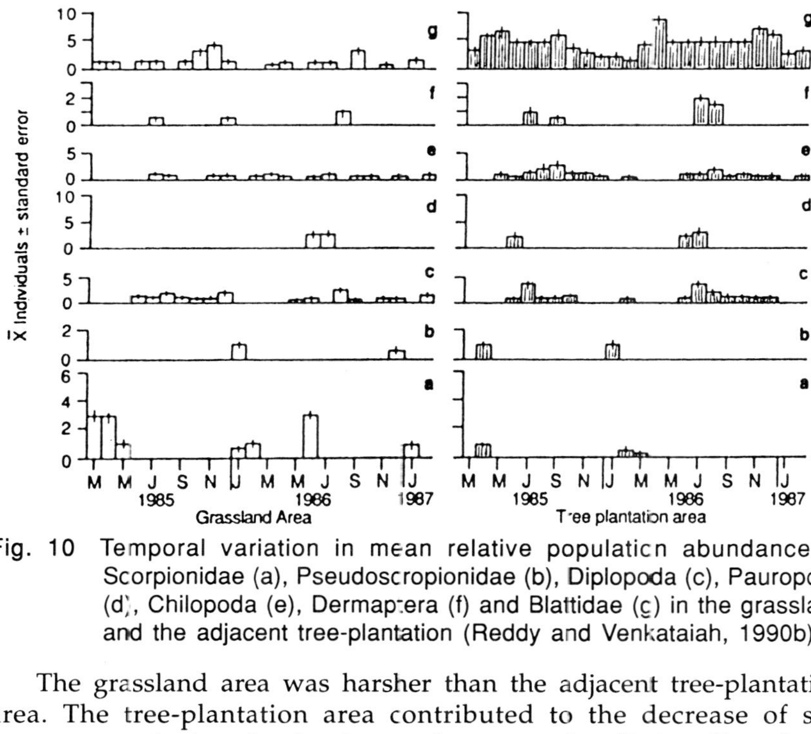

Fig. 10 Temporal variation in mean relative population abundance of Scorpionidae (a), Pseudoscropionidae (b), Diplopoda (c), Pauropoda (d), Chilopoda (e), Dermaptera (f) and Blattidae (g) in the grassland and the adjacent tree-plantation (Reddy and Venkataiah, 1990b).

The grassland area was harsher than the adjacent tree-plantation area. The tree-plantation area contributed to the decrease of soil temperature during the daytime and possessed a distinct litter layer. These changes provided favourable environment for the soil fauna that increased in density, biomass and diversity. In contrast, in ecosystems where abiotic factors are ameliorated by features inherent to the system, these variables should be less effective (Odum, 1971). Soil moisture levels influenced seasonal population abundance of the maximum number of arthropod taxa and thus was the key factor in both grassland and tree-plantation areas. Moreover, more arthropod taxa showed correlations of greater significance with the abiotic variables in the grassland than the tree-plantation area (Reddy and Venkataiah, 1990a).

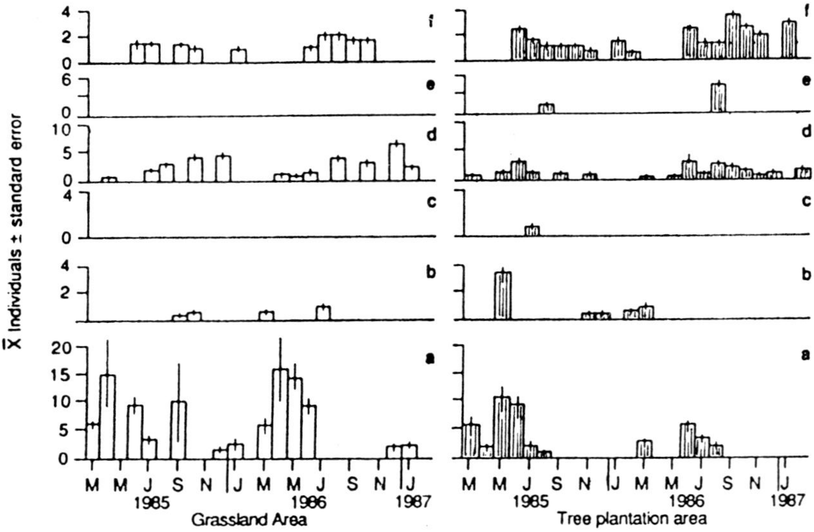

Fig. 11 Temporal variation in mean relative population abundance of Odontotermes sp. (a), Heteroptera (b), *S. hirtus* (c), Orthoptera (Gryllidae), (d) Psocoptera (e) and Lepidoptera larvae (f) in the grassland and the adjacent tree-plantation (Reddy and Venkataiah, 1990b).

Thus, these variables probably were relatively more important limiting factors in the grassland.

SOIL-INHABITING MICROARTHROPODS

Qualitative and Quantitative Composition

The soil microarthropods were sampled by soil-core samples taken from grassland and tree-plantation areas and processed through dry funnel extractors. Collembola, Formicidae, Pseudoscorpionidae, Isoptera and Coleoptera and their larvae were recorded only from its 0–5 cm layer of soil, while Cryptostigmata, Mesostigmata and Prostigmata were recorded in both 0–5 and 5–10 cm layers of the soil of grassland area. In tree-plantation area, Pseudoscorpionidae, Acarina and Collembola (except Onychiuridae), Isoptera, Coleoptera and their larvae were found in both the layers of the soil whereas Formicidae, lepidopteran larvae and Onychiuridae were recorded only in its 0–5 cm soil layer. In the grassland at the 0–5 cm layer of soil, eleven common species of microarthropods were recorded whereas only four species were found in 5–10 cm layer of soil, while in the tree-plantation area, 12 species were recorded at the 0–5 cm layer and nine species at 5–10 cm layer of soil, which is probably because of the well-developed soil and litter system. A maximum diversity of 17 representative species of microarthropods was recorded in *Eucalyptus globulus* litter, which decreased by six species at 1300m MSL (Ananthakrishnan *et al.*, 1993).

Vertical distribution of microarthropods was more or less similar between sites. About 81% of total number of microarthropods was recorded in the upper layer of grassland. In the tree-plantation area, about 76% of the total microarthropods were recorded in the surface layer, which was most probably because of the presence of litter layer. Collembola and Acarina constitued 76.6% of the total arthropods in grassland area and 71.1% in tree-plantation area. It showed the occurrence of more number of taxa belonging to miscellaneous groups of arthropods in the latter area. Quantitatively, the numbers of different microarthropods were higher in the tree-plantation area, and this may be due to the availability of plant litter, organic matter and humus in the area. The fact that the tree-plantation area harboured greater diversity and density of microarthropods compared to that of the grassland is consistent with earlier findings (Kuhnelt, 1963; Szujeki *et al.*, 1980; Reddy, 1984a; Niijima and Akiomi, 1991), that tree-plantations cause significant increase of microarthropods both qualitatively and quantitatively.

Similarly microarthropods inhabiting the soil-litter system of pine (*Pinus kesiya*) plantations in north-east India which is included as one of the hot-spots, Indo-Burma of the globe (Myers *et al.*, 2000), were grouped into categories such as Collembola, Acarina and Miscellaneous group, the latter including Isopoda, Thysanoptera, Hymenoptera, Araneidae, Protura, Diplura, Chilopoda, Diplopoda, Symphyla, Pauropoda, micro-coleopteran adults and their larvae and Dipteran larvae. Collembola being the most dominant group among the microarthropods in pine-plantation soil-litter system, was represented by five families (Isotomidae, Entomobryidae, Onichiuridae, Sminthuridae and Hypogastruridae) and dominated by *Isotoma trispinata* (MacGillivray, 1896). Acarina was represented by Prostigmata, Mesostigmata and Cryptostigmata and was dominated by the former. The author has recorded species belonging to eleven genus of oribatid mites and seven genus of Collembola in the soil-litter system of these pine plantations (unpublished data). Of the total microarthropod population 63.16% was Collembola, 31.79% was Acarina and only 5.05% was miscellaneous species. The Collembola group together with the Acarina group comprised 95% of the total microarthropods in the pine plantations (Reddy, 1984b). Similar findings were reported for summer population of microarthropods in the tea plantations in northeast India (Reddy, 1981b), and of microarthropod communities of *E. globulus* and *Tectona grandis* monoculture stands (Ananthakrishnan, 1996). However, Ananthakrishnan *et al.* (1993) reported higher densities of Dermaptera which were next in abundance to Collembola and mites in *E. globulus* stands.

Seasonal Fluctuation

Soil microarthropods including Collembola, Acarina and Coleoptera showed similar trends of seasonal fluctuation in their population densities during the annual cycles in both the grassland and tree-plantation areas (Reddy and Venkataiah, 1990c). This suggested that they had regular seasonal patterns of abundance influenced by abiotic factors (Levings and Windsor, 1985), and that fluctuations in Acarina populations are largely seasonal (Steinberger and Whitford, 1984). Their seasonal fluctuations at the surface layers were more or less similar to that of whole 0–10 cm layer of soil, as the total arthropod densities of the surface layer constituted a major portion (>75% in grassland area and >68% in tree-plantation area) of the entire 0–10 cm layer. This phenomenon has been reported in many earlier studies (Mitchell 1978; Takeda, 1978; Reddy, 1984a).

Total Microarthropods

The abundance of total microarthropods was minimum in summer, being absent in May and increasing to maximum in the rainy season in both the layers of the grassland. In the sub-layer, they were absent during the entire summer. In tree-plantation they were less abundant during the summer and increased to maximum density during early rainy season. They were higher in number compared to that of the grassland. However, at the sub-layer, low density was recorded in early summer being absent in summer (Fig. 12). The abundance decreased during post-monsoon period in both the areas (Reddy and Venkataiah, 1990c).

The decrease in abundance, or absence during the summer was probably because of dessication of the soil surface due to the higher temperature in absence of rainfall. The difference in microarthropod abundance between the grassland and tree-plantation areas was probably due to the presence of vegetation and litter cover that provide not only suitable microhabitat but also food in the later (Stanton, 1979; Reddy and Venkataiah, 1990c). Their vertical distribution may be due to their response to environmental conditions, which are subjected to seasonal changes. The differences in the pattern of abundance during the seasons in both the grassland and tree-plantation areas are probably due to the influence of abiotic environmental factors such as rainfall, temperature soil moisture and soil organic matter etc.

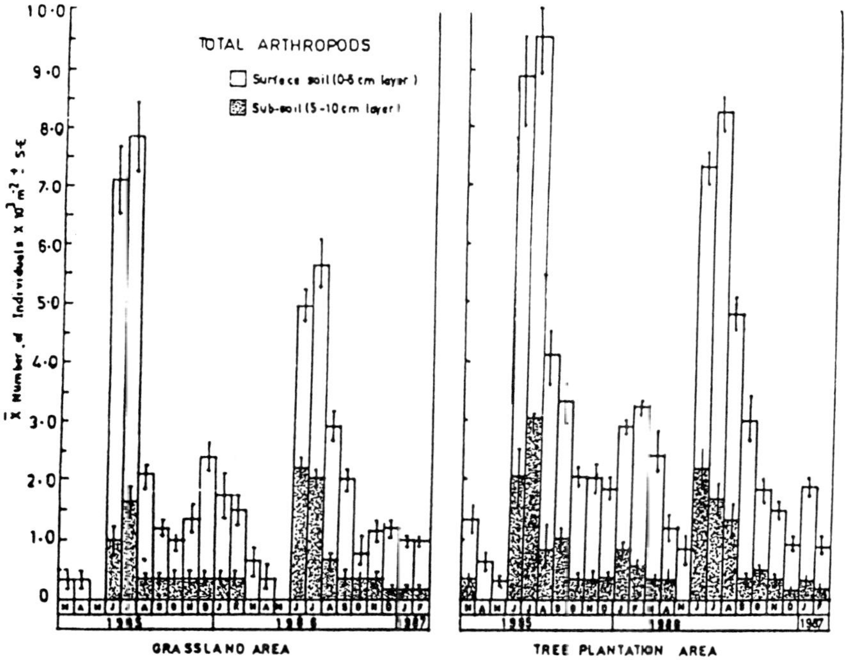

Fig. 12 Temporal variation in mean population densities of total soil microarthropods in the grassland and the adjacent tree-plantation (Reddy and Venkataiah, 1990c).

Collembola

The population densities of Collembola and its dominant species, *Lepidocyrtus* sp. occurred in highest numbers during the monsoon season. As they are abundant only in the wet-season and in moist soils, they may be the indicator species of soil moisture conditions (Reddy, 1984b; Niijima and Akiomi, 1991). They were found only at the surface layer of grassland area and at both the layers in tree-plantation area, and were recorded in higher number in the later (Fig. 13). Collembola were tend to persist in relatively higher densities in the humus soil interface in *E. globulus* stands (Ananthakrishnan *et al.*, 1993). It was probably because of the presence of more microhabitats in the upper layers, and the organic debris colonized by microflora on which the Collembola feed (Ferreira *et al.*, 1994). The Collembola were less abundant during the post-monsoon period, minimum in winter and absent in summer (Reddy and Venkataiah, 1990c).

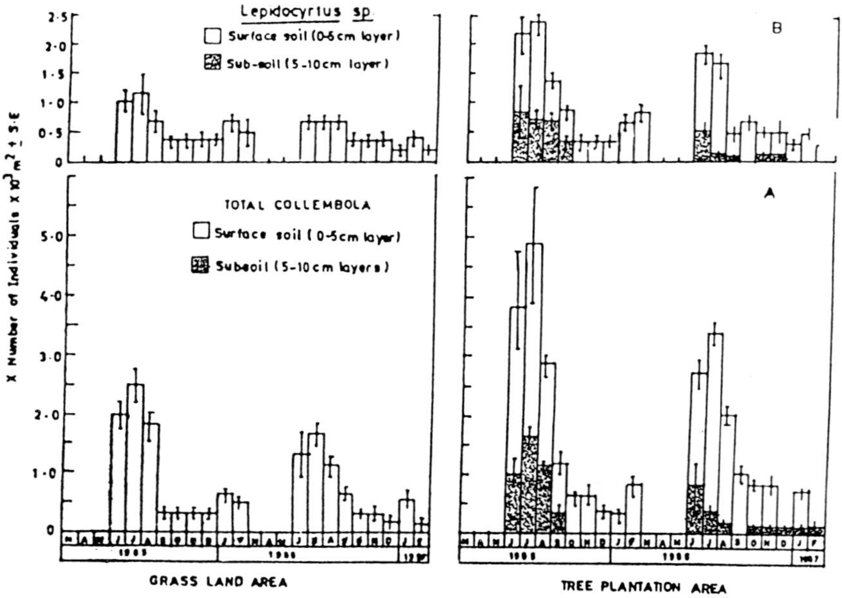

Fig. 13 Temporal variation in the mean population densities of total Collembola (A) and *Lepidocyrtus* sp. (B) of the grassland and the adjacent tree-plantation (Reddy and Venkataiah, 1990c).

Isotoma trispinata, Sphaeridia sp. nr. *pumilis* and Onychiuridae were absent during post monsoon, winter and summer (Fig. 14). However, they were recorded for a longer period of time in the year and at relatively higher numbers in tree-plantation area. Population of Onychiuridae was recorded in meagre numbers in both the grassland and tree-plantation areas (Fig. 14C) (Reddy and Venkataiah, 1990c). Sminthuridae have been reported to disappear during the dry season as they cannot penetrate into the soil during dry conditions, and probably undergo a quiescent stage (Price, 1973).

Studies from Portugal suggested that reafforestation with *Eucalyptus globulus* Labill caused a rupture on the composition of Collembola population. Collembola community of holm-oak, *Quercus rotundifolia* Lam. forest was much richer both qualitatively and quantitatively than that of *E. globulus* forest. The disruption in the community composition of Collembola was probably partially due to changes in soil surface environment derived from the establishment of the eucalyptus plantation and partly due to different forestry practices. (Sousa and da Gama, 1994). The difference in Collembola community between the *Quercus* stands and *Eucalyptus* plantation with different management practices,

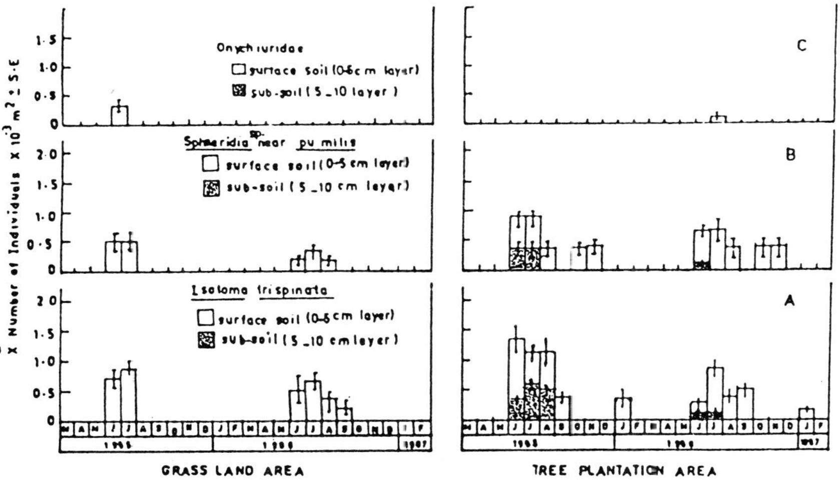

Fig. 14 Temporal variation in the mean population densities of *Isotoma trispinata* (A), *Sphaeridia* sp. nr. *pumilis* (B) and Onychiuridae (C) of the grassland and the adjacent tree plantation (Reddy and Venkataiah, 1990c).

was most probably due to the changes in habitat configuration, particularly at the organic horizon and vegetation cover. Both the nature of plant species, i.e., exotic species as well as the installation and management practices such as periodic removal of herb and shrub layers, and cellulose paste production etc. were probably responsible for disruption in the Collembola communities i.e., being greater under oak stand compared to that of eucalyptus plantation (Sousa *et al.*, 1997). They have concluded that any reafforestation activity especially with exotic tree species induced a change in the biocenotic equilibrium of soil communities. However, eucalyptus plantation, have a much higher number of collembolan individuals than pine, *P. pinaster* Aitom wood, though there was no clear distinction between both communities because of existence of homogeneity between the two biotopes. This difference in Collembola density was attributed to higher water content of eucalyptus soil particularly in the litter layer (Ferreira *et al.*, 1994), as moisture is one of the most important ecological factors controlling the distribution of Collembola in the soil-litter system (Reddy, 1984a,b; Rusek, 1989). Reafforestation with eucalyptus probably has caused changes in density and diversity of microflora and availability of trophic niches etc., and physico-chemical conditions in soil, which affected the soil communities leading to a rupture in their balance (Ferreira *et al.*, 1994).

Soil invertebrates under pine plantations were generally poor in diversity and density when compared to the natural forest soil; also there was lesser macrofauna under pine plantations (Watts, 1951; Armstrong, and Van Hensbergen, 1996). They were more numerous and evenly distributed in the top-soil under native forests than under pine plantations probably because of more acidic nature of the soil and lower concentration of nitrogen and organic carbon. The process of podsolization of the acid soils under pine vegetation made the soil less suitable for soil fauna. However, more fungivore fauna occured in organic layers of soil under these plantations (Watts, 1951). Soil fauna were even absent in some areas of pine stands where fungi were present (De Ronde, 1993). Similarly, reforestation with coniferous trees, *P. halepensis* and *P. nigra* in northern Spain produced a rupture on the structure and dynamics of soil faunal population such as oribatids, Collembola and nematodes. These soil invertebrates became poorer both qualitatively and quantitatively and few species appeared as exclusives from pine forest soils and only some species could adapt to new conditions. Probably, reforestation with *P. halepensis*, which led to closed environment compared to open land with Mediterranean bush did not produce and increase in species diversity and density of the edaphic fauna (Jordana *et al.*, 1987).

Acarina

Population densities of Acarina were maximum during early monsoon and decreased in the post-monsoon period; their abundance increased during winter and decreased to minimum density in early summer. They were absent in late summer in the surface soil layer of both grassland and tree-plantation areas (Figs. 15 and 16) (Reddy and Venkataiah, 1990c). They occurred in higher numbers in tree-plantation. In grassland area, they were recorded in greater numbers in surface layer compared to the sub-layer. In contrast, in tree-plantation areas, they were recorded in higher numbers in sub-layer compared to the surface layer (Fig. 15). The increase in Acarina abundance in the sub-layer was probably due to the presence of palatable decomposed litter fragments and suitable microenvironmental conditions. Cryptostigmata were absent during summer and increased in abundance during rainy season in the sub-layer of the tree-plantation (Fig. 15) (Reddy and Venkataiah, 1990c). Ananthakrishnan *et al.* (1993) reported the preponderance of mites (48%) among various groups of microarthropods in the fresh litter layers, which increased in very large numbers in F layers in *E. globulus* monoculture stands.

Lancetoppia willmanni Hammer decreased to minimum during post-monsoon (October) in Grassland area, while in tree-plantation area, its

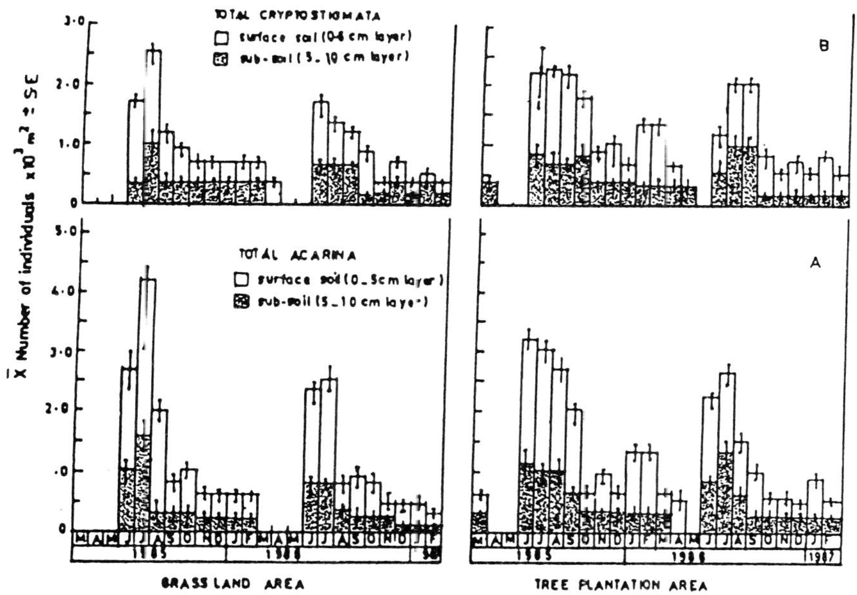

Fig. 15 Temporal variation in the mean population densities of total Acarina (A) and total Cryptostigmata (B) of the grassland and the adjacent tree-plantation (Reddy and Venkataiah, 1990c).

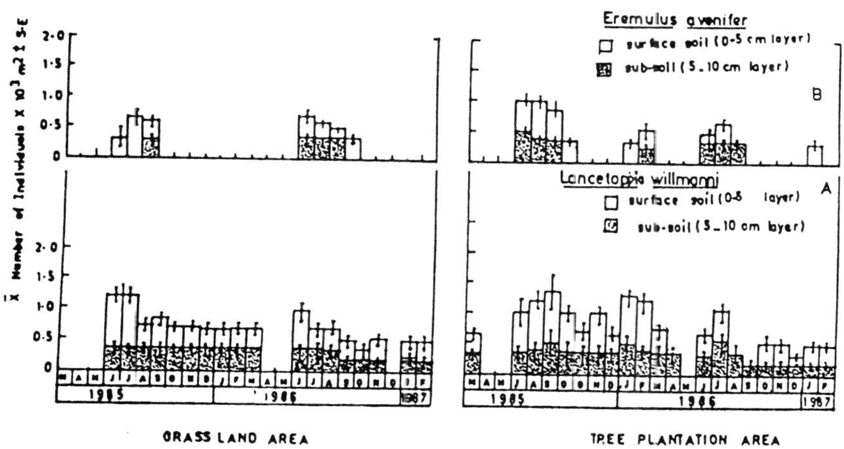

Fig. 16 Temporal variation in the mean population densities of *Lancetoppia willmanni* (A) and *Eremulus avenifer* (B) of the grassland and the adjacent tree-plantation (Reddy and Venkataiah, 1990c).

abundance decreased to low ebb during the period (September) (Fig. 16A). Similarly, the abundance of *E. avenifer* decreased to a low ebb in September in the former area. In surface layer of soil, its abundance decreased to minimum in (rainy season August). In the sub-layer of soil, the population was recorded only during the rainy season. In tree-plantation area, it decreased to minimum number in winter at the surface layer (Fig. 16B) and at the sub-layer of soil, *L. willmanni* and *E. avenifer* were recorded in rainy season, decreased in post-monsoon period and winter, which increased to maximum number during early summer but absent in summer (Reddy and Venkataiah, 1990c). Higher densities of oribatids were recorded in autumn with smaller peaks in winter-spring and summer (Schenker, 1986). Schenker related the minimum abundance with mild and very wet winter followed by a cool and dry spring.

Coleoptera

Population of total Coleoptera and that of *Pachycera* sp. and coleopteran larvae were recorded in the surface layer of soil of the grassland. In the surface layers of both the areas, the total Coleoptera and their larvae were abundant during the rainy season, decreased during winter and were absent during summer (Figs. 17A and C). Their density was more in tree-plantation compared to the grassland area. Population of *Pachycera* sp. was recorded for a longer period in the tree-plantation

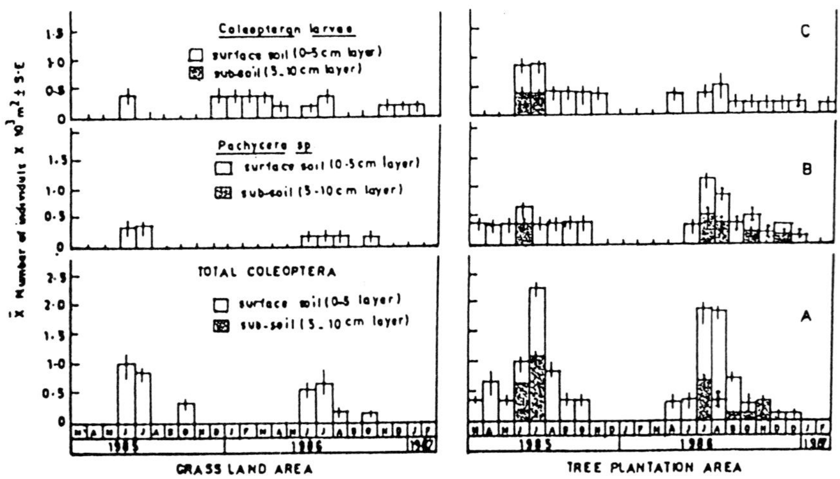

Fig. 17 Temporal variation in the mean population densities of total Coleoptera (A), *Pachycera* sp. (B) and the coleopteran larvae (C) of the grassland and the adjacent tree-plantation (Reddy and Venkataiah, 1990c).

area (Fig. 17B). This was probably because of the presence of more vegetation and litter on soil surface of tree-plantation area (Reddy and Venkataiah, 1990c). Scarabaeids and Carabids were recorded in decomposing *E. globulus* litter (Ananthakrishnan *et al.*, 1993). Tree-plantations, like forested ecosystems, have a closed canopy and different array of microclimatic conditions than an open grassland, and these differences are reflected in their respective beetle communities (Thiele, 1977).

SOIL MICROARTHROPOD COMMUNITIES, LITTER DECOMPOSITION AND NUTRIENT CYCLING IN PLANTATION FORESTS

Soil fertility is a determining factor in the growth and development of tropical forests and tree plantations which play a key role in regulating the global climate, soil erosion, flood control and water quality. Soil fertility in these ecosystems, particularly the natural biological fertility, depends on plant litter decomposition and nutrient dynamics, i.e., the release of nutrients from decaying litter through the integrated action of soil microorganisms and soil invertebrate fauna in the soil-litter system. Thus, soil faunal action is one of the important biological functions that affect nutrient cycling in these ecosystems.

Litter Decomposition Rate at Different Plantation Forests

Decomposition and mineralization of plant litter varies from species to species and from site to site, depending on a number of abiotic and biotic environmental factors (Swift *et al.*, 1979; Reddy, 1995). Some of the plant litter has low decay rates, while the litter of other species shows faster decomposition rates. The time required for complete decomposition and disappearance of the original biomass ranged from even less than two months to more than one year. An analysis of the litter decomposition rates in different plantation forests in the tropics and sub-tropics showed that the mass loss of needle litter in pine (*Pinus kesiya* Royle) plantation at Shillong in eastern India was 72.4% after one year, the annual average being 38% to 40.3% (Reddy, 1981a; Ramakrishnan and Das, 1983) while that of the oak-conifer forest in the western Himalayas ranged 54 to 55% (Pandey and Singh, 1981). The mass loss in bamboo litter in degraded forest fallows has been reported >97% (Toky and Ramakrishnan, 1984; Reddy and Toky, 1990) whereas the loss of litter mass in deciduous leaf litter in the humid sub-tropical forests in Meghalaya was 56–78%. (Singh and Ramakrishnan, 1982).

The mass loss rates of litter of *Dalbergia sissoo* in 5 mm mesh bags was 93% and in 2 mm mesh bags it was 84% and in 90 μm mesh bags, it was 60% after 354 days (Gupta and Rajvanshi, 1991). The mass loss of a teak litter in teak plantation was 90% after one year near Varanasi in Uttar Pradesh (Singh and Ambasht, 1980) while at Thailand and Nigeria it was 95% and 100% per year respectively (Aksornkoae *et al.*, 1972; Egunjobi, 1974). However, the mass loss of teak and eucalyptus litter was recorded as 96% and 64% respectively after 18 months at Peechi in Kerala (Sankaran, 1993), and it was 50.4% for eucalyptus litter in coarse (5 mm) mesh bags and 44.5% in fine (1 mm) mesh bags after one year at Warangal (Reddy and Venkataiah, 1989). The decomposition rates in different forest ecosystems have been reported to be very high. Brazilian tropical lowland rain forest, it was 93% (Klinge and Rodrigues, 1968) and that of the tropical lower montana forest of Puerto Rico was 94% (Wiegert, 1970), in the lowland rain forests in Nigeria it was 240% (Hopkins, 1966), 164% in Trinidad (Cornforth, 1970), and 120% in the tropical lower montane rain forests of New Guinea (Edwards, 1977).

The large differences in decomposition rates at different sites even with the same type of litter for example eucalyptus litter at Peechi in Kerala and Warangal in Andhra Pradesh in India were probably due to changes in some abiotic factors such as temperature, moisture and type of soil, and the biotic factors such as decomposer communities (particularly the micro- and macro-arthropods including termites). It has been suggested that the variation in temperature and moisture, and their profound combined influence on the decomposer communities can explain, to a large extent, the variation in the litter decomposition rates and mineralization (Williams and Gray, 1974; Reddy, 1995). Besides, the physical quality of litter (leaf thickness and leaf hardness or toughness) and the biochemical quality (lignin, nitrogen, and tannin contents and their ratios in the litter) influenced the decomposition rates and mineralization (Edwards, 1977; Melillo *et al.*, 1982; Takeda *et al.*, 1984). A high proportion of structural substances such as lignin and cellulose lowered the decomposition rates, while high levels of soluble components such as carbohydrates, fats and waxes increased the decomposition rates (Takeda *et al.*, 1984). All these biotic and abiotic environmental factors could be responsible for the very little or no accumulation of litter in tropical natural forests and plantation forests, which could imply a fast turnover of organic matter in the soil litter system. However, available information on tropical litter decomposition and nutrient dynamics particularly pertaining to plantation forest ecosystems is very limited and not sufficient for a complete holistic understanding of the functioning of the detritus system of these tropical ecosystems (Takeda *et al.*, 1984; Reddy, 1995).

LITTER INHABITING MICROARTHROPOD COMMUNITIES

The estimation of population densities of microarthropods inhabiting the soil-litter systems of plantation forests depends on the methods of sampling, extraction and enumeration. The published information generally showed underestimates of the natural population size of these arthropods (Seastedt, 1984). Their densities showed a wide range of variation in the tropics depending not only on the abiotic and biotic environmental factors of soil-litter system, but also on the physico-chemical qualities of the litter, and were influenced by both the amount and quality of detritus (Wallwork, 1976; Seastedt, 1984; Reddy, 1995). More information is needed on the relationship between microarthropod densities in habiting the litter system and their abiotic and biotic microenvironment, and the food resources in tropical plantation forests (Reddy, 1995).

Litter microarthropods, dominated by Collembola and Acarina in some ecosystems making up to 95% of the total microarthropods (Reddy, 1984a; Gupta and Rajvanshi, 1991; Reddy and Alfred, 1989; Ananthakrishnan *et al.*, 1993), constitute one of the major components of the biotic environment of the soil-litter system. Among them, isotomids and prostigmatids are the most abundant groups (Reddy, 1984a, 1995). Other microarthropods such as oribatids, mesostigmatids and astigmatids in Acarina, and sminthurids, podurids and entomobryids in Collembola are rarely dominant. Collembola and Acarina, and other taxa such as small isopods, pseudoscropions, diplopods (Polyxenidae, Polydesmidae, and Glomeridae), Pauropoda, Chilopoda, Symphyla, Protura, Diplura, Thysanura, Psocoptera, Isoptera,Thysanoptera, Homoptera, Hymenoptera (Formicidae), and coleopteran and dipteran larvae constitute the whole guild of microarthropods in the litter system (Reddy, 1995).

These organisms play a significant role in litter breakdown and communition leading to the release of mineral nutrients by microbial action and leaching (Vohland and Schroth, 1999). The introduction of mesofauna (mainly Collembola) in microcosm with deciduous forest litter increased the release of nitrogen (NO_3 and NH^{-+}) about three folds compared to that of the control without these animals (Anderson *et al.*, 1983). Other dominant groups of such litter-feeding mesofauna such as oribatid mites and small millipedes inhabiting the soil litter system of tropical plantation forests may enhance such nutrient release considerably. The microarthropods physically disintegrate plant litter with their mandibles and other mouthparts and expose a greater surface area for microbial colonization, decomposition and leaching (Swift *et al.*, 1979; Reddy, 1995). Although it has been reported earlier that bacteria and fungi are responsible for most of the litter decomposition in tropical forests (Lee, 1974), the exclusion experiments studying the effects of soil microarthropods on litter decomposition showed that these invertebrates played an indispensable role in the litter decomposition in plantation forest ecosystems (Madge, 1965; Reddy and Alfred, 1989; Gupta and Rajvanshi, 1991; Reddy, 1995). Vohland and Schroth (1999) are of the view that an abundant and active litter fauna may ensure rapid recycling of nutrients from plant litter. However, the role of microarthropods in litter decomposition and mineralization in different tropical plantation forest ecosystems is not adequately documented. Therefore, this review has attempted to examine the importance of soil and litter-inhabiting microarthropod communities during the process of litter decomposition across different plantation forests in the tropics, particularly in the indian subcontinent.

Microarthropods occupy all the trophic levels within the detritus food web on the floor of plantation forests (Ananthakrishnan, 1996). Some of them such as oribatids, collembolans, isopods and diplopods have well-developed mouthparts to fragment the litter, and feed on the microflora associated with the decayed litter, too. Such microarthropod activity creates new surface areas for enhanced microbial colonization and activity (Reddy, 1995) and in this manner, microarthropods and microorganisms play a synergistic role during the processes of litter decomposition and nutrient dynamics. Other microarthropods such as Prostigmata and Mesostigmata have piercing mouth parts, and being predators, along with other predatory microarthropods such as pseudoscropions and chilopods, regulate collembolan and other meso- or meio-faunal population densities and microbial biomass (Wallwork, 1976), which in turn influences the decomposition process.

Decomposition Rate in Relation to Microarthropod Abundance

Coniferous Litter

Pine needle litter is less palatable to soil fauna under pine plantation than native leaf litter (Watts, 1951; Armstrong and Van Hensbergen, 1996). Thus, there is less mixing of organic matter with mineral soil in these ecosystems. Population densities of microarthropods and the mass loss of pine needle litter from the litter bags in pine plantation at Shillong was observed to be low during dry period starting from November to March. On the other hand, the mass loss was higher when the microarthropod populations were relatively denser during the monsoon months (Reddy, 1989; Reddy, 1995). However, the higher mass-loss during the initial two rainy months was probably due to leaching of soluble organic compounds (Swift *et al.*, 1979; Reddy, 1995). The association of microarthropods with the communition of the needle litter was evident from the disintegrated litter recovered from the litter bags. Moreover, statistical significant positive correlations between mass loss and the densities of Collembola, Acarina and other miscellaneous groups of microarthropods were recorded. The microarthropod densities along with abiotic factors such as rainfall, litter moisture and temperature showed significant combined influence on the mass loss of needle litter, and they were found responsible for 75% variation in mass loss (R^2 = 0.75; $P<0.05$) (Reddy, 1989).

In another experiment using litter bags of different mesh sizes (3 mm, 1 mm, and 0.3 mm), the difference in mass loss of needle litter across the bags was recorded marginal, the average mass loss being maximum (38%) in the coarse (3 mm) mesh bags and minimum (34%) in the fine (1 mm) mesh bags. The difference in dominant micro-arthropods such as Collembola and Acarina inhabitating the three types of litter bags was also negligible, being 95.5%, 97.1% and 97.3% of the total microarthropods. In the needle litter of coarse mesh bags, the groups of Collembola, Protura, Pauropoda, Isopoda, Hemiptera and coleopteran adults and larvae, dipteran larvae and juvenile earthworms were recorded in large numbers (Reddy and Alfred, 1989). The presence of these faunal groups with larger populations probably made a greater contribution to mass loss during litter decomposition process of the needle litter in these bags (Santos and Whitford, 1983).

Deciduous Litter

In the mixed tree plantation forests at Warangal, the mass loss in eucalyptus leaf litter was significantly higher (50.4%) in the coarse mesh bags compared with the mass loss in fine mesh bags (44.5%) ($P<0.05$). The fine mesh bags showed about six times higher litter mass loss than in suspended bags (7.4%) ($P<0.01$). Interestingly, there were no microarthropods in the litter of suspended litter bags. However, the microarthropods were more abundant in the coarse mesh bags (Reddy and Venkataiah, 1989; Reddy, 1995). Similarly, in the case of neem leaf litter, the percentage of mass loss in the coarse mesh bags was found significantly higher than that in the fine mesh bags. The higher decomposition rates in coarse mesh bags may be related to the high population densities of microarthropods present in these bags (Reddy, 1992).

In another exclusion experiment, teak (*Tectona grandis* L.) leaf litter was treated with an organochlorine insecticide (Endosulfan) in order to keep away the microarthropods from the decomposed litter and was compared with that of normal leaf litter, i.e., without any chemical treatment, allowing access to all the microarthropods. The mass loss of the insecticide-treated teak leaf litter was significantly less (<54%) compared to that of normal leaf litter where the mass loss was more than 80% after 330 days ($P < 0.01$). It indicated the significant role of microarthropods in mass loss of teak litter during the process of litter decomposition (Reddy, 1995).

The mass loss of leaf and twig litter of *D. sissoo* was higher (93.0%) in 5 mm mesh bags, 83.9% in 2 mm mesh bags, and 60.2% in 90 μm mesh bags. The mass-loss, as reported in the eucalyptus leaf litter (Reddy and Venkataiah, 1989), was minimum (27.8%) in suspended aerial bags, which offered no access to microarthropods (Gupta and Rajvanshi, 1991). Collembola and Acarina were dominant groups of microarthropods in the decaying litter in the coarse mesh bags, and their activity along with termite activity caused significant mass loss in these mesh bags. These findings clearly showed the direct and significant role of microarthropod densities in mass loss during litter decomposition process in different tropical plantation forests in the Indian subcontinent.

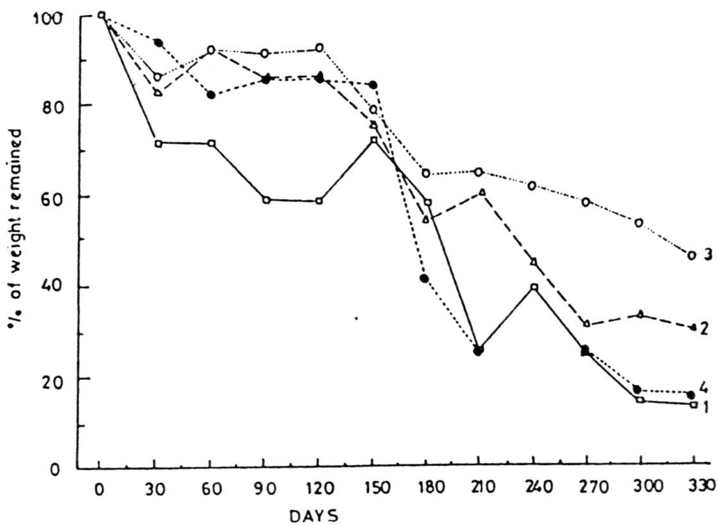

Fig.18 Percentage of mass-remained in the litter bags containing teak leaf litter treated with insecticide in relation to that of normal decomposed leaf litter (without any pesticide treatment) (Reddy, 1995).

1–Litter bags treated with bactericide, 2–with fungicide
3–With insecticide and 4–decomposed leaf litter without any pesticide treatment.

NUTRIENT DYNAMICS DURING LITTER DECOMPOSITION

Nutrient release from the decaying litter is the net effect of changes in nutrient concentrations and litter mass. When the densities of fauna are generally high per g of litter, the fragmentation and communition, and grazing effects appear to be the dominant factors influencing nutrient dynamics.

CONIFEROUS NEEDLE LITTER

Seasonal patterns of nitrogen (N), phosphorus (P), potassium (K), calcium (Ca) and magnesium (Mg) concentrations in the decomposing pine needle litter in the pine plantations near Shillong showed much less variation than could be expected from the bimodal pattern of rainfall there. Nitrogen concentrations slowly increased throughout the first year of decay, while Mg exhibited no clear pattern (Reddy, 1989). Calcium exhibited a sharp decline during the initiation of the dry period, while K showed an increase in concentration during the dry period. This increase is correlated with an apparent increase in litter mass (Reddy, 1989), suggesting deposition of dry particulate matter on the litter (Seastedt and Crossley, 1980). Although microarthropods have often been reported to influence the net mineralization of the chemical elements of decaying litter (Seastedt, 1984), there was no significant correlation between the abundance of any of the group of microarthropods and the N and Ca concentrations of the decomposed needle litter (Table 4) (Reddy, 1989). The lack of a relationship supports the report of Macauley (1975) that microarthropods have little effect on the N concentration of the decomposed needle litter. Similarly, microarthropods have been reported to have little effect on the concentration of Ca (Seastedt, 1984). However, Anderson *et al.*, (1983) reported that Collembola have a considerable effect on the leaching of N from the decomposed litter than any other taxa of soil microarthropods. Collembolan grazing has been reported to increase the leaching of ammonium, nitrate and calcium from decomposed oak litter (Ineson *et al.*, 1982). Potassium concentrations of the decomposed needle litter, while not showing any correlation with the densities of Collembola and Acarina, was significantly negatively correlated with the abundance of miscellaneous groups of microarthropods (Table 4). However, Ineson *et al.* (1982) reported that the concentration of K was little affected by the microarthropods. The Mg concentrations of litter showed significant positive correlations with the abundance of Acarina in the little bags (Table 4). In contrast, Seastedt (1984) reported that microarthropods have no consistent effect on Mg concentration of the decayed litter.

Table 4 Linear regression equations between the abundance of different taxa of microarthropods and the mass loss and elemental concentrations (% of dry mass) of the decaying leaf litter

Microarthropod taxa	*Mass loss*	*N*	*P*	*K*	*Mg*
Coniferous litter: (Pine needle)					
Total Collembola	y=5.58+0.007 X t=2.0*	–	–	–	–
Total Acarina	y=5.9+0.002x t=2.9**	–	–	–	y=0.2+0.004x t=2.08*
Miscellaneous taxa	y=6.17+0.28x t=4.0**	–	–	y=0.07-0.014x t=2.19*	
Deciduous litter: (eucalypt leaf) Fine-mesh bags					
Total Collembola	–	y=-1.29-0+008x t=2.48	y=0.31+0.003x t=2.57*	–	–
Lepidocyrtus sp.	–	y=-1.29-0.01x t=2.43*	y=0.31+0.004x t=2.49*	–	–
Sminthuridae	–	y=-1.27-0.03x t=2.28*	y=0.31+0.01x t=2.34	–	–
Coarse-mesh bags: Total Collembola	–	y=-1.36-0+009x t=2.81*	y=0.28+0.003x t=2.45*	–	–
Lepidocyrtus sp.	–	y=-1.36-0.001x t=2.65*	y=0.28+0.004x t=2.31*	–	–
Sminthuridae	–	Y=-1.34-0.04x t=2.71*	y=0.29+0.001x t=2.31*	–	–
Total Microarthropods	–	y=1.61+0.08x t=2.55*	y=0.28+0.01x t=2.22		

*P<0.05; **P<0.01

Deciduous Litter

Seasonal variation in the concentrations of N and Ca and that of P and K in the decomposed leaf litter of eucalyptus indicated that the concentration of N increased from 0.56% to 1.56% while the increase in Ca concentration was negligible (Fig. 19). The concentration of P decreased from 0.6% to 0.2%, and that of K from 0.18% to 0.04% over one year (Fig. 20) (Reddy and Venkataiah, 1989). However, there were negligible changes in the concentration of these mineral elements in the litter of suspended bags that has no access to microarthropods. The concentration of N in the litter was higher in the coarse mesh bags (1.21%) than in the fine mesh bags (1.16%) and more than double

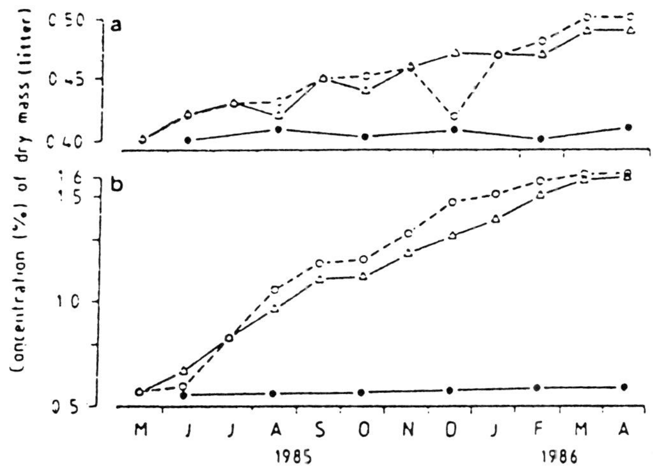

Fig.19 Temporal changes in concentration of Ca(a) and N(b) (Percentage of dry mass) of leaf litter in fine (Δ) and coarse (O) mesh bags and in suspended (•) bags in a eucalyptus plantation (Reddy and Venkataiah, 1989).

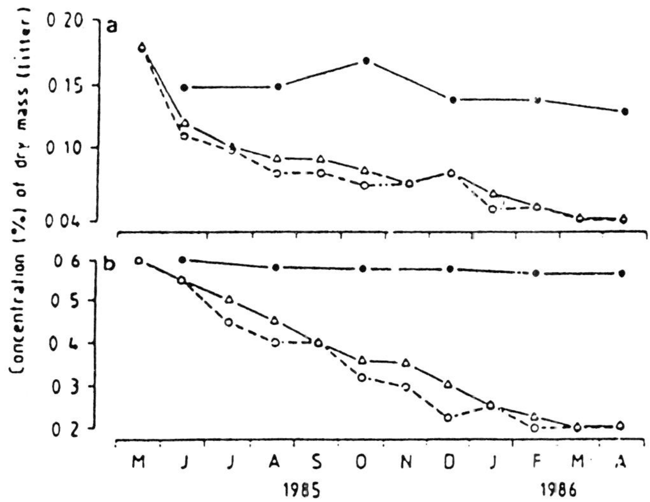

Fig. 20 Temporal variation in the concentration of K(a) and P(b) (Percentage of dry mass) of litter in fine (Δ) and Coarse (O) mesh bags and in suspended (•) bags in a eucalyptus plantation (Reddy and Venkataiah, 1989).

compared with its concentration in the suspended bags. The Ca concentration was identical in the litter of both the mesh bags and was higher than that of the suspended bags. The concentrations of P and K were slightly lower in the coarse mesh bags than in the fine mesh bags and far lower than the concentrations in the suspended bags. The almost total absence of K leaching and changes in concentrations of other nutrients from the suspended bags suggests that the litter must be fragmented or otherwise physically altered by the fauna in order to promote nutrient leaching (Reddy and Venkataiah, 1989).

The measurements of mass loss and elemental concentrations, and the population densities of microarthropods are consistent with the hypothesis that the abundance of microarthropods plays a significant role in the seasonal dynamics of nutrients in decaying litter during the process of litter decomposition (Seastedt and Crossley, 1980). The seasonal variation in the concentration of N in decaying eucalyptus leaf litter showed significant positive correlation with the total microarthropods in the coarse mesh bags (Table 4), which may imply that the abundance of total microarthropods contributes to the release of N (Seastedt, 1984; Reddy, 1995). The variation in concentration of Ca and K did not show any correlation with the changes in the abundance of any of the microarthropods. The seasonal variation in the concentration of P showed positive linear correlation with the abundance of total Collembola, with that of *Lepidocyrtus* sp. and of Sminthuridae in both the mesh bags, and with the total microarthropod abundance in the litter of the coarse mesh bags (Table 4) (Reddy and Venkataiah, 1989). Therefore, the increase in microarthropod density appears to enhance P concentration, whereas the decrease in their abundance during winter and summer reduced its concentration in the eucalyptus leaf litter during decomposition.

These studies on litter decomposition suggest that microarthropods play an indispensable role in mass loss and nutrient dynamics of coniferous as well as deciduous litter during decomposition in plantation forests. The net result of rapid mass loss with no significant increase in the elemental concentration of the decaying needle mass indicates a very rapid mineralization effect of the microarthropods. Both the increase and decrease in the mineral nutrients of the decomposed litter are determined by the relative balance of inputs minus the outputs (Seastedt and Crossley, 1980). The outputs are most probably directly increased by consumption of litter by the microarthropods and other meso- and macro-fauna in soil litter system and indirectly by the fragmentation and stimulation of more microflora or by some other mechanisms (Crossley, 1977). The various positive and negative correlations of microarthropod densities with N and P concentrations of decomposed leaf litter of eucalyptus, if not spurious, could suggest that the cumulative effect of microarthropod fauna on mass loss and nutrient concentrations of decaying litter may differ from the responses of the individual faunal groups.

REFERENCES

Aksornkoae, S., Khenmark, C., and Kaewla-lad, Raewla-lad, T., 1972. Study on organic matter in teak plantations, *Kasetsart Univ. For. Res. Bull.*, No. 23, 36 pp.

Ananthakrishnan, T.N., 1996. *Forest Litter Insect Communities: Biology and Chemical Ecology*, Oxford & IBH Publ. Co. Pvt. Ltd., New Delhi.

Ananthakrishnan T.N., Gopichandran, R., and Subramaniam, V.R., 1993. Dynamics of microarthropod communities in relation to the chemical environment of decomposing *Eucalyptus globulus* litter. *International J. of Ecology and Environmental Sciences*, **19:** 83–92.

Anderson, J.M., Ineson, P., and Huish, S.A., 1983. The effects of animal feeding activities of element release from deciduous forest litter and soil organic matter, In: *Proceedings of the VII Int. Colloq. of Soil Zoology*, (P.H. Lebrum, H.M. Andre, A. DE MEDTS, C. Gregoire Wibo, and G. Wauthy, eds.), Dieu-Brichart, Ottignies – Louvain – la- Neuve, 87–100 pp.

Armstrong, A.J., and Van Hensbergen, H.J., 1996. Impact of afforestation with *Pinus* on assemblages of native biota in South Africa, *South Afrikaanse Bosboutydsksif*, **175:** 35–42.

Choudhuri, D.K., and Banerjee, S., 1975. Qualitative and quantitative composition of Acari and Collembola in relation to soil organic matter and microbe complex, *Orient Insect*, **9:** 313–316.

Cornforth, I.S., 1970. Leaf fall in a tropical rain forest, *J. Appl. Ecol.*, **7:** 603–608.

Crossley, D.A. Jr., 1977. The role of terrestrial saprohagous arthropods in forest soils. Current status of concepts, In: *The role of arthropods in forest ecosystems*, (W.J. Mattson, ed.), Springer, Berlin Heidelberg, New York, 49–54 pp.

Deharveng, L., 1996. Soil Collembola diversity, endemism and reforestation: a case study in Pyrenees (France), *Conserve. Biol.*, **10:** 74–84.

Dennison, D.F., and Hodkinson, I.D., 1984a. Structure of the predatory beetle community in a woodland soil ecosystem III. Seasonal activity pattern as revealed by pit-fall trapping, *Pedobiologia*, **26:** 45–56.

Dennison, D.F., and Hodkinson, I.D., 1984b. Structure of the predatory beetle community in a woodland soil ecosystem V. Summary and conclusions, *Pedobiologia*, **26:** 171–177.

De Ronde, C., 1993. Spatial Variation of forest floor loading in pine stands of the Tsitsikamma, *S. Afr. For. J.*, **166:** 1–7.

Dindal, D.L., 1977. Influence of human activities on oribatid mite communities, In: *Biology of oribatid mites, SUNY* Coll. Environ. Sci. Forest., (D.L. Dindal, ed.), Syracuse, USA, 105–20 pp.

Donnelly, D., and Giliomae, J.H., 1985. Community structures of epigaeic ants in a pine plantation and a newly burnt fynbos, *J. ent. Soc. S. Afr.*, **48:** 259–265.

Edwards, P.J., 1977. Studies on mineral cycling in a montane rain forest in New Guinea. II. The production and disappearance of litter, *J. Ecol.*, **65:** 971–992.

Egunjobi, J.K., 1974. Litter fall and mineralization in a teak stand, *Oikos.*, **25:** 222–226.

Elton, C.S., 1966. *The Pattern of Animal Communities*, John Wiley and Sons, New York, USA.

Evans, J., 1982. *Plantation Forestry in the Tropics*, ELBS and Clarendon Press, Oxford, England.

Felker, P., 1986. *Tree Planting in Semi Arid Regions*, Elsevier, Amsterdam.

Ferreira, C.S., da Gama, M.M., Sousa, J.P., and Vasconcelos, T.M., 1994. Comparative study of the collembola population from a *Pinus pinaster* Aitom forest and *Eucalyptus globulus* Labill. plantation, *Cienc. Biol. Ecol. Syst.*, (Portugal), **14**(1/2): 19–34.
Grenslade, P.M., 1964. Pit-fall trapping as a method for studying populations of Carabidae (Coleoptera), *J. Anim. Ecol.*, **33**: 301–310.
Grum, L., 1959. Seasonal changes of activity of the Carabidae, *Ekol. Pol. Ser.*, **A7**: 255–268.
Gupta, S.R., and Rajvanshi, R., 1991. The effect of biotic and abiotic factors on litter decomposition in a tropical *Dalbergia sissoo* forest ecosystem, In: *Advances in Management and Conservation of Soil Fauna*, (G.K. Veeresh, D. Rajagopal, and C.A. Viraktamath, eds.), Oxford & IBH Publ. Co. Pvt. Ltd., New Delhi, 491–495 pp.
Hopkins, B., 1966. Vegetation of the Olekemeji Forest Reserve, Nigeria. IV. The litter and soil with special reference to their seasonal changes, *J. Ecol.*, **54**: 637–703.
Ineson, P., Leonard, M.A., and Anderson, J.M., 1982. Effects of collembolan grazing upon nitrogen and cation leaching from decomposing leaf litter, *Soil Biol. Biochem.*, **14**: 601–605.
Jordana, R., Arbea, J.I., Moraza, L., Montenegro, E., Mateo, M.D., Hernandez, M.A., and Herrera, L., 1987. Effect of reafforestation by conifers in natural biotopes of middle and South Navarre (Northern Spain), *Revue Suisse Zool*, **94**(3): 491–502.
Klinge, H., and Rodrigues, W.A., 1968. Litter production in an area of Amazonian terra firma forest, *Amazonian*, **1**: 287–310.
Kuhnelt, W., 1963. Soil inhabiting Arthropods, *Ann. Rev. Entomol*, **8**: 115–136.
Lamotte, M., 1979. Secondary Production: Consumption and decomposition, In: *Tropical Grazing Land Ecosystems, UNESCO Natural Resources Res.*, **16**: 146–208.
Larmuth, J., 1979. Aspects of plant habitat as a thermal refuse for desert insects, *J. Arid Environ.*, **2**: 323–327.
Lavelle, P., Dangerfield, M., Fragoso, C., Eschenbrenner, V., Lopez-Hernandez, D., Pashani, B., and Brussaard, L., 1994. The relationship between soil fauna and tropical soil fertility, In: *The Biological Management of Tropical Soil Fertility*, (P.L. Woomer, and M.J. Swift, eds.), TSBF, J. Wiley & Sons. Chi Chester, 137–169 pp.
Lee, K.E., 1974. The significance of soil animals in organic matter decomposition and mineral cycling in tropical forest and savanna ecosystems. Trans. 10th Int. Congr. Soil Sci., Moscow, Volume III, 43–51 pp.
Levieux, J., 1982. A comparison of ground dwelling ant populations between a Guinea savanna and an evergreen rain forest of the Ivory Coast, In: *The biology of Social insects*, (M.D. Breed, C.D. Michener, and H.E. Evans, eds), Westview Press, Boulder, USA, 48–52 pp.
Levings, S.C., and Windsor, D.M., 1985. Litter arthropod populations in a tropical deciduous forest: relationships between years and arthropods groups, *J. of Animal Ecology*, **54**: 61–69.
Luff, M.L., 1973. The annual activity pattern and life cycle of *Pterostichus mandidus* (F.). (Col., Carabidae). *Entomol, Scand*, **4**: 259–273.
Macauley, B.J., 1975. Biodegradation of leaf litter in *Eucalyptus pauciflora* communities. 1. Techniques for comparing the effect of fungi and insects. *Soil Biol. Biochem*, **7**: 341–344.
Madge, D.S., 1965. Leaf fall and litter disappearance in a tropical forest, *Pedobiologia*, **5**: 273–288.
Mboukou-kimbatsa, I.M.C., Bernhard-Reversat, F., and Loumeto, J.J., 1998. Changes in soil macrofauna and vegetation when fast growing trees are planted on savanna soils, *Forest Ecology and Management*, **110**: 1–12.
Melillo, J.M., Aber, J.D., and Muratore, J.F., 1982. Nitrogen and lignin control of hardwood leaf litter decomposition dynamics, *Ecology*, **63**: 621–624.
Mitchell, M.J., 1978. Vertical and horizontal distributions of oribatid mites (Acari: Cryptostigmata) in an aspen woodland soil, *Ecology*, **59**: 516–525.

Moeed, A., and Meeds, M.N.J., 1985. Seasonality of pit-fall trapped invertebrates in three types of native forest, Orongorongo valley, New Zealand, N.Z. *J. Zool.*, **12:** 17–53.

Myers, N. Mittermeier, R.A., Mittermeier, C.G., de Fonseca, G.A.V., and Kent, J., 2000. Biodiversity hotspots for conservation priorities, *Nature*, **403:** 853–858.

Niijima, K., and Akiomi, Y., 1991. The effects of reforestation on soil fauna in the Philippines, *The Philippine J. of Science*, **120**(1): 1–19.

Odum, E.P., 1971. *Fundamentals of ecology*, 3rd edition, Saunders WB Com., Philadelphia, USA.

Ostbye E., Hagver, S., Hande, K.M., Hande, P.S., Melaen, J., and Skortrett, A., 1978. Pit-fall catches of surface-active arthropods in some high mountain habitats at Finse, South Norway II. General results at group level, with emphasis on Opiliones, Araneida and Coleoptera, *Norw. J. Ent.*, **25:** 195–205.

Pandey, U., and Singh, J.S., 1981. A quantitative study of the forest floor, litter fall and nutrient return in an oak conifer forest in Himalayas. II. Pattern of litter fall and nutrient return, *Acta Oecologia*, (Oecol. Gener.), **2:** 83–99.

Parameter, R.R., and MacMahan, J.A., 1984. Factors influencing the distribution and abundance of ground beetles (Coleoptera) in shrub-steppe ecosystem: The role of shrub architecture, *Pedobiologia*, **26:** 21–34.

Perfecto, I., and Vandermeer, J., 1996. Microclimatic changes and indirect loss of ant diversity in a tropical agroecosystem, *Oecologia*, **108:** 577–582.

Price, D.W., 1973. Abundance and vertical distribution of microarthropods in the surface layers of a Californian pine forest soil, *Hilgardia*, **42:** 121–148.

Ramakrishnan, P.S., and Das, A., 1983. Studies on pine ecosystem function in Meghalaya, *Trop. Pl. Sci. Res.*, **1:** 15–24.

Reddy, M.V., 1981a. Microarthropods and the rate of needle litter disappearance in a pine plantation ecosystem of North Eastern India, *Pedobiologia*, **22:** 339–344.

Reddy, M.V., 1981b. Faunal composition of summer population of soil microarthropods of tea nursery beds at Nayabunglow, North Eastern India, *Indian J. Ecol.*, **8:** 108–111.

Reddy, M.V., 1984a. Ecology of soil and litter inhabiting arthropods, *Indian Review of Life Sciences*, **4:** 169–217.

Reddy, M.V., 1984b. Seasonal fluctuation of different edaphic microarthropod population densities in relation to soil moisture and temperature in a pine, *Pinus kesiya* Royle plantation ecosystem, *International Journal of Biometeorology*, **28:** 55–59.

Reddy, M.V., 1989. Mass loss and concentrations of nutrients in relation to microarthropod abundance, during needle decomposition in a pine plantation, *Proc. Indian Acad. Sci.* (Anim. Sci.), **98:** 285–289.

Reddy, M.V., 1992. Effects of microarthropod abundance and abiotic variables on mass-loss, and concentration of nutrients during decomposition of *Azadirachta indica* leaf litter; *Trop. Ecol.*, **33:** 89–44.

Reddy, M.V., 1995. *Soil Organisms and Litter Decomposition in the Tropics*, Oxford & IBH Publ. Co. Pvt. Ltd., New Delhi and Westview Press, Colorado, 276 pp.

Reddy, M.V., and Alfred, J.R.B., 1989. Seasonal abundance of microarthropods of needle-litter during decomposition in a pine plantation in relation to litter mass-loss, moisture and temperature, *J. Soil. Biol. Ecol.*, **9:** 104–117.

Reddy, M.V., and Toky, O.P., 1990. Decomposition of grass, bamboo and tree litter in relation to microarthropod densities and abiotic factors in a "Jhum" fallow of northeast India, *J. Soil. Biol. Ecol.*, **10:** 27–35.

Reddy, M.V., and Venkataiah, B., 1989. Influence of microarthropod abundance and climatic factors on weight loss and mineral nutrient contents of eucalyptus leaf litter during decomposition, *Biol. Fertil. Soils*, **8:** 319–324.

Reddy, M.V., and Venkataiah, B., 1990a. Effect of tree plantation on qualitative and quantitative composition of soil arthropods of a semi-arid tropical savanna, *Environment and Ecology*, **8**(1): 361–367

Reddy, M.V., and Venkataiah, B., 1990b. Seasonal abundance of soil surface arthropods in relation to some meteorological and edaphic variables of the grassland and tree-planted areas in a tropical semi-arid savanna, *Int. J. Biometeorol*, **34:** 49–59.

Reddy, M.V., and Venkataiah, B., 1990c. Effects of tree-plantation on seasonal community structure of soil microarthropods in a tropical semi-arid savanna, *Tropical Ecology*, **31**(1): 96–105.

Remmiert, I.L., 1981. Body size of terrestrial arthropods and biomass of their population in relation to the abiotic parameters of their milieu, *Oecologia*, (Berlin), **50:** 12–13.

Rusek, J., 1989. Ecology of Collembola, In: *Proceedings of 3rd Intern. Sem. Apterygota, Sciena*, (R. Dallai, ed.), 271–281 pp.

Sankaran, K.V., 1993. Decomposition of leaf litter of albizia (*Paraserianthes falcataria*), eucalypt (*Eucalyptus tereticornis*) and teak (*Tectona grandis*) in Kerala, India, *Forest Ecol. and Manag.*, **56:** 225–242.

Santos, S.F., and Whitford, W.G., 1983. The influence of soil biota on decomposition of plant material in a gypsum sand dune habitat, *South West Natur.*, **28:** 423–427.

Schenker, R., 1986. Population dynamics of oribatid mites (Acari: Oribatei) in a forest soil ecosystem, *Pedobiologia*, **29:** 239–246.

Seastedt, T.R., 1984. The role of microarthropods in decomposition and mineral processes, *Ann. Rev. Entomol.*, **29:** 25–46.

Seastedt, T.R., and Crossley, D.A. Jr., 1980. Effects of microarthropods on the seasonal dynamics of nutrients in forest litter, *Soil Biol. Biochem.*, **12:** 337–342.

Singh, A.K., and Ambasht, R.S., 1980. Production and decomposition rate of litter in a teak (*Tectona grandis L.*) plantation at Varanasi (India). *Rev. Ecol. Biol. Sol.*, **17:** 13–22.

Singh, J.S., and Ramakrishnan, P.S., 1982. Structure and function of a subtropical humid forest of Meghalaya. II. Litter dynamics and nutrient cycling, *Proc. Indian Acad. Sci. (Plant Sci.)*, **91:** 255–268.

Sousa, J.P., and da Gama, M.M., 1994. Rupture in a Collembola community structure from a *Quercus rotundifolia* Lam. forest due to the reafforestation with *Eucalyptus globulus* Labill, *Eur. J. Soil. Biol.*, **34**(2): 71–78.

Sousa, J.P., Vingada, J.V., Barrocas, H., and da Gama, M.M., 1997. Effects of introduced exotic tree species on Collembola communities: the importance of management techniques, *Pedobiologia*, **41:** 145–153.

Stanton, N., 1979. Patterns of diversity in temperate and tropical litter mites, *Ecology*, **60:** 295–304.

Steinberger, Y., and Whitford, W.G., 1984. Spatial and temporal relationships of soil microarthropods on a desert watershed, *Pedobiologia*, **26:** 275–284.

Swift, M.J., Heal, O.W., and Anderson, J.M., 1979. *Decomposition in Terrestrial Ecosystems*, Blackwell, Oxford, 372 pp.

Szujeki A., Szusszko, J., Mazur, S., Perlinski, and Tracz, A., 1980. Studies on requirements and possibilities of zooamelioration of afforested arable land, In: *Soil biology as related to land use practices*, (D.L. Dindal, ed.), EPA, Washington, DC, USA, 444–449 pp.

Takeda, H., 1978. Ecological studies of collembolan populations in a pine forest soil. II. Vertical distribution of Collembola, *Pedobiologia*, **18:** 22–30.

Takeda, H., Prachaiyo, B., and Tsutsmi, T., 1984. Comparison of decomposition rates of several tree leaf litter in a tropical forest in the northeast Thailand, *Jap. J. Ecol.*, **34:** 311–379.

Thiele. H.U., 1977. *Carabid Beetles in their Environments: A study of habitat selection by adaptions in physiology and behaviour*, Springer-Verlag, New York, 369 pp.

Toky, O.P., and Ramakrishnan, P.S., 1984. Litter decomposition related to secondary succession and species type under slash and burn agriculture (Jhum) in northeast India, *Proc. Indian Nat. Sci. Acad.*, **50:** 57–65.

Venkataiah, B., 1988. *The Effects of Tree Plantation on the Structure and Function of the Soil Arthropod Populations of a Semi-arid Tropical Savanna*, Ph.D. thesis, Kakatiya University, Warangal, India.

Vohland, K., and Schroth, G., 1999. Distribution patterns of the litter macrofauna in agroforestry and monoculture plantations in central Amazonia as affected by plant species and management, *Applied Soil Ecology*, **13**(1): 57–68.

Wallwork, J.A., 1976. *The distribution and diversity of soil fauna*, Academic Press, London, 355 pp.

Watts, J.C.D., 1951. *Some comparative studies of the fauna in soils developed under natural forest, pine and blue gum*, M.Sc. thesis, Rhoes University, Grahamstown.

Wiegert, R.G., 1970. Effects of ionizing radiation on leaf fall, decomposition and litter microarthropods of a montane rain forest, In: *A tropical rain forest*, (H.T. Odum and R.F. Pigeon, ed.), USAEC, Oak-Ridge, Tennessee, 89–100 pp.

Whitford, W.G., 1978. Structure and seasonal activity of Chihuahua desert ant communities, *Insect Soc.*, **25**: 79–88.

Williams, S.T., and Gray, T.R.G., 1974. Decomposition of litter on the soil surface, In: *Biology of Plant Litter Decomposition*, (C.H. Dickinson, and G.J.F. Pugh, eds.), Vol. 2, Academic Press, London, 611–632 pp.

Young, A.M., 1982. *Population biology of tropical insects*, Plenum Press, New York, USA.

CHAPTER 10

TREE PERFORMANCE, SOIL PLANT-LITTER CHARACTERISTICS AND FAUNAL ACTIVITY IN SOME TROPICAL PLANTATIONS

F.K. Akinnifesi[1], G. Tian*[2] and B.T. Kang[3]

[1]*Universidade Estadual do Maranhão (UEMA), Caixa Postal 3004, São Luis-MA, Brazil*

[2]*International Institute of Tropical Agriculture (IITA), P.M.B. 5320, Ibadan, Nigeria*

[3]*15 Beechwood Drive, Haverhill, MA 01832, USA*

**Present address: SADC-ICRAF Agroforestry Project Makoka Agricultural Research Station, P.O. Box 134, Zomba, Malawi*

ABSTRACT

Plantation forestry forms an important alternative means of wood production in the tropics. The area under forest plantations is expanding rapidly in many countries in the tropics. However, the productivity at many sites is below the potential, and opportunities exist for increasing wood production by adopting appropriate soil and management practices. An important influence that vegetation has on soil is the deposition of dead plant debris on the forest floor. Plant litter deposition and decomposition affect the soil organic matter build-up and nutrient cycling at the soil-litter interface, which may also modify both the soil physico-chemical properties and soil biota. Information on litter production and its effects on soil properties and/or soil biota is therefore needed for better management of the long-term productivity of forest plantations in the tropics. We conclude that available data specific to tropical plantation tree species is generally scarce. In this chapter, we combine the available information from the literature, with new data obtained from plantation stands located in southern Nigeria and northeast Brazil, regarding the differential effects of species, age and site conditions on litter production, soil properties and faunal activities. Tree growth and wood volume were affected by stand density and age. Rate of radial growth was highest during juvenile phase and reduced at stand maturity. The stand wood volume increased with stocking in both the Brazilian and Nigerian sites. Tree performance was better at the prehumid than sub-humid tropical conditions. In addition, litter production depends on tree species, size, stand age and stocking. The phylloclades component constituted greater (> 60%) of the total litters for most tropical plantation species. Forest plantations showed differential effects on soil physico-chemical properties. Most species increased the soil organic matter status of surface soil for the initial 10 years, with declining effect at 16 years after tree establishment. Trees effects on soil pH was more pronounced on the acid than high base status soils. Plantation species depressed the extractable soil P levels in low P status soil while increased same in the high P status soils. Exchangeable K, Ca and Mg levels are maintained at high levels under most plantation species, particularly in *Gmelina arborea*. Forest plantation species differ in litter production and quality, thus may have direct and indirect effects on the activities of earthworms and microarthropods. Activities of these soil fauna are higher under plantation forests than grass control.

Key Words: Tropical tree plantations, plant-litter, soil faunal activity, earthworms, soil fertility, soil organic matter.

The establishment of forest plantations is an alternative form of wood production that has grown rapidly in the tropics in recent years (FAO, 1994), with the major goal of expanding the production of wood and non-wood products and/or services. According to FAO (1995), the area of plantations in the tropics had increased from 11.5 million hectares in 1980 to about 44 million in 1990. Brown *et al.* (1997) recently reviewed the global demand and productive capacity of wood from tropical plantation forests, and suggested that although greater portion of the wood produced from tropical forests came from plantations, however, the productivity of plantations at many sites is below the potential, and that there are opportunities for increasing production on a wide scale by adopting appropriate soil and stand management practices.

Performance of trees species, provenance, progenies, and/or clones could vary with sites, depending on edaphic and climatic conditions (Jackson and Ojo, 1973; Ladipo and Leakey, 1986; Kang *et al.*, 1994a). MacDicken and Brewbaker (1989) reported significant location and species interactions in their work on five tropical fuelwood plantation species. Site-soil properties are considered to have a direct effect on the performance of woody species and also on their ability for soil fertility restoration (Challinor, 1968; Lundgren, 1978; Chijike, 1980). One of the most obvious effects that vegetation has on soil under it is the deposition of dead plant debris, and are strongly related to the soil organic matter built-up in the soil-litter interface on the forest floor. However, the long-term productive capacity of tropical plantations depend on availability of judicious and efficient management of soil, nutrients and water resources.

Litter deposition and decomposition constitute a fundamental process of nutrient transformations in plantation forest ecosystems. Litterfall, which include falling leaves, twigs or branchlets, bark peeling, and reproductory structures, is a major avenue for additions of organic matter and nutrients from trees. These processes can provide basic information for estimation of productivity, energy flow and nutrient cycling (Misra and Nisanka, 1997). The litter component of tropical plantations serve as an important resource through which trees modify soil properties and microsite conditions. These benefits include, soil erosion control (reduction of run-off), increase in nutrient cycling (i.e. removal and return of nutrients by trees to the soil), maintenance or enhancement of soil organic matter status, improvement of soil physico-chemical properties, weed suppression and enhancement of activities of soil biota.

There is increasing evidence of site specificity with litter production and its quality (composition) and decomposition in varying eco-regions of the tropics (Misra and Nisanka, 1997; Barbosa and Fearnsides, 1996; Gill *et al.*, 1987; Das and Ramakrishnan, 1985). Therefore, the rate of nutrient cycling depend on the quantity and patterns of litter deposition, humification and decomposition, which in turn are affected by soil faunal activities and litter quality. Although data on litterfall dynamics has continued to accumulate in recent years, especially for temperate and tropical natural forests, however, information on the soil-litter dynamics of tropical plantations are relatively scanty and much dispersed. This paper aims at combining the review information with field data to describe the differential effects of tree growth on soil-plant litter characteristics in some tropical plantations.

CHARACTERISTICS OF STUDY SITES

The field data variously presented in this study were obtained from four different sites in southern Nigeria and northeast Brazil (Table 1). The Ibadan, Gambari and, Onne sites are located in southern Nigeria, with latitudinal range from 4° 43′ to 7° 30′ N, while the Brazilian site is located at latitude of 3° 13′ S.

These are lowland sites with altitudes ranging between 10 m and 240 m. Annual rainfall is highest at Onne in southeast Nigeria, a perhumid forest zone with strongly acidic Ultisol (Typic paleudult), and is characterized by low base saturation (Kang *et al.*, 1994a). Ibadan is located in a subhumid, forest-savanna transition zone, and soil is classified as an Alfisol (Oxic paleustalf) with high base saturation (Moorman *et al.*, 1975). The vegetation at Gambari is a dry

Table 1 Description of study sites in southern Nigeria and northeast Brazil.

Site	*Location*		*Elevation*	*Annual rainfall*	*Soil type*	*Vegetation type*
	Latitude	*Longitude*	(m)	(mm)		
Ibadan, SW Nigeria	$7^0 30'$ N	$3^0 54'$ E	150	1280	Alfisol	Forest-savanna transition zone
Gambari,SW Nigeria	$6^0 50'$ N	$3^0 50'$ E	120	1500	Alfisol	Semi-deciduous forest
Onne, SE Nigeria	$4^0 43'$ N	$7^0 01'$ E	50	2480	Ultisol	Humid forest
Urbano Santos, NE Brazil	$3^0 13'$ N	$43^0 24'$ W	28	1300	Podzol	Dry-wet savanna

semi-deciduous forest, and the soil is an Alfisol, with high base saturation (Smith and Montgomery, 1962). It has a temperature range from 20° to 32° C (Agbede *et al.*, 1987). The Urbano Santos site is located on Cerrado (dry-wet savanna) ecosystem in northeast Brazil. Soil is acidic and is classified as Podzol (Podzolico Vehmelho Amarelho) (EMBRAPA, 1986).

The plantations at Ibadan and Onne were established in 1979 as arboreta at the main and sub research stations of the International Institute of Tropical Agriculture. The species reported for this study are represented at both Ibadan and Onne sites and include: *Cordia alliodora, Dactyladenia barteri, Dialium guineense, Gmelina arborea, Irvingia gabonensis, Leucaena leucocephala* (K28), *Pterocarpus soyauxii and Treculia africana.* Spacing at Ibadan is 2 × 4 m, while the Onne site include a spacing trial at 2 × 2 m, 2 × 4 m, 2 × 6 m and 2 × 8 m. The plantations varied in plot sizes.

The Gambari site comprised *G. arborea* (Roxb.) plantation stands which were established at different age sequences on 40 ha per age series by the Oyo State Forestry Department, Nigeria. Four age series were selected for the present study (8, 10, 13 and 16 years old) and the detailed criteria for selection and sampling were described by Akinnifesi and Akinsannii (1995). Tree spacing was 3 × 4 m for all age series reported for this study.

The plantation at Urbano Santos-MA, Brazil was established by the Comercial Agricola Paineiras (a division of the Susana company Ltd-a leading pulp and paper producer in Brazil), comprising two *Eucalyptus camaldulensis* clones, BX-05 and UTL-34 respectively established in 1989 and 1993, as well as a stand of *Acacia mangium* established in 1989. The spacing for *A. mangium* is 3 × 4 m. Each of the two clones of *Eucalyptus* occupies an area of 10 ha. The experiment for both *E. camaldulensis* clones include the varying spacing treatments. The methodology for all tree dendrometric data reported in this study has been detailed in Akinnifesi and Akinsanmi (1995). Soil and litter sampling and analyses is reported in Kang *et al.* (1994 a, b). Litter mass on the forest floor at Urbano Santos was evaluated from 1 × 1 m quadrats at the end of 1998 dry season in January of 1999. An average of 36 samples per density class were presented in this paper.

CHARACTERISTICS OF TREE SPECIES

DESCRIPTION OF SPECIES

The plantation species grown in the different sites have the following characteristics:

G. arborea is widely grown in Nigeria, and constitutes one of the two most planted exotic plantation species in the country (FORMECU, 1984; FORMECU/BEAK, ADfB-support Forest Resource Study project report, in preparation). Through the Taungya system, *G. arborea* and Teak (*Tectona grandis*) were reported to have been planted at 20,000 ha annually (Akinnifesi and Akinsannii, 1995). A recent survey showed that this rate has declined (FORMECU/BEAK report, in preparation). *G. arborea* has also been grown on a wide scale in the Brazilian amazon under the huge Jari project (Nock, 1985; Brown *et al.*, 1997). This species is valued for its good quality as industrial raw material for pulp and paper production (Akinnifesi and Akinsannii, 1995), and for its tremendous coppiceability that enhanced its value in fuelwood plantations (Kang *et al.*, 1994a). Currently it is being explored by the furniture industry in some countries.

C. alliodora is an important timber species, and has been reported to play the significant role of shade providing trees in agroforestry systems in Central America (Beer, 1988; Beer *et al.*, 1990). It is a valued timber species in the Brazilian amazon and in Nigeria.

I. gabonensis (wild mango) provides fruits for use as food, juice and highly valued condiments (Okafor, 1983; 1990; Kang *et al.*, 1994a) in West Africa. It is ranked among the top five multipurpose food trees preferred by farmers in West and East Africa (ICRAF, 1994). Fruit productivity of two *I. gabonensis* species are reported by Kang *et al.*, (1994a). When managed for timber production, it can provide well sought softwood for furniture.

D. guineense (velvet tamarind) a local species, in addition for use as fruit drink, fodder, craft wood, fuelwood and charcoal, is also known for fallow improvement (Kang *et al.*, 1994a). The dietary importance of *T. africana, I. gabonensis* and *D. guineense* for food, as sources of protein, minerals and vitamins are well known (Okafor, 1990).

Leucaena leucocephala (Lam) de Wit. is a commonly grown exotic, N_2-fixing, multipurpose tree and has been well described by Brewbaker *et al.* (1985). It is a source of fuelwood and charcoal, pulp for paper and rayon, leaves for fodder and green manure, timber for building and furniture and poles and crafts, shade for trees like coffee and cacao, seeds for crafts, and gum for glues (FACT, 1997).

Senna siamea is a widely grown non N_2-fixing, multipurpose tree in Nigeria and many African countries.

A. mangium is a fast-growing, N_2-fixing, multipurpose tree suited for recuperation of degraded areas. In Indonesia, *A. mangium* was planted as a source of small-diameter timber in about 47,000 hectares (Brown *et al.*, 1997).

P. soyauxii, a traditional N_2-fixing legume is described as a leafy vegetable with high potential in the home-garden systems of the humid zone of west Africa (Okafor, 1983; 1990).

T. africana (African breadfruit) is widely used for human food and animal feed (Okafor, 1990), and has been reported to have a high runkel ratio (Okafor, 1983). For food production, Okafor (1990) has recommended using budded plants of the large-fruited *T. africana var. africana* (as done in the two sites reported here), in order to reduce the tree bole, and height of fruit set, while the smaller-fruited var. *mollis* and the *inversa*, which are better suited for pulp production can be grown from unbudded planting materials.

D. barteri is an important fallow improvement species in the traditional fallow systems of southern Nigeria (Getahun *et al.*, 1982). Like industrial timber species, the natural variations within non-industrial multipurpose trees and shrubs can offer similar opportunities for conserving, developing and utilizing plant resources.

Eucalyptus spp. represent more than 50% of all species used in Brazilian afforestation programmes, and was established at the rate of 350,000 ha per annum (Nock, 1985). In 1988 pulp and paper production had reached 3.8 millions tons in Brazil, of which 67% were derived from Eucalyptus and the remaining mainly from Brazilian pines (FAO, 1994). Thus, Eucalypts from afforestation has made Brazil one of the largest producer and exporter of pulp and paper in the world (Dubois *et al.*, 1996). According to FAO (1995) Eucalypts, pines, teak and acacias together account for about 50% of all plantations in the tropical countries. Although, the area of planted forests is relatively small in Brazil (7 million ha) compared to the huge area of native forest (566 million ha), plantations accounted for about 60% of total wood production in the country (FAO, 1995; Brown *et al.*, 1997).

Effect of Site and Age on Tree Growth

There is increasing evidence in the tropics regarding the site and age effects on tree growth characteristics (Jackson and Ojo, 1973; Kang *et al.*, 1994a, MacDicken and Brewbaker, 1988). Table 2 showed the radial growth characteristics of some tropical plantation species. Regardless of sites the radial growth in DBH increased in most species with tree age. *A. mangium* increased by 43.8% during 3 to 10 years after establishment (YAE) at Urbano Santos. However, radial growth in DBH was not important for *C. alliodora* during IO to 16 YAE at both Ibadan and Onne, for *G. arborea* at Gambari, and for *P. soyauxii* at Ibadan indicating reduced rate of radial growth after trees have approached maturity. *D. guineense* increased by 53.1% at Onne compared to 16.5%

Table 2 Influence of age and site on growth characteristics of some plantation species in the tropics (n ≥30)

Species	*Site*	*Stand ages (years)*	*Dbh (cm)*	*Tree height (m)*	*Basal area (m^2/ha)**	*Source**
Acacia mangium	U/Santos	3	8.70 ± 0.1	10.5 ± 0.1	7.2 (11.9 ± 0.6)	(1)
Acacia mangium	U/Santos	10	15.5 ± 0.9	15.7 ± 1.1	27.8 (21.8 ± 0.9)	(1)
Cordia alliodora	Ibadan	10	19.9 ± 1.3	23.3 ± 2.4	38.9 (23.3 ± 2.1)	(2)
Cordia alliodora	Ibadan	16	20.7 ± 0.6	25.9 ± 1.0	42.1 (25.9 ± 1.0)	(2)
Cordia alliodora	Onne	10	19.1 ± 2.1	22.3 ± 1.9	35.8 (22.3 ± 1.4)	(2)
Cordia alliodora	Onne	16	19.6 ± 4.0	25.2 ± 5.6	37.7 (25.2 ± 3.6)	(3)
Dialium guineense	Ibadan	10	9.1 ± 0.9	8.4 ± 0.5	8.1 (11.7 ± 1.5)	(2)
Dialium guineense	fbadan	16	10.9 ± 0.8	9.5 ± 0.5	11.7 (15.7 ± 1.1)	(3)
Dialium guineense	Onne	10	9.1 ± 1.0	21.2 ± 1.3	8.1 (21.2 ± 1.3)	(2)
Dialium guineense	Onne	16	19.4 ± 4.0	22.5 ± 4.1	36.9 (22.5 ± 4.1)	(3)
Eucalyptus camaldulensis	U/Santos	5	11.3 ± 0.9	18.4 ± 1.9	12.1 (20.4 ± 0.8)	(1)
Eucalyptus camaldulensis	U/Santos	10	14.8 ± 1.1	20.9 ± 2.2	20.8 (28.7 ± 1.4)	(1)
*Gmelina arborea**	Ibadan	10	26.8 ± 0.8	23.3 ± 1.2	70.5 (23.3 ± 1.2)	(2)
Gmelina arborea	Ibadan	16	27.0 ± 0.6	22.0 ± 1.8	71.6 (46.8 ± 3.4)	(3)
Gmelina arborea	Onne	10	28.9 ± 1.3	27.1 ± 1.0	82.0 (37.1 ± 2.2)	(2)
Gmelina arborea	Onne	16	29.1 ± 0.8	25.5 ± 1.4	83.1 (39.6 ± 3.0)	(3)
Gmelina arborea	Gambari	8	21.9 ± 1.0	20.8 ± 0.7	33.4 (27.8 ± 1.8)	(4)
Gmelina arborea	Gambari	10	23.8 ± 0.9	23.5 ± 2.4	37.1 (30.2 ± 2.0)	(4)
Gmelina arborea	Gambari	13	25.4 ± 1.0	24.2 ± 2.1	42.2 (31.7 ± 1.9)	(4)
Gmelina arborea	Gambari	16	26.7 ± 1.1	23.9 ± 1.9	46.6 (35.6 ± 1.4)	(4)
Irvingia gabonensis	Ibadan	10	12.4 ± 0.7	8.3 ± 0.3	15.0 (15.4 ± 1.0)	(2)
Irvingia gabonensis	Ibadan	16	14.2 ± 1.0	11.8 ± 0.6	19.8 (25.2 ± 1.9)	(3)
Irvingia gabonensis	Onne	10	16.6 ± 1.3	12.2 ± 0.8	27.0 (18.7 ± 1.2)	(2)

Contd.

Table 2 Contd.

Species	*Site*	*Stand ages (years)*	*Dbh (cm)*	*Tree height (m)*	*Basal area (m^2/ha)**	*Source**
Irvingia gabonensis	Onne	16	21.1 ±1.5	15.6 ± 0.5	43.7 (35.0 ± 2.3)	(3)
Leucaena leucocephala	Ibadan	10	23.8 ±2.1	20.4 ± 3.5	55.6 (28.2 ± 2.4)	(2)
Leucaena leucocephala	Ibadan	16	24.3 ±1.9	18.8 ± 2.9	58.0 (35.4 ± 1.9)	(3)
Pterocarpus soyauxii	Ibadan	10	20.4 ±1.0	18.1 ± 0.8	40.9 (27.0 ± 1.2)	(2)
Pterocarpus soyauxii	Ibadan	16	21.6 ±2.4	14.6 ± 1.2	45.8 (28.6 ± 1.5)	(3)
Pterocarpus soyauxii	Onne	10	13.7 ±0.9	9.2 ± 0.7	18.4 (17.7 ± 1.5)	(2)
Pterocarpus soyauxii	Onne	16	17.8 ±0.4	12.3 ± 0.9	31.1 (23.4 ± 2.6)	(3)
Senna siamea	Ibadan	10	17.4 ±0.9	17.8 ± 0.5	29.7 (24.2 ± 1.0)	(2)
Senna siamea	Ibadan	16	20.4 ±1.0	16.9 ± 0.8	40.9 (35.7 ± 2.2)	(3)
Senna siamea	Onne	5	9.0 ±0.4	11.5 ± 0.3	8.0 (12.5 ± 0.6)	(3)
Senna siamea	Onne	10	19.9 ±1.0	18.0 ± 0.7	38.9 (25.2 ± 2.0)	(2)
Treculia africana	Ibadan	10	18.5 ±1.1	15.5 ± 0.4	33.6 (27.9 ± 1.0)	(2)
Teculia africana	Ibadan	16	21.1 ±1.4	15.6 ± 0.4	43.7 (35.3 ± 2.3)	(3)
Treculia africana	Onne	10	21.7 ±1.6	24.2 ± 2.8	46.7 (39.0 ± 1.8)	(2)
Treculia africana	Onne	16	29.9 ±3.8	19.1 ± 2.3	87.8 (44.9 ± 1.1)	(3)

*Data from: (1) Akinnifesi, F.K. and J.P Linhares, unpublished; and (2) Kang *et al.*, 1994a; Akinnifesi F.K., B.T. Kang., G.O Kolawole and O. Ezekiel, unpublished.

at Ibadan during 10 to 16 YAE. This is because *D. guineense* is a slow growing species. Although, DBH for *D. guineense* was same at both Onne and Ibadan at 10 YAE, trees were taller at the Onne site at 16 YAE. *S. siamea* also increased rapidly in DBH by 54.8% at Onne during 5 to 10 years. More growth is generally observed during juvenile phase than after maturity in most species, and explained why older Senna plantation at Ibadan had less radial growth during 10 to 16 YAE.

G. arborea performed well at both Ibadan and Onne, but trees were superior in radial growth and taller at the Onne site, and growth characteristics were lowest at Gambari for each comparable age series. This suggests better vigour of *G. arborea* on humid than subhumid conditions. At Ibadan, development of epicormic branching and neiloidal stump was pronounced at 16 YAE than other sites. Stump diameter growth during 10 to 16 YAE was thrice (50.2%) as much at Ibadan than Onne (16.3%). These are probably indications of quicker over-maturity, and/or need for pruning to maintain clear boles in the drier site.

Growth in DBH was not significantly different for *C. alliodora* between Onne and Ibadan regardless of age, however, superior height was estimated at Ibadan. *I. gabonensis* and *T. africana* performed better at Onne than at Ibadan for all growth parameters measured.

Height growth was small for most species from 10 to 16 YAE, especially for *D. guineense* and *G. arborea* at Onne, and for *T. africana* at Ibadan. Shorter trees were even observed in stands of *Gmelina* during 16 than 13 YAE for Gambari, and during 16 than 10 YAE at Onne and Ibadan. Similar trend was observed for *C. alliodora* at Onne, *L. leucocephala* and *P. soyauxii*, *S. siamea* and *T. africana* at Ibadan. These reduced height growth in these species may be attributed to reduced apical growth following tree maturity and steady declining state of activity (Lundgren, 1978).

Effect of Stocking on Tree Performance

Tree spacing (density) is often considered a feasible strategy by foresters, to accommodate short and long-term operational needs of the forest estate owners. Even where there is acceptable planting distance, spacing may also vary according to site quality and the objective of the plantation (Nwoboshi, 1982; Nyland, 1996). Data presented in Table 3 showed the effect of spacing or tree density on dendrometric parameters of some plantation species. At both Onne and Urbano Santos sites, trees in stands with wide spacing grow more in diameter than those at close spacing (Table 3). Similarly at Onne during 16 YAE, *C. alliodora* planted with wide spacing or low stocking (625833 trees/ha) are larger in diameter by 24.5% than those planted at narrow spacing or high stocking rate (1250–2500 trees/ha). For *G. arborea* a tree density range of 625–2500 tree/ha did not significantly affect radial growth at 16 YAE. Tree height was however superior at narrower spacing. Such trend was not consistent in *T. africana.*

Although, tree density has no consistent effect on individual standing tree wood volume (Table 3), the stand wood volume per hectare seemed to decline with wider spacing for *G. arborea* and *T. africana.* Both the basal area and wood volume per hectare were consistently superior at highest tree density and (2 × 2 m) for *C. alliodora, G. arborea* and *T. africana* at the Onne site. Nwoboshi (1982) also reported that at any given age, the plantation of Opepe (*Nauchlea diderrichi*) at Sapoba in the rainforest zone southern Nigeria, had larger diameter and greater taper for individual trees under a wider (4.2 × 4.2 m) than narrow (1.8 × 1.8 m) spacing, whereas, superior wood volume was reported for the stand wood volume. According to Nwoboshi (1982), these characteristics give wide-spaced trees somewhat shorter appearance than those of densely stocked plantation. The observation that radial and height growths in *T. africana* were not significantly altered by spacing up to 16 YAE, was similar to the observations reported by Benneth (1960) for *Pinus radiata* and Nwoboshi (1982) for *E. grandis* and *N. diderrichi.* As in earlier works, bifurcation and mortality were however less in narrower spacing (data not shown). There was low mortality at Urbano Santos, due to better management of the plantations by the private firm.

For *E. camaldulensis* planted at Urbano Santos, the mean diameter and wood volume per tree increased linearly with decreasing tree density or increasing spacing (Table 3). Similar trend was also shown by Nwoboshi (1982) in Zambia for *E. grandis.* On the other hand, both the basal area and wood volume per hectare increased initially as tree density decreased from 1208 to 1890 tree/ha, then these parameters decreased linearly with decreasing density or wider spacing. Although, the spacing of 2 × 2 m was superior in basal area and wood volume per hectare at Onne for *C. alliodora, G. arborea* and *T. africana* at 16 YAE, the spacing of 2.3 × 3.6 m seemed optimal for *E. camaldulensis* at Urbano Santos-MA during 10 YAE. Similar result was also presented by Ricardo de Souza (1996), who reported for two 14 years old Pine species (*Pinus caribaea var. hondurensis* and *P. oocarpa*) plantations grown at Anhembi SP Brazil. These results agreed with the views of Sattoo *et al.* (1955) and Nyland (1996), that gross production by stands planted at close spacing is greater than those raised under wide spacing. Close planting ensures early canopy closure and also promotes timber quality, by reducing the stem taper, branch size and bifurcation, and encouraging quick self-pruning. Whereas, this may involve higher expense of establishment and management, while each tree has less

Table 3 Effect of stocking on breast height diameter, height and stand basal area and standing wood volume of *Cordia alliodora*, *Gmelina arborea* and *Treculia africana* at Onne Nigeria, and *Eucalyptus camaldulensis* (BX-05) at Urbano Santos-MA, Brazil.

Species	*Spacing* (*m*)	*Density* (*stem/ha*)	*DBH* (*cm*)	*Basal area* (m^2/*ba*)	*Tree height* (*m*)	*Wood volume*	
						m^3/*tree*	m^3/*ha**
			(Onne, Nigeria)**				
Cordia alliodora	2 × 2	2500	19.5 ± 1.9	74.66 (100)*	20.3 ± 1.3	0.277	691
	2 × 4	1250	19.6 ± 4.0	37.71 (100)	21.7 ± 3.4	0.248	310
	2 × 6	833	25.9 ± 4.4	43.89 (86)	28.4 ± 1.3	0.519	368
	2 × 8	625	25.8 ± 1.9	32.67 (83)	24.3 ± 2.4	0.453	235
Gmelina arborea	2 × 2	2500	26.1 ± 5.5	133.76 (100)	20.0 ± 3.5	0.511	1276
	2 × 4	1250	29.1 ± 1.4	83.13 (100)	25.5 ± 1.4	0.643	804
	2 × 6	833	28.3 ± 4.5	52.40 (100)	18.9 ± 2.3	0.606	505
	2 × 8	625	28.9 ± 6.0	41.00 (83)	19.3 ± 2.4	0.634	329
Treculia africana	2 × 2	2500	27.0 ± 3.9	143.14 (92)	18.9 ± 1.9	0.422	971
	2 × 4	1250	29.9 ± 3.8	87.77 (85)	19.1 ± 2.3	0.524	556
	2 × 6	833	25.9 ± 2.9	43.89 (80)	19.7 ± 2.2	0.413	275
	2 × 8	625	30.0 ± 1.4	43.89 (32)	18.4 ± 2.3	0.554	111

Contd.

Table 3 Contd.

Species	*Spacing* (*m*)	*Density* (*stem/ha*)	*DBH* (*cm*)	*Basal area* (m^2/*ba*)	*Tree height* (*m*)	*Wood volume*	
						m^3/*tree*	m^3/*ha**
			(Urbano Santos-MA, Brazil) ***				
Eucalyptus camaldulensis (BX-05)	2.30 × 2.30	1890	11.7 ± 0.6	19.77 (99)	19.5 ± 0.5	0.075	140
	2.30 × 3.60	1208	14.8 ± 0.6	20.78 (100)	20.9 ± 0.3	0.141	170
	2.30 × 4.50	966	16.1 ± 0.6	19.67 (98)	20.9 ± 0.3	0.173	164
	2.30 × 5.62	773	17.2 ± 0.7	17.96 (100)	21.1 ± 0.3	0.202	156
	3.60 × 3.60	772	16.8 ± 0.5	17.11 (100)	20.8 ± 0.3	0.191	148
	3.60 × 4.50	617	17.9 ± 0.7	15.53 (100)	21.3 ± 0.3	0.222	137
	3.60 × 5.62	496	18.9 ± 0.6	13.92 (100)	21.2 ± 0.3	0.251	125
	4.50 × 4.50	494	19.7 ± 0.7	15.06 (98)	21.2 ± 0.3	0.276	134
	4.50 × 5.62	395	20.1 ± 0.7	12.53 (100)	21.7 ± 0.3	0.288	114
	5.62 × 5.62	317	20.9 ± 0.7	10.88 (99)	21.5 ± 0.3	0.349	111

*Standing wood volume per hectare was corrected for tree form and species mortality rate; Figures in parenthesis indicate % mortality.

**Species at Onne were evaluated at 16 years after establishment (n– 30) (Akinnifesi F.K., B.T. Kang., G.O Kolawole and O. Ezekiel, unpublished).

****Eucalyptus* sp. at Urbano Santos was evaluated at 10 years after establishment (n = 90) Akinnifesi, F.K. and J.P Linhares, unpublished).

space below and above ground for development, especially on harsh environments.

At wider spacing (lower stocking) however, the trees initially grow bigger and faster, reach minimum merchantable sizes sooner, and thus may provide the landowners with early revenues to repay or off-set part of initial investment (Smith, 1986). However, our experience showed that widening the spacing below optimum leads to development of large branches, strongly tapering and poorly formed boles. It is therefore important to find spacing or stocking level that best satisfy the goal of afforestation, and must fit the silvical characteristics of the component trees. According to Nyland (1996), in plantations intended for commodity production, the desired spacing or density should 1) ensure a maximum volume and highest quality of sought after products in the shortest time; 2) minimize the cost of establishment that is consistent with goals, and 3) facilitate eventual harvesting during intermediate tending, and for the reproduction method desired at the end of rotation.

SOIL FERTILITY CHANGES UNDER PLANTATION FORESTS

Effect on Soil Organic Matter

The site soil properties have been reported to have a direct effect on the performance of woody species and also on its ability for soil-nutrient restoration (Challinor, 1968; Lundgren, 1978; Chijike, 1980; Kang *et al.*, 1994a, b). In tropical plantations, as in natural forests, carbon and nutrient dynamics are achieved through two major processes; (1) uptake from soil by vegetation and (2) removal from vegetation and return to the soil as litter, rainwash and root excretion and decay. The actual amounts involved in these processes depend on the species, soil types, climate, tree age and management factors.

There is evidence that some plantation species improved the soil organic carbon after 5, 10 and 16 YAE (Table 4). Under most tree species soil organic carbon level was higher than the control treatment up to 10 YAE, suggesting their positive influence on soil organic matter during the growth phase, which according to Lundgren (1978), has a similar effect as the fallow phase of shifting cultivation. At Ibadan, even the no-tree control fallow which showed a steady decline in organic carbon up to 10 years after land preparation (baseline data), had surprisingly, restored the original organic carbon level after 16 years of grass fallow (Table 4). For all species evaluated at Onne and Ibadan the levels of soil organic carbon were higher than control at 5 and 10 YAE. Measured at 10 YAE, *L. leucocephala*, *C. alliodora*, *D. barteri* and *G. arborea* increased

Table 4 Soil chemical characteristics under some plantations species in southern Nigeria in Urbano Santos-MA, Brazil sampled at 0–15 cm depth.

Species/clone	*Stand age* (*years*)	*pH* (H_2O)	*Org. C.* (*g/kg*)	*Extr. Bray* P_1 (*mg/kg*)	*Exchangeable cations* *K*	*Ca* (*c mol/kg*)	*Mg*
Ibadan, Nigeria*							
Baseline (at plot establishment)	0	5.0	11.6	8.4	0.15	1.88	0.17
Control (no-tree)	5	6.2	9.1	2.7	0.23	0.40	0.11
Control (no-tree)	10	5.6	6.6	7.1	0.10	1.57	1.25
Control (no-tree)	16	6.4	11.6	3.4	0.29	1.72	1.45
Cordia alliodora	10	6.6	13.2	3.5	0.48	1.94	0.55
Cordia alliodora	16	6.6	13.0	5.1	0.44	1.95	0.63
Dactyladenia barteri	5	6.1	13.6	3.3	0.28	1.42	0.27
Dactyladenia barteri	10	6.0	12.0	6.8	0.27	2.07	0.26
Dactyladenia barteri	16	5.8	11.9	2.8	0.19	1.22	0.61
Dialium guineense	10	6.4	8.4	2.5	0.23	2.32	0.43
Dialium guineense	16	6.2	9.7	6.9	0.19	1.31	0.33
Gmelina arborea	5	6.3	14.2	2.2	0.16	3.95	0.30
Gmelina arborea	10	6.4	12.0	2.2	0.27	4.23	1.00
Gmelina arborea	16	6.7	12.6	2.9	0.23	2.06	0.41
Irvingia gabonensis	10	5.8	10.6	3.3	0.19	1.41	0.37
Irvingia gabonensis	16	6.1	6.6	2.3	0.16	1.27	0.34
Leucaena leucocephala (K28)	5	6.2	14.6	1.6	0.34	2.51	0.16
Leucaena leucocephala (K28)	10	6.2	14.8	3.9	0.27	2.08	0.35
Leucaena leucocephala (K28)	16	6.4	11.0	3.4	0.20	1.59	0.34
Pterocarpus soyauxii	10	6.2	6.8	4.6	0.10	0.86	0.27
Pterocarpus soyauxii	16	5.6	8.3	2.8	0.19	1.26	0.40
Treculia africana	5	5.8	11.4	2.7	0.18	1.42	0.27
Treculia africana	10	6.1	9.6	2.7	0.28	2.07	0.26
Treculia afticana	16	6.7	9.7	2.9	0.26	1.22	0.61
Gambari, Nigeria							
Gmelina arborea		7.4	13.8	20.7	0.49	4.25	0.55

Contd.

Table 4 Contd.

Species/clone	*Stand age* (*years*)	*pH* (H_2O)	*Org. C.* (*g/kg*)	*Extr. Bray P_1* (*mg/kg*)	*Exchangeable cations* *K*	*Ca* (*c mol/kg*)	*Mg*
Treculia afticana	16	6.7	9.7	2.9	0.26	1.22	0.61
Gambari, Nigeria							
Gmelina arborea	–	7.4	13.8	20.7	0.49	4.25	0.55
Onne, Nigeria*							
Control (no-tree)	10	4.0	10.2	48.1	0.07	0.31	0.06
Control (no-tree)	16	4.6	9.9	46.8	0.07	0.49	0.13
Cordia alliodora	10	4.6	12.0	75.6	0.06	0.29	0.08
Cordia alliodora	16	4.4	12.6	99.2	0.12	0.24	0.07
Dactyladenia barteri	16	4.5	16.9	82.0	0.09	0.33	0.33
Dialium guineense	10	4.5	12.0	54.4	0.07	0.38	0.13
Dialium guineense	16	4.5	11.0	68.8	0.05	0.28	0.05
Gmelina arborea	10	4.4	11.5	84.7	0.08	0.28	0.06
Gmelina arborea	16	4.5	12.3	81.8	0.06	0.33	0.06
Irvingia gabonensis	10	4.5	12.2	87.7	0.10	0.37	0.08
Irvingia gabonensis	16	4.5	9.3	74.3	0.06	0.24	0.07
Pterocarpus soyauxii	10	4.5	11.6	66.5	0.09	0.23	0.10
Pterocarpus soyauxii	16	4.2	14.3	61.2	0.08	0.28	0.06
Treculia africana	10	4.4	12.0	58.4	0.09	0.42	0.19
Treculia africana	16	4.5	12.8	46.0	0.10	0.26	0.11
Urbanos Santos-MA, Brazil**							
Eucalyptus camaldulensis (BX-05)	0	4.1	14.2	71	5.10	6.00	2.45
Eucalyptus camaldulensis (BX-05)	5	4.2	13.5	82	4.95	5.73	2.19
Eucalyptus camaldulensis (BX-05)	10	4.4	16.7	73	3.90	7.40	2.30
Eucalyptus camaldulensis (UTL-34)	0	4.3	18.1	96	6.40	10.0	3.70
Eucalyptus camaldulensis (UTL-34)	5	4.4	19.2	101	9.55	6.70	2.66

*Akinnifesi F.K., Kang B.T., G.O. Kolawole and O. Ezekiel (unpublished data); Plantations at Ibadan and Onne were unfertilized.
**Akinnifesi, F.K. and J.P. Linhares (unpublished data); Eucalypts plantations fertilized at establishment with 400 kg/ha NPK 14-07-28, and 240 kg/ha ofS;

soil C levels by 81.8% and 124.2 over the control. *T. africana* and *D. guineense* showed lower effects (27.3 to 37.7 %), and *P. soyauxii* had little effect on soil organic carbon. Both *D. barteri* and *G. arborea* maintained higher levels of soil organic carbon at 5, 10 and 16 YAE than the original level at plot establishment. Positive effect of *D. barteri* on soil organic matter built-up had earlier been reported by Kang *et al.* (1990).

Apart from *C. alliodora, D. barteri* and *G. arborea* the tree species lower the levels of soil organic carbon were observed at 16 YAE compared to 10 YAE. In addition, *D. guineense, I. irvingia,* and *P. soyauxii* had lower soil organic C than the baseline at 10 and 16 YAE. These trends generally suggest that at 5 to 10 years after tree establishment the trees form closed canopy, and thereafter approach a maximum growth (Sanchez *et al.*, 1985). In the report of Lundgren (1978), plantations of *P. patula* and *Cupressus lusitanica* in Tanzania had initial improvement on soil organic matter during the growth phase (4 to 8 years), followed by a period of deterioration during 10 to 20 years, which corresponded to maximum plantation growth. A decrease in soil organic matter during the last phase of tree growth occurred because trees have approached a steady or decline state of activity (Lundgren, 1978), more sequestering of C in biomass, greater root crowding in the surface soil, and possibility of lower undergrowth cover due to shading by mature tree canopy, that can increase surface soil runoff and erosion.

It is not known why *T. africana* had inferior soil organic C level than baseline at 5, 10 and 16 YAE on the Alfisol at Ibadan, while the performances was different on the Ultisol at the Onne site. In general, most species at the Onne site, improved the soil organic carbon at 16 YAE. Similar positive effect of the two *E. camadulensis* clones were observed on soil organic content at Urbano Santos at 10 YAE. The effects are attributed to litter deposition and partial fallow effect. However, effects of woody species on soil organic carbon seemed to be more pronounced on Alfisol at Ibadan than both on Ultisol at Onne and on Podzol at Urbano Santos.

Effect on Soil Acidity, Phosphorus and Exchangeable Bases

In the present study, all the species generally depressed the soil extractable P levels at Ibadan in relation to the original soil P-level at tree establishments (Table 4). However, effect of trees seemed favorable at the Onne site during 16 YAE. *Cordia* and *Gmelina* had more pronounced increase in P-levels compared to the control plot without trees during the interval of 10 to 16 YAE. Effect of the two *Eucalyptus* clones were positive on soil P at 5 and 10 YAE for BX-05 and 5 YAE for

UTL-34. Soil P was generally higher in this particular Ultisol at Onne due to the marine origin of parent material, and in Podzol at Urbano Santos than in Alfisol at Ibadan. Kang *et al.* (1994a) suggested that the lower extractable P level under plantation species at Ibadan may be partly due to the uptake by trees in this zone, but noted that some of the P is held in organic form which is not extractable by the Bray P-1 extractant.

Effect of trees on soil pH seemed to be more pronounced at Ibadan than Onne and Urbano Santos. These effects are attributed to the recycling of bases in tree litters particularly on the high base status region. Sanchez and Russell (1987), showed that soil pH under *G. arborea* in the Brazilian Amazon at Para, increased from pH 4.8 at land clearance to 5.1 points at 8 YAE.

The effect of trees on soil exchange Ca differed among species and with time at the Ibadan site. Although the no-tree control also showed a steady increase in Ca, but the initial soil Ca level was not restored to original level after 16 years under no-tree grass fallow. Like for soil organic carbon, most species attained peak levels of Ca at 10 YAE, and declined thereafter. *G. arborea* seemed to be most efficient at maintaining higher Ca levels on the Alfisol at Ibadan. *G. arborea* has been reported to be a Ca and Mg accumulator by Sanchez *et al.* (1985), and has particular unique potential in recycling exchangeable cations (Young, 1989). For two plantations of *G. arborea* evaluated by Chijike (1980) in Nigeria and Brazil, high Ca (171 to 161 kg/ha) was reported to be returned to the soil via litterfall. Similar increase of 67 % in Ca levels compared to original level was observed under *G. arborea* plantation established 8 years earlier on an Acrisol in Para (Brazilian Amazon) (Sanchez and Russell, 1987). The trees did not improve, and in fact reduced soil exchange Ca at Onne (Ultisol) at 16 YAE (Table 4), and is partly attributed to Ca uptake by trees, and the inherently low base saturation of the soil. Morris (1995) estimated 56% reduction in top soil exch. Ca at 13 years after afforestation with *P. patula* on Usutu Forest, Swaziland. Effects of the two *Eucalyptus* clones on soil Ca were also contrasting on Podzol at Urbano Santos.

At Ibadan, the soil exchangeable K was generally maintained at higher levels than the level at trees establishment (Table 4). *Cordia alliodora* had the most pronounced positive effect on soil exch. K at both Ibadan and Onne at 16 YAE. At Ibadan, *Dactyladenia barteri, Irvingia gabonensis* and *Leucaena leucocephala* showed less improvement on soil exchange K. Whereas, *Gmelina arborea* and *Treculia africana* showed an initially increasing trend up to 10 YAE, then decline at 16 YAE. Conversely, the trend in soil exch. K at Onne for *Treculia africana* indicates a sharp increase during 10 to 16 YAE. The two *Eucalyptus* clones also had contrasting effects on soil exch. K at Urbano Santos. The trend at Ibadan, in exch. Mg in no-tree control is consistent with Ca levels, except that the trees generally had lower Mg levels than the control plot, but higher than baseline value regardless of the tree age. At Onne, presence of trees did affect soil exch. Mg. Substantial removal of soil Ca and Mg by trees was also reported by Nwoboshi (1972).

LITTER PRODUCTION IN TROPICAL PLANTATIONS

ANNUAL LITTERFALLS IN TROPICAL PLANTED FORESTS

In plantation forest ecosystems, a large proportion of the net primary production is shed annually as litter. The greatest contribution to soil humus layer is from the litter, mainly from the detritus falling from the trees on the soil surface. This plays a fundamental role in cycling of nutrients and transfer of energy between plants and soil. The amount of nutrients delivered by litterfall to the forest floor is of great importance to sustainable wood production and can provide an index of plantation forests productivity (Madeira *et al.*, 1995; Ho-Taek *et al.*, 1998).

The range of annual litter production in plantation forests varied widely between sites and species (Table 5). At Ibadan Nigeria, Kang *et al.* (1994b) reported litter production for 10-year old tree stands of *Cordia alliodora, Gmelina arborea Irvingia gabonensis, Leucaena leucocephala* (K28), *Pterocarpus soyauxii* to have ranged between 4.64 t/ha and 9.67 t/ha (Table 5). Seven to ten years old *Pinus carribaea* was also reported to have yielded mean litterfall of 5.88 t/ha at similar environment (Egunjobi and Onwuluzo, 1979), and this compared with 8-years old *Pinus patula* in India (Jeeva and Ramaskrishnan, 1997). However, rate of litterfall from 19-year old *Pinus patula* plantations in Tanzania (Lundgren, 1978), was comparable to plantations in India, of the same species but 7 to 10 years old (Das and Ramakrishnan, 1985) and 10-year old (Jeeva and Ramakrishnan, 1997) in India. The reported rates of litterfall for different fast-growing plantation species commonly exceed 2 t/ha except for juvenile plantations such as for *E. tereticormis* (Gill *et al.*, 1987), and 2-year old plantations of *E. globulus* and *Pinus patula* in India (Jeeva and Ramakrishnan, 1985).

Exceptionally high litter production levels (8.69 to 29.62 t/ha) were however, estimated for stands of varying age sequences of *Casuarina equisetifolia* in India (Misra and Nisanka, 1997). This high litter production may be linked to the good soil properties, and rainfall (2000 mm) at the site.

Although, it is generally assumed that latitudinal position can influence tree performance and hence the litter productivity, available

Table 5 Litterfall under some plantation species in the tropics.

Species	*Latitude*	*Litterfall* (t/ha)		*Source*
		Leaf/Needle	*Total*	
Mexico (Humid woodland)				
Pinus radiata	38° 20'S	2.8–3.21	–	Baker (1993)
Eucalyptus obliqua	38° 20'S	2.59–2.67	3.88–5.18	Baker (1993)
Eucalyptus requas	38° 20'S	2.59	6.86	Baker (1993)
Eucalyptus sieberi	38° 20'S	2.67	5.37	Baker (1993)
India				
Eucalyptus tereticornis	76° 56'N	–	1.03	Gill *et al.* (1987)
Acacia nilotica	76° 56'N	–	2.50–5.75	Gill *et al.* (1987)
Alnus nepalensis	27° 7'N		7.73–8.71	Rao *et al.* (1972)
Tectona grandis	30° 24'N	5.01	6.95	Pande & Sharma (1986)
Tectona grandis	26° 47'N	6.06–17.3	–	Singh (1982)
Tectona grandis	27° 7'N	–	6.97–8.72	Rao *et al.* (1972)
Pinus patula	26° 47'N	–	3.75–11.40	Singh (1982)
Pinus kesiya	24° 45'N	6.50–7.02	6.66–8.98	Das & Ramakrishnan (1985)
Eucalyptus globulus	11° 22'N	–	1.93–5.97	Jeeva and Ramakrishnan (1997)
Eucalyptus spp.	30° 24'N	5.91	7.07	Pande & Sharma (1986)
Casuarina equisetifolia	19° 16'N	8.32–22.32	8.69–29.62	Misra & Nisanka (1997)
Pinus ruxburgii	30° 24'N	4.78	9.67	Pande & Sharma (1986)
Pinus patula	11° 22'N	–	1.02–6.25	Jeeva and Ramakrishnan (1997)
Shorea robusta	30° 34'N	8.29	11.27	Pande & Sharma (1986)
Shorea robusta	27° 7'N	–	9.93–12.43	Rao *et al.* (1972)
Nigeria (Ibadan, Subhumid)				
Cordia alliodora	7° 30'N	–	9.67	Kang *et al.* (1994)
Gmelina arborea	7° 30'N	–	5.50	Kang *et al.* (1994)
Irvingia gabonensis	7° 30'N	–	6.03	Kang *et al.* (1994)
Leucaena leucocephala	7° 30'N	–	4.64	Kang *et al.* (1994)
Pterocarpus soyauxii	7° 30'N	–	7.42	Kang *et al.* (1994)
Treculia africana	7° 30'N	–	8.84	Kang *et al.* (1994)
Pinus carribaea	6° 30'N	–	5.88	Egunjobi and Onwuluzo (1979)
Tanzania				
Pinus patula	26° 47' N	–	6.22	Lundgren (1978)

data is not sufficient to validate such claim in tropical plantations. Various reports in the tropics indicate a strongly positive linear relationship between tree age and litter productivity for diverse sites (Singh, 1982; Das and Ramakrishnan, 1985; Jeeva and Ramakrishnan, 1997; Misra and Nisanka, 1997; Morris, 1995; Rao *et al.*, 1972; Sharma and Ambasht, 1987). Age series used in the studies ranged widely with yearly intervals, and up to maximum of 13 years old for *Casuarina* (Misra and Nisanka, 1997), two yearly intervals up to 10 years old for *E. globulus* and *P. patula* (Jeeva and Ramakrishnan, 1997), irregular intervals starting from 7 to 22 years old for *P. kesiya* (Das and Ramakrishnan, 1985). All these studies showed that the annual rates of litterfall increase linearly with plantation age. However, Singh (1982) compared estimates from a wider age series and found that litter production increased steadily at 10, 12, 17 and 25 years after tree establishment (YAE) with rates ranging from 3.75 to 11.4 t/ha. These annual litter production however declined to 8.75 t/ha at 35 YAE. Similar trend was also obtained by Sharma and Ambasht (1987) under young to old (7 to 56 years) stands of *Alnus nepalensis* in eastern Himalayas. The effect of tree age on litterfall is consistent with trend in tree growth, suggesting that the decline might be explained by reduced activity of tree growth as plantation matures as was also shown for organic matter changes at Ibadan. According to Rao *et al.* (1972) stand age and density are two important determinants of litter production in plantation forests. Rao *et al.* (1972) reported that in *Shorea robusta*, for the same age, the litter production was similar inspite of difference in stand density. But in Teak (*T. grandis*) of the same age, higher density gave higher litter production. However, a young plantation of *P. roxburghii* at higher density produced relatively less litter. These results suggest that influence of age and plant density on litter production may vary according to the tree species. Efforts are needed to relate rates of litterfall productivity with tree age and soil physico-chemical properties in tropical plantations.

Tree Stocking and Floor Litter Biomass

Rapid cycling of nutrients through the forest floor is generally accepted as an important factor in successful plantation growth on sites with low nutrient capital available (Morris, 1995). Quantification of litter accumulation on the forest floor is important in plantations of fast-growing tree species grown in short rotations coppiced stands. Proper understanding of litterfall, forest floor litter and nutrient cycling are important for sustained yield management of forest ecosystems.

The range of floor litter estimates of fast-growing plantations in the tropics are given in Table 6. For Pine plantations, floor litter mass are relatively low and ranged between 4 and 7 t/ha for *P. carribaea* in Nigeria (Egunjobi and Onwuluzo, 1979), and in South Africa (Vesveld, 1981).

Whereas, estimates in *A. mangium, E. camaldulensis clones* in Urbano Santos, Brazil (F.K. Akinnifesi and J.P. Linhares, unpublished data), *P. patula* in Tanzania (Lundgren, 1978) gave intermediate (15 to 35 t/ha) floor litter biomass. Relatively high (51 to 217 t/ha) floor litter mass estimates were differently reported by Singh (1982) in India, Morris (1995) in Swaziland, and Schutz *et al.* (1993) in South Africa.

Morris (1995) showed that forest floor biomass was linearly correlated with tree size ($r^2 = 0.59^*$), and also related to age, site elevation and site index. Several soil parameters especially Ca is correlated strongly with the litter biomass in that study. Similarly, in our study at Urbano Santos, Brazil, the mean litter biomass accumulated on the floor at the end of the dry season ranged from 15.2 to 18.9 t/ha in young *E. camaldulensis* (UTL-34) to 20.0 to 23.8 t/ha in older clone (BX-34) grown in the same site (Table 6). Floor litter biomass in UTL-34 increased steadily with lower stand density (Table 7). The increasing floor litter biomass with spacing in younger plantation (UTL-34) is related to the tree size. However, the intermediate stocking (617 to 773 trees/ha) was superior for the older BX-05, and this may suggest effect of tree age difference. Sharma and Ambasht (1987) showed that forest floor litter layer biomass increased with stand age from 14.1 t/ha in 7-year stand to 30.5 t/ha in 56-year stand. In the study at Urbano Santos, the branchfall, respectively accounted for 23.3 and 27.7% of the total floor litter biomass estimates for UTL-34 and BX-05. Whereas, the mean phyfloclades (leaf-litter), accounted for 65.7% of total floor litter mass in UTL-34 and 72.8% in BX-05. This may be partly genetic, and partly due to difference in age. The younger stand (UTL-34) is likely to achieve more intensive self-pruning involving branchfall than older stand which may be heavier in leaffall during the dry season. Most studies agreed that foliage litter forms the greater proportion of total litter production in tropical plantations (Rao *et al.*, 1972; Jeeva and Ramakrishnan, 1997). In Mexican lowland rainforests, Alvarez-Sanchez and Sada (1993) reported higher fractions of litter for leaves (67.9%), intermediate for branches (21.6%), and low for nonidentified fragment (6.8%) and reproductive structures (4.7%).

The rates of biological processes have been related to tree size (Adu-Bredu *et al.*, 1997), and is related to stand respiration, photosynthesis (Yokota and Hagihara, 1996; Yokota *et al.*, 1994) and stand age (Morris, 1995). Litter thickness was observed to differ between species, and was generally lower (3 to 5 cm) for *Eucalyptus* sp. than for *A. mangium* (6 to 12 cm). Nutrient production on the floor is related to litter biomass

Table 6 Forest floor biomass under some fast-growing plantations in the tropics

*Species**	*Country*	*Age (years)*	*Mass (t/ha)*	*References*
Alnus nepalensis	India	7	13.0–15.3	Sharma and Ambasht (1987)
Alnus nepalensis	India	17	17.2–21.0	Sharma and Ambasht (1987)
Alnus nepalensis	India	30	16.9–21.1	Sharma and Ambasht (1987)
Alnus nepalensis	India	46	22.5–25.8	Sharma and Ambasht (1987)
Alnus nepalensis	India	56	28.0–32.9	Sharma and Ambasht (1987)
Pinus patula	Swaziland	11–15	50.6–155.6	Morris (1995)
Pinus patula	South Africa	44	217	Schutz (1983)
Pinus patula	India	12–34	21–70	Singh *et al.* (1984)
Pinus patula	Tanzania	10–30	25–35	Lundgren (1978)
Pinus radiata	South Africa	37	7	Vesveld (1981)
Pinus carribaea	Nigeria	10	4	Egunjobi and Onwuluzo (1979)
Acacia mangium	Brazil*	10	47–51	Akinnifesi and Linhares (unpublished data)
Eucalyptus camaldulensis (UTL-34)	Brazil	5	15–19	Akinnifesi and Linhares (unpublished data)
Eucalyptus camaldulensis (BX-05)	Brazil	10	20–24	Akinnifesi and Linhares (unpublished data)

*Litter depth was 2.5-5 cm for the two *Eucalyptus camaldulensis* clones; and 6–13 cm for *Acacia mangium* in same site at Urbano Santos, Brazil.

**Calculated from Table 7.

Table 7 Effect of stand stocking on growth and litter mass accumulated on forest floors[a] under two clones of *Eucalyptus camaldulensis* in Urbano Santos-MA, Brazil (Akinnifesi, F.K. & J.F.P. Linheres, unpublished).

Clone	*Stand density* (*tree/ha*)[b]	*Dbh* (*cm*)	*Tree height* (*m*)	*Litterfall* (*t/ha*)[c] *Branchlets*	*Stem bark*	*Phylloclades*	*Rep. structure*	*Total*
BX-05								
	317–496	19.90 ± 0.69	21.53 ± 0.36	4.71 ± 0.08	0.87 ± 0.15	16.20 ± 1.76	0.22 ± 0.06	22.00
	617–773	17.30 ± 0.64	21.07 ± 0.28	5.21 ± 0.42	0.97 ± 0.24	17.48 ± 2.16	0.18 ± 0.06	23.84
	966–1890	14.17 ± 0.61	20.43 ± 0.27	4.73 ± 0.18	0.90 ± 0.16	14.26 ± 1.43	0.14 ± 0.06	19.98
	Mean	17.12	21.01	4.88	0.91	15.96	0.18	
UTL-34								
	317–496	16.09 ± 0.22	19.70 ± 0.24	4.18 ± 0.42	0.85 ± 0.03	13.17 ± 1.79	0.57 ± 0.05	18.78
	617–773	13.71 ± 0.31	19.20 ± 0.15	4.73 ± 0.43	0.84 ± 0.05	10.72 ± 1.54	0.37 ± 0.05	16.57
	966–1890	10.81 ± 0.33	18.04 ± 0.27	4.85 ± 0.08	0.61 ± 0.19	9.35 ± 1.39	0.34 ± 0.03	15.17
	Mean	13.54	18.98	4.59	0.77	11.08	0.43	

[a]UTL-34, five years old, BX-05, ten years old; Soil was fertilized at two years after establishment with 400 kg/ha NPK 14-07-28, and 240 kg/ha of S;
[b]Growth (n=270) and litter (n= 36) data were averaged for plots of close density ranges;
[c]Branchlets, twig or small branches <2cm diameter; Phylloclades, leaves and petioles; Stem bark, bark peeling from the stem; Reproductive structure, fruits and inflorescence; *Acacia mangium* of same age, accumulated 47–51 t/ha of floor litter mass.

Table 8 Nutrient yield* from the forest floor litter under two *Eucalyptus* clones and *Acacia mangium* in Urbano Santos-MA, Brazil.

Species/clones	*Maan Litter mass (t/ha)***	*Nutrient yield (kg/ha)*				
		N	*p*	*K*	*Ca*	*Mg*
Eucalyptus sp. (BX-05)						
Phylloclades	15.96	263.7	8.3	6.3	360.7	87.8
Branchlet	4.88	19.0	1.5	0.2	100.5	16.1
Bark	0.91	7.2	0.3	0.1	18.3	3.1
Reproductive structure	0.18	1.3	0.1	0.1	5.9	0.2
Total	21.93	297.2	10.2	6.7	484.7	107.2
Eucalyptus sp. (UTL-34)						
Phylloclades	11.08	231.6	8.3	4.2	292.5	33.2
Branchlet	4.59	26.6	1.4	1.1	101.0	14.7
Bark	0.77	3.2	0.1	0.2	16.6	2.7
Reproductive structure	0.43	4.9	0.2	0.3	9.3	1.9
Total	16.87	265.9	10.0	5.8	419.4	52.5
*Acacia mangium****						
Phylloclades	42.63	1700.9	21.3	17.1	1100.0	119.4
Branchlet	2.71	57.7	0.8	0.8	68.8	7.6
Reproductive structure	1.45	38.9	0.6	0.8	32.5	3.0
Total	46.79	1797.5	22.7	18.7	1201.3	130.0

*Leaf N content; *Acacia mangium,* 3.9 mg/kg; *Eucalyptus* sp. (BX-05), I.7 mg/kg; *Eucalyptus* sp. (UTL-34), 2.1 mg/kg;
**Calculated from Table 7
*** Bark peeling negligible (Adapted from Akinnifesi and Linhares, unpublished)

(Table 8), with leaf representing greater fraction of the nutrient accumulation on forest floor. *A. mangium* is relatively higher in N-concentration and biomass nutrient because it is an N_2-fixing leguminous tree (Table 8). Litter thickness and composition of litter layer on floor, may vary from season to season due to characteristic cycling including litterfall and decomposition. It is presumed that the measurements made at the end of the season may efficiently reflect the differences in spacing.

The processes of nutrient release from decomposed litter are complex and depend on loss in dry weight. A recent review by Misra and Nisanka (1997) showed that the rate of loss in litter biomass is linearly related to period of decomposition. The decomposition constant (k) calculated during the period of 12 to 24 months in India, ranged from 0.690 to 0.931 for *A. nepalensis* (Sharma and Ambasht, 1987), 0.392 to 0.931 for *C. equisetifolia* (Misra and Nisanka, 1997), and 0.460 to 0.780 for *P. kesiya* (Das and Ramakrshnan, 1985). Litter decomposition constant, turnover time for biomass and different nutrient elements generally decline with time in months. The release of nutrients depend on its availability to microbial needs, and the rate of litter biomass decomposition (Swift *et al.*, 1979).

EFFECT OF TREES ON SOIL FAUNAL ACTIVITY

Effect on Earthworms

Worm-casting activity is known to vary with species, ambient conditions, soil type and particularly vegetation cover. Investigations in southwestern Nigeria showed the presence of two major types of worm-casts: (1) the turret-shaped produced by *Hyperiodrilus africanus* (Figure 1), and (2) the granular casts produced by *Eudrilus eugeniae.* Between the two species, *H. africanus* shows higher casting activity (Madge, 1965; Cook *et al.*, 1979). While *H. africanus* casting activity is limited mainly to areas with good cover, that of *E. eugeniae* can also be observed in bare grounds. As results of investigations reviewed in this paper were mainly concerned with the effect of ground cover in soil management for sustainable agricultural production, only observations on *H. africanus* were reported.

The effects of ground cover on the casting activity of *H. africanus* are observed to be related to: (1) indirect effect of shading that maintains lower temperature and higher moisture retention of the surface soil, and (2) direct effect of plant litter as a food source (Hauser, 1993; Cook *et al.*, 1979; Kang *et al.*, 1994b; Tian *et al.*, 1993a). Kang *et al.*, (1994b) reported large differences in earthworm-casting activity under ten year old plantations of *D. barterii, Gliricidia sepium, L. leucocephala, S. siamea* and *T. africana* (Table 9). These differences were related to:

(1) Differences in canopy structure and density of the tree species, which affected the amount of incident radiation and surface soil temperature. *D. barteri* showed the highest canopy density and lowest incident radiation and surface soil temperature, while *G. sepium* with a more open canopy showed the highest incident radiation and surface soil temperature.

(2) The amount of litterfall which differs significantly between tree species, with *T. africana* showing the highest amount of litterfall (Table 10).

(3) The longetivity of the plant litter which serves as mulch. Tian *et al.* (1993b) showed that plant residue decomposition and nutrient release vary with their chemical composition. Low quality plant residues with high lignin and C/N ratios such as *D. barteri, T. africana* and *S. siamea* litters decomposed slowly. (For example *D. barteri* leaf prunings applied as mulch has half life of over 4 months). While quality plant residues characterized by low lignin content and C/N ratios such as *G. sepium* and *L. leucocephala* litters showed faster decomposition rates, with half life of less than 4 months. The slow decomposing materials have a longer duration effect as mulch, and can thereby retain higher surface soil moisture level for higher earthworm activity than fast decomposing materials. This explains the higher worm casting activities under *D. barteri* and *T. africana* (Table 9). It was observed (Kang, B.T., unpublished), that the mean diameter size of *H. hyperiodrilus* worm casts is affected by tree species. For example, under *D. barteri* wormcast size was smaller but more numerous than under *T. africana,* and

(4) Chemical composition of plant litter. The high quality plant materials because of their higher nutritional level are known to stimulate higher nutrient content in worm casts (Tian *et al.,* 1993a). This may explain the relatively high N-enriched worm casts produced under *G. sepium and L. leucocephala* (Kang *et al.,* 1994b).

The above observations showed the differential effects of woody species primarily through their combine tree canopy, litter quality and physical mulching effects on casting activity of *H. africana*. Similar differential effects of hedgerow tree species in alley cropping were also reported by Kang *et al.,* (1990) and Hauser (1993). They also observed that worm-casting activity of *H. africana* were higher under the tree hedgerows than in the alleys.

Table 9 Annual earthworm cast dry weight and earthworm population under canopies of five woody species grown on an Oxic paleustalf in the forest-savama transition zone of southwestern Nigeria (Adapted from Kang *et al.*, 1994b).

Species	*Wormcast dry weight** (*mg/ha/year*)	*Number of earthworm* (*No./m²*)	*Worm fresh weight* (*g/m²*)
Dactyladenia barteri	26.38	51	40.9
Gliricidia sepium	24.44	72	14.5
Leucaena leucocephala	18.64	52	25.6
Senna siamea	18.33	66	14.4
Treculia africana	22.89	57	37.4
LSD. (0.05)	0.36	15	12.2

*Moisture content of worm casts, 30.7 to 33.0%; Soil temperature at 8.00 a.m., 23.4 to 25.0°C; at 2.00 p.m., 24.5 to 28.9°C.

Table 10 Effect of tree species grown on an Oxic paleustalf in the forest - savanna transition zone of southwestern Nigeria on physico-chemical properties of worm casts and corresponding surface (0–15 cm) soils (Adapted from Kang *et al.*, 1994b).

Properties	*Dactyladenia barteri*		*Gliricidia sepium*		*Leucaena leucocephala*		*Senna siamea*		*Treculia africana*		*LSD* (0.05)
	Casts	*Soil*	*Casts*	*Soil*	*Casts*	*Soil*	*Casts*	*Soil*	*Casts*	*Soil*	
Bulk density (mg/m^3)	1.55	1.12	1.71	1.37	1.66	1.13	1.57	1.18	1.62	1.22	0.30
Clay content (g/kg)	140	120	220	120	240	140	140	100	160	120	–
pH-H_2O	6.4	6.0	6.6	6.0	6.6	6.2	6.7	6.4	6.7	6.3	0.1
Org. C. (g/kg)	51.7	12.0	47.7	11.4	44.4	14.8	49.5	11.9	47.8	9.4	4.9
Exch. K (cmol/kg)	0.69	0.27	0.84	0.22	0.86	0.29	0.72	0.26	0.80	0.25	0.10
Exch. Ca (cmol/kg)	8.02	2.05	7.48	1.91	8.06	1.86	7.24	2.03	8.51	2.05	0.62
Exch. Mg (cmol/kg)	1.36	0.45	1.58	0.42	1.66	0.43	1.97	0.44	1.67	0.38	0.19
Extr. P (mg/kg)	8.30	6.80	9.25	3.70	9.90	3.90	11.50	6.32	4.90	3.02	2.68

The improvement in surface soil characteristics of high base Alfisol (Table 4) under woody species can be partially attributed to the earthworm casting activity. Earthworms play an important role in recycling large quantities of soil annually (Table 9). By passing through their digestive system, they also enriched the soil with nutrients and organic carbon (Table 10). It also appears that the woody vegetation has a differential effect on the physicochemical properties of the *H. africanus* worm casts (Table 10). In addition to improving the soil aggregate stability (Kang *et al.*, 1994b), earthworm activity also increases soil porosity. On the high base status soil in the humid tropics, the combine interactive effects of woody species cover and earthworm activity can play an important role in differentially improving soil properties.

EFFECT ON MICROARTHROPODS

Soil microarthropods, comprised mainly soil mites (Acari) and springtails (Collembola) (Crossley *et al.*, 1992) with body width ranging from 0.08 mm to 0.5 mm. They are abundant in agricultural soils, and play an important role in nutrient cycling through their regulatory activities in decomposition of organic materials and nutrient turnover (Seastedt, 1984; Reddy, 1995). The springtails are largely detritus or fungal feeders, as are many of the mites, whereas the predatory mites subsist on other members of the soil animal community. Microarthropods control the distribution, abundance, and activity of some soil fungi and bacteria (Lussenhop, 1992). Most oribatid mites, free-living astigmatid mites and springtails have well-developed mouthparts capable of fragmenting plant materials (Seastedt, 1984). Tian *et al.* (1998) suggested that microarthropods have a "buffering effect" in regulating leaf decomposition and N release. Population density and composition of the fauna in soil indicate soil conditions (Haq, 1994; Reddy, 1986).

Badejo *et al.* (1998) carried out a study on the soil microarthropod dynamics under selected trees in the IITA arboretum at Ibadan. The plots selected for study are plantations of *D. barteri*, *T. africana*, *L. leucocephala* and *G. sepium*. For both dry and wet months, microarthropod populations were higher under the trees, as compared to the natural grass and secondary forest plots (Table 11). The effect was particularly pronounced in the dry months, despite the overall low density. The population under trees appeared to be correlated with the residue quality of tree litter, particularity during the dry months. The populations of microarthropods were highest under *G. sepium*, which produced litter with the highest quality among the species studied. Throughout a year, there was a positive correlation between soil moisture content and soil collembola in the experimental plots; and there was a negative correlation between soil temperature and numbers of collembola extracted from the soils under canopies of the different woody species.

Table 11 Soil microarthropod population (x 1000/m^{-2}) under various trees in a plantation at Ibadan, southwestern Nigeria. (Adapted from Badejo *et al.*, 1999).

Plot	*Dry months* (*Dec. 93–Feb. 94*)	*Wet months* (*Apr.–Jun. 94*)
Grass control	0.3	23
Forest	1.3	16
Leucaena leucocephala	8.4	38
Treculia africana	4.5	38
Dactyladenia barteri	5.1	37
Gliricidia sepium	13.0	68

Restoration of Soil Microarthropods by Trees

Adejuyigbe *et al.* (1999) monitored soil microarthropod populations from May 1996 to April 1997 under various planted woody species in a severely degraded soil at Ibadan, southwestern Nigeria. The soil was degraded due to depletion of soil organic matter and nutrients, loss of soil biota, soil erosion and compaction during a long period of cultivation. The woody species planted in 1989 for soil regeneration included *A. leptocarpa, S. siamea* and *L. leucocephala.* Natural regrowth vegetation was dominated by *Chromolaena odorata* and continuous cropping with maize.

Both natural and planted woody fallows improved populations of soil microarthropods during the dry and wet months (Table 12). Except for *L. leucocephala,* planted woody fallow showed a high ability in restoring soil microarthropod population, compared to the natural fallow. The highest effect was observed with *S. siamea,* which also adapted to

Table 12 Soil microarthropod population (x 1000/m^{2}) under various trees planted for soil regeneration at Ibadan, southwestern Nigeria. (Adapted from Adejuyigbe *et al.*, 1999).

Plot	*Dry months* (*Dec. 93–Feb. 94*)	*Wet months* (*Apr.–Jun. 94*)
Maize/cassava	1.5	2
Natural fallow	9.0	28
Acacia leptocarpa	9.6	40
Leucaena leucocephala	6.2	27
Senna siamea	14.0	37

the environment better than other species (Kang *et al.*, 1997). Positive correlation exist between soil microarthropod populations and soil moisture contents. Litter with high lignin contents favoured the restoration of soil microarthropods by planted woody species (Adejuyigbe *et al.*, 1999). The results from both studies suggested that trees could provide favorable habitat for soil microarthropods. The population density of soil microarthropods under trees is related to the litter quality and microclimate under trees.

ACKNOWLEDGMENTS

The authors wish to thank the International Institute of Tropical Agriculture Ibadan (IITA), Nigeria and the Comercial Agricola Paineiras (CAP) of Urbano Santos-MA, Brazil for supporting this research. In addition, Dr. Hans Krogh and Mauricio da Silva of CAP are thanked for their permission to use the Eucalypts plantations as well as for field supports. Messrs O. Kolawole, P. Ezekiel, P. Igboba, O. Ganiyu of IITA, J.P. Linhares of UEMA, Brazil are acknowledged for their assistance during field data collection and or laboratory analyses.

REFERENCES

Adejuyigbe, C.O., Tian, G., and Adeoye, G.O., 1999. Soil microarthropod populations under natural and planted fallows in the humid tropics, *Agroforestry Systems*, (in press).

Adu- Bredu, S., Yokota T., Ogawa, K., and Hagihara, A., 1997. Tree size dependence of litter production, and above-ground net production in a young Hinoki (*Chamae cyparis* Obtusa) stand, *J. of Forest Research*, **2:** 31–37.

Agbede, O.O., Adeola, A.O., and Oyeboade, O.A., 1987. Taungya in Nigeria: Tree planting time and fertilizer effects on growth and crop yield, *Forest Ecology and Management*, **20:** 291–298.

Akinnifesi, F.K., and Akinsannii, F.A., 1995. Linear equations for estimating the merchantable wood volume of *Gmelina arborea* in Southwest Africa, *Journal of Tropical Forest Science*, **7:** 391–397.

Alvarez-Sanchez, J., and Sada, S.G., 1993. Litter dynamics in a Mexican lowland tropical rainforest, *Tropical Ecology*, **14:** 127–142.

Badejo, M.A., and Tian, G., 1999. Abundance of soil mites under four agroforestry tree species with contrasting litter quality, *Biology and Fertility of Soils*, (acceptable with revision).

Badejo, M.A., Nathaniel, T.I., and Tian, G., 1998. Abundance of springtails (Collembola) under four agroforestry tree species with contrasting litter quality, *Biology and Fertility of Soils*, **27:** 15–20.

Barbosa, R.I., and Fearnside, P.S., 1996. Carbon and nutrient flows in an Amazonian forest: Fine litter production and composition at Apian, Roraima, Brazil, *Tropical Ecology*, **37:** 115–125.

Baker, T.G., 1983. Dry matter, nitrogen and phosphorus content in litterfall and branch in *Pinus radiata* and Eucalyptus forests, *New Zealand J. of Forest Science*, **13:** 205–221.

Beer, J., 1988. Litter production and nutrient cycling in coffee (*Coffea arabica*) or cacao (*Theobroma cacao*) plantations with shade trees, *Agroforestry Systems*, **7:** 103–114.

Beer, J., Bonneman, A. Chavez, W., Fassbender, A., Imbach, A.C., and Martel, I., 1990. Modeling agroforestry systems of cacao with *Cordia alliodora* and *Erithrina poepigiana* in Costa Rica: V. Productivity indices, organic matter models and sustainability over ten years, *Agroforestry Systems*, **12:** 229–249.

Benneth, F.A., 1960. Spacing and early growth of planted slash pine, *J. of Forestry*, **58:** 966–967.

Brown, A.G., Narnbiar, E.K.S., and Cossalter, C., 1997. Plantations for the tropics—their role, extent and nature, In: *Management of soil, nutrient and water in tropical plantation forests*, (E.K.S. Nambiar, and A.G. Brown, eds.), Australian Center for International Agricultural Research (ACIAR), Monograph No. 43, 1–23 pp.

Brewbaker, J.L., 1985. Nitrogen fixing trees for fodder and browse in Africa, In: *Alley farming in the humid and subhumid tropics*, (B.T. Kang, and L. Reynolds, eds.), IITA Ibadan.

Challinor, D., 1968. Alterations of surface soil characteristics four tree species, *Forest Ecology Management*, **49:** 286–290.

Chijike, E.O., 1980. Impact on soils of fast-growing species in lowland tropics. FAO Forestry Paper No. 21, FAO Rome, 111pp.

Cook, A.G., Critchley, U., Perfect, T.J., and Yeadon, R., 1979. Effects of cultivation and DDT on earthworm activity in a forest soil in the sub-humid tropics, *J. Applied Ecol.*, **17:** 21–29.

Crossley, D.A. Jr., Mueller, B.R., and Perdue, J.C., 1992. Biodiversity of microarthropods in agricultural soils: relation to process, *Agriculture, Ecosystem and Environment*, **40:** 37–46.

Das, A.K., and Ramakrishnan, P.S., 1985. Litter dynamics in Khasi pine (*Pinus kesiya*) of North East India, *Forest Ecology and Management*, **10:** 135–153.

Dubois, J.C.L., Viana, V.M., and Anderson, A.B., 1996. *Manual agroflorestal para a Aniazônia*, volume 1, 227 pp.

Egunjobi, J.K., and Onwuluzo, B.S., 1979. Litterfall, mineral turnover and litter accumulation in *Pinus caribaea* L. stands in Nigeria, *Biotropica*, **11:** 251–255.

EMBRAPA (Empresa Brasileira de Pesquisa Agropecuaria), 1986. *Servicos nacional de levantainento e conservacao do solos do Estado do Maranhao*, Boletim de pesquisa No. 35, Rio de Janeiro.

FACT, 1997. *Forest, farm, and community tree research reports*, volume 2, 1997.

FAO (Food and Agricultural Organization), 1994. *Readings in sustainable forest management*, FAO Forestry paper 122, 266 pp.

FAO (Food and Agricultural Organization), 1995. *Forest Resources Assessment 1990: tropical forest plantation resources*, FAO Technical paper 128, 81 pp.

FORMECU (Forest Monitoring Evaluation and Coordination Unit), 1984. *Nigeria Forestry Statistics*, Federal Department of Forestry, Ibadan, Nigeria. Research Paper No. 2.

Getahun, B., Wilson, F., and Kang, B.T., 1982. The role of trees in farming systems in the humid tropics, In: *Agroforestry systems prospects for development*, (H. MacDonald, ed.), United Nations, 28–32 pp.

Gill, H.S., Abrol, I.P., and Sarnia, J.S., 1987. Nutrient recycling through litter production in young plantations of *Acacia nilotica* and *Eucalyptus tereticornis* in a highly alkaline soil, *Forest Ecology and Management*, **22:** 59–69.

ICRAF (International Centre for Research in agrogorestry Institute), 1994. Annual report, 52 pp.

Haq, M.A., 1994. Roles of oribatid mites in soil ecosystem, In *Ecology and Biology of Soil Organisms*, (S.C. Bhandari, and L.L. Somani, eds.), Agrotech Publishing Academy, Udaipur, Nepal, 143–177 pp.

Hauser, S., 1993. Distribution and activity of earthworms and contribution to nutrient recycling in alley cropping, *Biol. Bert. Soils*, **15:** 16–20.

Ho-Teak, P., Hattor, S., and Tanaks T., 1998. Development of a numerical model for evaluating the effect of litter layer on evaporation, *Journal of Forest Reasearch,* **3:** 25–33.

Jackson, J.K., and Ojo, G.A. 1973. *Provenance trials of Eucalyptus camaldulensis in the Savannah region of Nigeria,* Forestry Research Station series No. 14, Samaru Nigeria. Forestry Research Station series; No. 14, Samaru, Nigeria.

Jeeva, V., and Ramakrishnan, P.S., 1997. Studies impact of plantation forestry in Nilgiri Hills of the Ghats on soil quality and nutrient cycling, *Tropical Ecology,* **38:** 215–235.

Kang, B.T., Akinnifesi, F.K., and Ladipo, D.O., 1994a. Performance of selected woody agroforestry species grown on an alfisol and alfisol in the humid lowland of West Africa, and their effects on soils properties, *Journal of Tropical Forest Science,* **7:** 303–312.

Kang, B.T., Akinnifesi, F.K., and Pleysier, J.L., 1994b. Effect of agroforestry woody species on earthworm activity and physicochemical properties of worm casts, *Biol. Fert. Soils,* **18:** 193–199.

Kang, B.T., Reynolds, L., and Atta-Krah, A.N., 1990. Alley farming, *Adv. Agron.,* **43:** 315–350

Kang, B.T., Salako, F.K., Akobundu, I.O., Pleysier, J.L., and Chianu, J.N., 1997. Amelioration of a degraded Oxic Paleustalf by leguminous and natural fallows, *Soil Use and Management,* **13:** 130–136.

Ladipo, D.O., and Leaky, R.R.B., 1986. Progeny and clone site interaction in Triplochiton scleroxylon K. Schum, Research paper No. 61, Forestry Research Institute of Nigeria, Ibadan.

Lundgren, B., 1978. *Soil conditions and nutrient cycling under natural and plantation forests in Tanzanian highlands,* Reports Forestry Ecological and Soils, Swedish University of Agricultural Science, Uppsala, Sweden.

Lussenhop, J., 1992. Mechanisms of microarthropod-microbial interactions in soil, *Advances in Ecological Research,* **23:** 1–33.

MacDicken, K.G., and Brewbaker J.L., 1989. Growth rates of five tropical leguminous fuelwood species, *Journal of Tropical Forest Science,* **1:** 85–93.

Madeira, M., Araujo, J., and Pereira, S., 1995. Effects of water and nutrient supply on amount and nutrient concentration of litterfall and forest floor litter in *Eucalyptus globulus* plantations, In: *Developments in plant and soil sciences: Nutrient uptake and cycling in Forest Ecosystem,* (L.O. Nilson, R.F. Huttl, and U.T. Johansson, eds.), Kluwer Academic Publisher, 287–295 pp.

Madge, D.S., 1965. Field and laboratory studies on the activity of two species of tropical earthworms, *Pedobiologia,* **9:** 188–214.

Misra, M.K., and Nisanka, S.K., 1997. Litterfall, decomposition and nutrient release in *Casuarina equisetifolia* plantations on the sandy coast of Orissa, India, *Tropical Ecology,* **38:** 109–119.

Moorman, F.R., Lal, R., and Juo, A.S.R., 1975. *The soils of IITA,* International Institute of Tropical Agriculture, Ibadan, Nigeria.

Morris, A.R., 1995. Forest floor accumulation, nutrition and productivity of *Pinus patula* in the Usutu Forest, Swaziland, *Plant and Soil,* **168–169:** 271–278.

Nock, H.P., 1985. The Brazilian afforestation programme, *Plant Research* and *Development,* **16:** 61–73.

Nwoboshi, L.C., 1972. Responses of teak (*Tectona grandis* L.F.), Idigbo (*Terminalia ivorensis A.* Chev.) and Opepe (*Nauchlea diderrichii* Merrill) seedlings to various light intensities, *Nigerian Journal of Forestry,* **2**(2): 48–53.

Nwoboshi, L.C., 1982. *Tropical Silviculture. Principles and techniques,* Ibadan University Press, ch. 14, 323 pp.

Nyland, R.D., 1996. *Silviculture: Concepts and Application,* McGraw-Hill Publishers, ch. 6, 633 pp.

Okafor, J.C., 1983. Horticulturally promising indigenous wild plant species of Nigerian Forest Zone, *Acta Horticultural*, **123:** 165–176.

Okafor, J.C., 1990. Development and selection of commercially viable cultivars from forest species for fruit. Proc. 12th plenary meeting of AEFAT, mitt. Allg. Bot. Hamburg, Germany Symposium, 81–97 pp.

Pande, P.K., and Sharma, S.C., 1986. Seasonality and pattern in leaf-fall and litter accretion on the forest floor in plantations of demonstration area. Forest Research Institute and College, Dehra Dun (India), *Indian Forester*, **112:** 328–341.

Rao, B.K.S., Dabral, B.G., and Pande, S.K., 1972. Litter production in forest plantations of Chir (*Pinus roxburghii*) and Teak (*Tectona grandis*) and Sal (*Shorea robusta*) at new forest, Dehra Dun, In: *Simposium on Tropical Ecolgy with emphasis on organic productivity*, (P.M. Golley, and F.B. Golley, eds.), Athens, 235–243 pp.

Reddy, M.V., 1986. Soil-inhabiting arthropods as indicators of environmental quality. Acta Biologica Hungarica, **37:** 79–84.

Reddy, M.V., 1995. *Soil Organisms and litter Decomposition in the Tropics*, Oxford & IBH Publ. Co. Pvt. Ltd., New Delhi, India and Westview Press, Colorado, USA, 277 pp.

Ricardo de Souza, D., 1996. *Efeito do espaçamanto na produtividade volumetrica de madeira em povoamento de Pinus oocarpa* schiede e *Pinus caribaea var. hondurensis bar*, Et Golf ate os 14 anos de idade. *Quarta* Simpósio Intemacional sobre ecossistemas florestais, Minas Gerais 13–6 de Agosto de 1996; 60 pp.

Sanchez, P., and Russell, C.E., 1987. Contrasting effect of *Pinus caribaea* and *Gmelina arborea* on soil properties, In: *Tropsoils: technical report*, (N. Candle, and C.B. McCants, eds.), North Carolina state University, 29–39 pp.

Sanchez, P.A., Palm, C.A., Davey, C.B., Szott, L.T., and Russell, C.E., 1985. Tree crops as soil improvers in the humid tropics?, In: *Attributes of trees as a crop plants. Attributes of trees as crop plants*, (M.G.R. Cannell, and J.E. Jackson, eds.), Abbots Ripton, UK: Institute of Terrestial Ecology, 327–358 pp.

Sattoo, T., Nakamura, T., and Senda, N., 1955. Material for study of growth in stands I. Young stands of Japanese red pine of various density, *Bulletin of Tokyo University Forests*, **48:** 65–80.

Schutz, C.J., Bredenkamp B.V., and Herbert, M.A., 1993. Stand density and litter depth of *Pinus patula South African Forest, J.*, **124:** 43–49.

Seastedt, T.R., 1984. The role of microarthropods in decomposition and mineralization processes, *Ann. Rev. Entom.*, **29:** 25–46.

Sharma, E., and Ambasht, R.S., 1987. Litterfall, decomposition and nutrient release in an age sequence of *Alnus nepalensis* plantation stands in the Eastern Himalayas, *J. of Ecology*, **25:** 997–100.

Singh, H., 1982. Nutrient content of a standing crop and biolcgical cycling in *Pinus patula* ecosystem. *Forest Ecology Management*. **4:** 317–332.

Singh, R.P., Mathur H.N., and Gupta, M.K., 1984. Litter production and release of nutrients in coniferous forests of Himachal Pradesh, *Indian Forester*, **110:** 1014–1022.

Smith, A., and Montgomery, R.F., 1962. *The soils and land use of central Western Nigeria*, The Government Printers, Ibadan, Nigeria, 246 pp.

Smith, D.M., 1986. *The Practice of Silviculture*, 8th ed., John Wiley & Sons Inc., New York, 578 pp.

Smith, W.D., and Strub, M.R., 1991. Initial spacing: How many trees to plant, In: *Forest Regeneration Manual*, (M.L. Durye, and P.M. Dougherty, eds.), ch. 15, 281–289 pp.

Swift, M.J., Heal, O.W., and Anderson, J.M., 1979. *Decomposition in Terrestial Ecosystems*, Blackwell Sci. Publishers, Oxford, 372 pp.

Tian, G., Adejuyigbe, C.O., Adeoye, G.O., and Kang, B.T., 1998. Role of soil microarthropods in leaf decomposition and N release under various land-use practices in the humid tropics, *Pedobiologya*, **42:** 33–42.

Tian, G., Brussaard, L., and Kang, B.T., 1993b. Mulching effect of plant residues with chemically contrasting composition on maize growth and nutrient accumulation, *Plant and Soil*, **153:** 179–187.

Tian, G., Brussaard, L., and Kang, B.T., 1993a. Biological effects of plant residues with contrasting chemical composition under humid tropical conditions: Effects on soil fauna, *Soil Biol. Biochem.*, **25:** 731–737.

Vesveld, D.B., 1981. Litterfall and decompositin in stands of native *Pinus radiata*. *South African Forestry Journal*, **116:** 40–50.

Young, A., 1989. *Agroforestry or Soil conservation*, CAB International, BPC wheatons Ltd, Exerter, ch. II, 275 pp.

Yokota, T., and Hagihara, A., 1996. Dependence of the aboveground CO_2 exchange rate on tree size in field-grown hinoki cypress (*Chamaecyparis obtusa*), *J. of Plant Research*, **109:** 177–184.

Yokota, T., Ogawa, K., and Hagihara, A., 1994. Dependence of the above-ground respiration of hinoski cypress (*Chamaecyparis obtusa*) on tree size, *Tree Physiology*, **14:** 467–479.

CHAPTER 11

EARTHWORMS IN TROPICAL TREE-PLANTATIONS: EFFECTS OF MANAGEMENT AND RELATIONS WITH SOIL CARBON AND NUTRIENT USE EFFICIENCY

Xiaoming Zou[1] and Grizelle González[2]

[1]*Institute for Tropical Ecosystem Studies, University of Puerto Rico, P.O. Box 363682, San Juan, PR 00936, USA*

[2]*International Institute of Tropical Forestry, USDA Forest Service, P.O. Box 25000, San Juan, PR 00928, USA*

ABSTRACT

With the vast amount of abandoned tropical land due to non-sustainable farming practices, tropical tree-plantations become an effective means in restoring soil productivity and preserving ecosystem biodiversity. Because earthworms are the dominant soil fauna in moist tropical regions and play an important role in improving soil fertility, understanding the mechanisms by which forest management practices affect the abundance and community structure of earthworms will be crucial in designing future reforestation programmes. Forest management practices include site preparation, tree species selection, fertilization, and harvesting. While native earthworms are often negatively affected by using exotic tree species, they can often be preserved in plantations using native tree species. Conventional practices of site preparation and harvesting often favor exotic endogeic earthworms and impose negative impact on native epigeic earthworms. The effects of chemical fertilization on earthworms vary with soils and fertilizers. These management practices alter soil properties such as soil C and N levels, and change the quantity and quality of plant litter. Under similar climate and soil conditions, earthworm abundance is positively correlated with soil C and N levels. High earthworm density is associated with high litter quality and low nutrient use efficiency of tree-plantations. We conclude that forest management practices can drastically affect earthworm populations and that maintaining a healthy population of earthworms can further promote forest nutrition in tropical tree-plantations.

Key Words: Earthworms, tropical tree plantations, forest management, forest fertilization, soil carbon, nutrient use efficiency.

Human deforestation has nearly claimed half of the world's native forest, and human demand for wood will increase 50% in the next decade. For the last century, deforestation has expanded at an accelerating rate throughout tropical regions. Deforested tropical lands have reached 1.9 billion ha by late 1980s (Grainger, 1988). Because of non-sustainable farming practices, these deforested lands have been abandoned at a rate of 12×10^6 ha y^{-1} (Lal and Stewart, 1990). These vast amounts of abandoned lands have drawn attention from land managers and ecologists to search for strategies to restore land productivity and reserve biological species diversity (Shukla *et al.*, 1990; Saunders *et al.*, 1991; Schulze and Mooney, 1993). One such strategy is to establish tropical tree-plantations. By late 1980s, tropical tree plantations accounted for 1% of total tropical forested areas.

Earthworms are recognized to accelerate the decomposition of soil organic matter and to increase soil fertility (Lee, 1985; Lavelle, 1988; Edwards and Bohlen, 1996; Bohlen *et al.*, 1997). Under moist climatic conditions, they often dominate soil macrofauna biomass (Fragoso and Lavelle, 1992; Hendrix, 1995). However, few studies have examined earthworms in tropical tree-plantations. This chapter intends to show how forest management practices affect the abundance and community structure of earthworms in tropical tree-plantations, and the chapter will also examine the correlation between soil properties and earthworm abundance.

EARTHWORMS AND FOREST MANAGEMENT

Forest management practices can influence earthworm communities through changing physical and chemical properties of soil, net primary productivity, or plant litter chemistry (Fig. 1a,b). These practices typically include site preparation, tree species selection, fertilization, and harvesting (Nyland, 1996). Site preparation can alter soil structure and plant communities. Tree species differ in above- and below-ground net productivity and in plant litter chemistry, thus vary in decomposition and nutrient cycling processes. Fertilization can directly change plant nutrition and soil properties, and consequently plant production and

Fig. 1a Wormcast of *Hyperiodrilus africanus* under *Leucaena leucocephala* hedgerows grown on high base status Afisol in southwestern Nigeria.

litter chemistry. Harvesting may impose direct effects on soil structure, plant species composition, and plant production.

Responses of earthworm to forest management practices may differ among various worm communities. Earthworms are classified into three categories (Bouche, 1977): the endogeic worms, the epigeic worms, and the anecic worms. Endogeic earthworms live and predominantly feed within the mineral soil layer. The epigeic earthworms live and feed within the surface litter layer. The anecic worms live in the mineral soil layer but predominantly feed on organic materials in the surface litter layer. Changes in physical and chemical properties in the mineral soils can have drastic impact on the endogeic worms, whereas changes in the quantity and chemistry of plant materials in forest floor mass may predominantly affect the epigeic worms. Changes either in mineral or plant litter layers can affect the anecic worms.

Site Preparation

Site preparation includes clearing vegetation and plowing soil. Clearing vegetation has direct effects on the net primary productivity and plant species composition, thus can affect earthworm communities. Because epigeic earthworms live and feed in plant litter layer, clearing vegetation may impose immediate effects on epigeic and anecic worms. Plowing

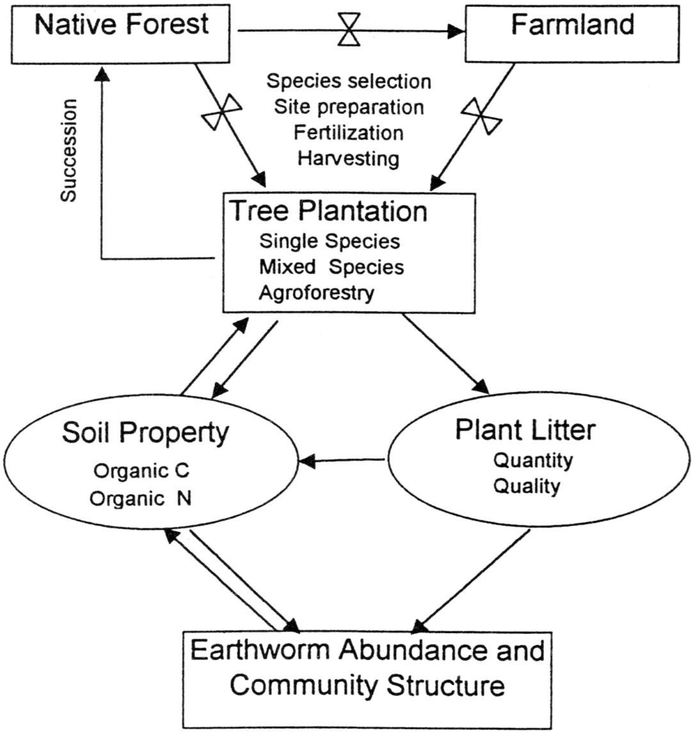

Fig. 1b Forest management practices (species selection, site preparation, fertilization, and harvesting) that affect earthworm abundance and community structure in tropical tree plantation. Native forest refers to forests that are not disturbed by human activities. Farmland includes crop fields, pastures, and other land use types like mine fields. Soil properties include chemical, physical, and biological factors such as soil C and N level, pH, microbial activity (Modified from González *et al.,* 1996).

disturbs soils and causes earthworm mortality due to injuries from machinery and predation by animals after worms are exposed to soil surface. In agricultural fields, plowing can reduce earthworm density by as much as 20%. Additionally, plowing buries plant litter into sub-soils layers, thus improves soil aeration, accelerates soil erosion, mixes mineral soils with plant materials, and redistributes plant seeds (Lee, 1985; Edwards and Bohlen, 1996). Plant productivity will be reduced for a short period of time ranging from a few months to a few years. Soil bulk density will often decrease by about 10% after plowing. Changes in soil bulk density can affect soil moisture conditions, often by increasing effective soil water storage capacity (the difference between field water holding capacity and plant wilting point). Increase in the effective soil water storage capacity improves soil water conditions for the survival and growth of earthworms, especially during dry seasons.

The effect of plowing on soil erosion is the balance of two processes: reduction in soil surface runoff water and increase in soil erodibility. Plowing increases large soil pore space and consequently soil water infiltration rates. However, the newly exposed soils are not protected by vegetation and litter, thus are under direct impact of raindrops and surface runoff. Effects of plowing on soil erosion are rarely studied in tropical tree-plantations. But many studies from agricultural fields indicated a general pattern of increasing soil erosion after plowing (Soil Conservation Service, 1988). Accelerated soil erosion can have long-term impact on earthworm communities. It may reduce worm species diversity and population abundance.

Much of the forest floor mass can be buried into sub-soil layers after plowing. The incorporation of plant materials into soil may favor endogeic worms, but may reduce the abundance of anecic and epigeic earthworms.

By redistributing plant seeds in soil profile, plowing affects the survival and gemination rates of plants, thus alters the species composition of plant community. Changes in composition of plant species can have drastic impact on earthworm communities (Zou, 1993; Zou and González, 1997; Zou and Bashkin, 1998; González and Zou, 1999a; González *et al.*, 1999). Furthermore, plowing may suppress infectious diseases and this may have positive effect on worms, but little information is available.

Species Selection

Plant species differ in above- and below-ground carbon allocation patterns. They also differ in the chemistry of plant litter. In tree-plantations at Hawaii, fine litterfall was 8.25 mg ha^{-1} y^{-1} in an *Albizia falcataria* (L.) Fosb. plantation as compared to 5.04 mg ha^{-1} y^{-1} in an *Eucalyptus saligna* Sm. plantation (Binkley *et al.*, 1992). Nitrogen content of Albizia litter is much higher than the Eucalyptus litter. A high nitrogen content in litter materials was correlated with high earthworm density (Zou, 1993). In another study in the tropical island of Puerto Rico, above-ground net primary productivity was higher in tree plantations of *Pinus caribaea* Morelet and *Swietenia macrophylla* King than in adjacent secondary forest (Lugo, 1992). Litter nitrogen content was higher in the secondary forest than in the plantations. Earthworm density was found higher in the secondary forests (González *et al.*, 1996). Standen (1988) also showed a reduction of earthworm abundance in pine plantations as compared to pastures.

Tree plantations with exotic species are often not favored by native earthworms. In Puerto Rico, exotic worm species occurred in plantations of both *P. caribaea* and *S. macrophylla* (exotic species) as well as in secondary forests of native species, but native worm species occurred only in the secondary forests (González *et al.*, 1996). In New Guinea, establishment of pine plantations on grassland reduced the number of native earthworms and introduced exotic species (Standen, 1988). In contrast, the establishment of *Araucaria* did not show a strong negative effect on the native earthworms. The negative effects of pine plantations on earthworms were pronounced as the plantations get older (Yeates, 1988).

Changes in composition of plant species often occur during secondary succession (Horn, 1974; Reiners *et al.*, 1994; Zou *et al.*, 1995). In Hawaii, secondary plant communities developed from abandoned sugarcane fields supported a higher earthworm population and diverse community structure (Zou and Bashkin, 1998). In Puerto Rico, changes in composition of plant species from grass dominated pasture to woody species corresponded to a reduction of earthworm density from 850 to 200 worms per square metre. On the other hand, earthworm diversity increased from endogeic *Pontoscolex corethrurus* community in the pasture to both endogeic and anecie worm community in the secondary forests.

Variation in plant litter chemistry and decomposition may also affect soil chemical properties such as pH. A greater uptake pattern of cations than anions by plants can also enhance soil acidification. Decrease in soil pH in acid soil is not favored by most earthworms (Lee, 1985; Edwards and Bohlen, 1996).

Forest Fertilization

Forest fertilization refers to the application of organic and inorganic fertilizers in forests. Organic fertilizers include plant and waste materials such as plant residues, farmyard manures, animal slurries, municipal sewage wastes, and industrial wastes from breweries, paper pulp, or potato processing. In most studies, organic fertilizers increase earthworm density and biomass (Marshall, 1977). Earthworm abundance can be doubled or tripled within a year after the application of organic materials. However, exceptions exist in the cases of applying fresh animal manure that has not been aged or composted, or in the cases where soil inorganic chemistry such as pH has drastically changed due to the application of organic materials (Curry, 1976). Information on using organic materials in tropical tree-plantation is limited.

Chemical fertilization may increase or decrease earthworm abundance and biomass. Application of nitrate fertilizers often acidifies soil pH, thus may have negative effect on earthworm communities. However, addition of N without decrease in soil pH may have positive effect on earthworm communities through increasing plant production and improving litter quality. In a tropical wet forest of Puerto Rico, the application of complete inorganic fertilizers (N, P, K, Ca, Mg, and micronutrients) decreased soil pH values and earthworm density (Zou, unpublished data).

HARVESTING

Logging of trees can reduce forest productivity and change plant species composition. The machinery used may also compact soils and injure earthworms. Because epigeic worms inhabit forest floors, they are most likely be affected by logging activities (Camilo and Zou, 1999). Harvesting can be divided into selective logging, strip clearing, and large-scale clearing. In a tropical wet forest of Puerto Rico (Zou *et al.*, 1995), epigeic, anecic, and endogeic worms all exist fifty years after selective logging (González and Zou, 1999a; González *et al.*, 1999). However, selective logging also introduced exotic earthworms *P. corethrurus* and *Amynthas rodericensis* into the forest. Strip harvesting may impose a stronger effect on epigeic earthworms than selective logging. Epigeic earthworms were not found in strip logged sites, whereas both anecic and endogeic worms survived the disturbance in the same wet forest in Puerto Rico (Camilo and Zou, 1999). Large-scale clearings often show the strongest effect on earthworms. Converting forests to pasture reduced worm species from five (3 native and 2 exotic) to one exotic species (Zou and González, 1997), and no epigeic or anecic worms survived the disturbance.

In Australia, Abbott (1985) also reported that exotic earthworm species occurred in pasture or orchards after forests were cleared. These data support the conclusions from studies in the temperate region that human disturbance reduces the abundance of native earthworms and increase the abundance of exotic earthworms (Kalisz and Dotson, 1989).

EARTHWORMS IN RELATION TO SOIL CARBON AND NUTRIENT USE EFFICIENCY

SOIL CARBON AND NITROGEN

Converting forest to agricultural lands often reduces soil C and N levels (Post and Mann, 1990), although pasture lands are known to remain the same level of soil C (Lugo *et al.*, 1986; Reiners *et al.*, 1994). Soil organic C often recovers after establishment of tree plantations on abandoned agricultural lands (Makeschin, 1994). In Hawaii, C and N levels in native forest soils reached 15.9% and 0.57%, respectively. Fifty years of sugarcane cultivation reduced soil C and N levels to 6.45% and 0.32%, respectively. The subsequent establishment of *Eucalyptus* plantations increased surface soil C (0–25 cm) and N by 30% (Zou and Bashkin, 1998). Meanwhile, earthworm population increased from 0 in the sugarcane fields to 398 individuals per square metre in the 10 years old *Eucalyptus* plantation. There is a positive correlation between earthworm abundance and soil C and N levels (Fig. 2). However, this correlation

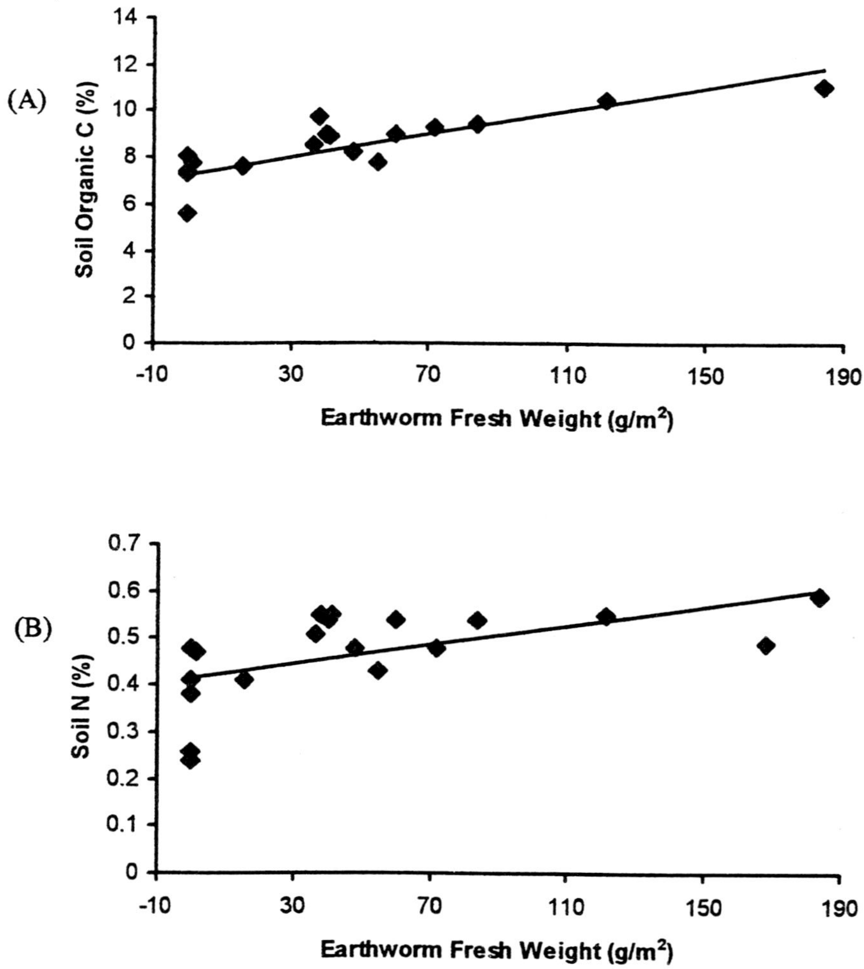

Fig. 2 Correlation between soil C (A) and N (B) levels and earthworm fresh weight in tropical tree plantations of *Eucalyptus* in Hawaii.

may not exist when data are derived from plantations with different tree species or soil types because tree species and soils themselves alone can affect earthworms.

Nutrient Use Efficiency

Nutrient use efficiency is practically defined as the ratio of above-ground net primary productivity to the quantity of nutrients in annual litterfall (Vitousek, 1982). It differs among tree species. In Hawaiian plantations, both nitrogen and phosphorus use efficiencies were the highest in the pure *Eucalyptus* stands, lowest in the pure *Albizia* stands, and medium in the mixed stands. Earthworm density was negatively and exponentially correlated with both N and P use efficiencies of these stands (Fig. 3), suggesting either a competition for nutrients between trees and earthworms or earthworm increased soil nutrient availability to trees through accelerated decomposition of soil organic matter. Although field experiments showed an increased decomposition rate when earthworms were excluded (Bohlen *et al.*, 1997), little information is available to address the competition hypothesis in tropical tree-plantations. James and Seastedt (1986) reported that there were no increase in above-ground plant biomass but there was a reduction in root biomass when earthworms were excluded. González and Zou (1999b) also demonstrated that reduction in earthworms reduced soil N availability but did not increase the growth of *Cecropia scheberiana,* a tropical pioneer tree species. Edward and Lofty (1980) with barley and Stockdill (1982) in pasture also showed that exclusion of earthworms actually reduced plant growth. These evidences tend to support that earthworms may not compete for nutrients with trees, but increase soil nutrient levels by accelerating decomposition.

FUTURE PERSPECTIVES

With vast abandoned agricultural lands, tropical tree-plantations are an effective strategy for the restoration of ecosystem productivity and preservation of biodiversity. However, little is known about the role which earthworms play in tropical tree-plantations, or vice versa. Much research is needed in order to establish sound management practices.

Among research priorities, relationship between tree species and earthworms appears to be essential. There are two approaches to examine this relationship. One is to manipulate either earthworm populations (e.g. James and Seastedt, 1986; González and Zou, 1999a) or tree species (e.g. Zou, 1993, González *et al.*, 1996; Zou and Bashkin, 1998). The other approach is to examine earthworm populations in areas surrounding

(A)

(B)

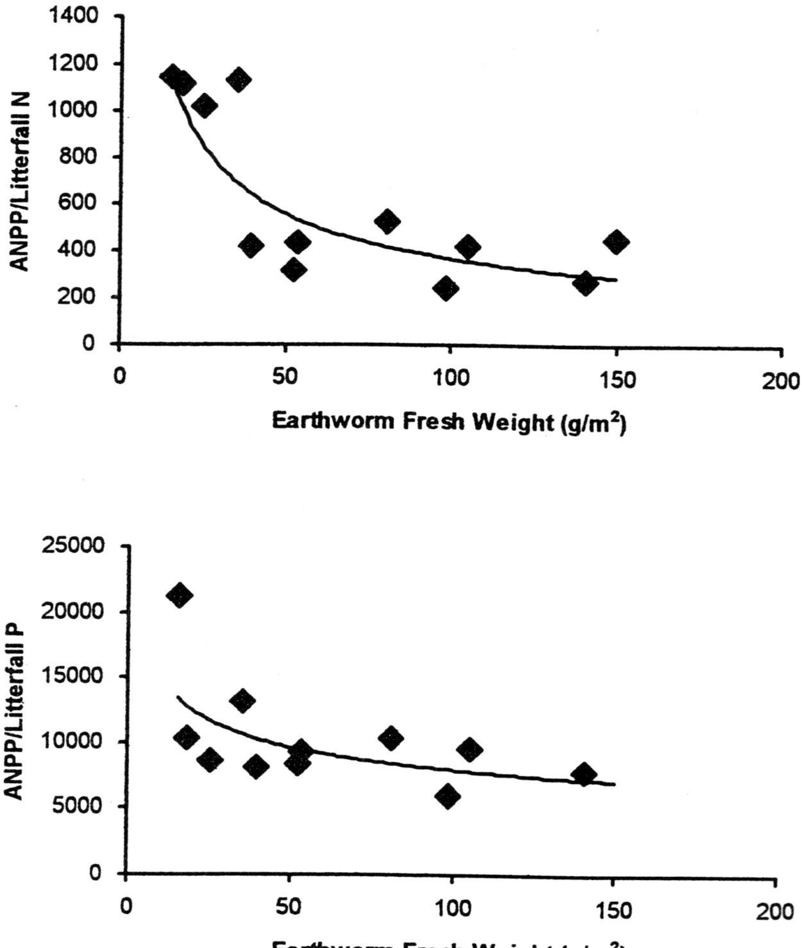

Fig. 3 Relations between nitrogen and phosphorus use efficiency and earthworm fresh weight in tropical tree plantations of *Albizia* and *Eucalyptus* in Hawaii.

one particular species and area outside the influence of this plant species (e.g. González and Zou, 1999a).

The second need is to examine the role that earthworms play in forest production and nutrient cycling in field experiments. Many early studies were conducted in agricultural fields in the temperate region.

These studies showed promising results. For example, an earthworm manipulation experiment in cornfields of Ohio showed worms can accelerate the decomposition of corn residues and increase soil nitrogen availability for plant growth (Bohlen *et al.*, 1997; Bohlen *et al.*, 1999). Research topics can expand to include the earthworm effect on green house gas production, soil inorganic and organic C fluxes, soil erosion, and water balance.

Thirdly, there is a need to assess earthworm biodiversity and its role in regulating soil processes and soil food web dynamics. Does earthworm diversity matter? How does earthworm diversity affect soil food webs and the flow of energy and elements? How do earthworm species interact with each other? This information is crucial in making management policies for tropical tree plantations, but little is available. These questions are most likely to be addressed in green-house experiments with incubation chambers. Although field experiments to manipulate earthworm populations are possible, they would be expensive.

ACKNOWLEDGEMENTS

This research was partially funded through grants from U.S. National Science Foundation (HRD 9353549) and National Aeronautics and Space Administration (NAGW 4059) to the University of Puerto Rico.

REFERENCES

Abbott, I., 1985. Distribution of introduced earthworms in the northern jarrah (*Eucalyptus marginata*) forest of Western Australia, *Australian J. of Soil Research*, **23:** 263–270.

Binkley, D., Dunkin, K.A., DeBell, D., and Ryan, M.G., 1992. Production and nutrient cycling in mixed plantations of *Eucalyptus* and *Albizia* in Hawaii, *Forest Science*, **38:** 393–408.

Bohlen, P. J., Parmelee, R.W., Allen, M.F., and Ketterings, Q.M., 1999. Differential effects of earthworms on nitrogen cycling from various nitrogen-15-labeled substrates, *Soil Science Society of America Journal*, **63:** 882–890.

Bohlen, P. J., Parmelee, R.W., McCartner, D.A., and Edwards, C.A., 1997. Earthworm effects on carbon and nitrogen dynamics of surface litter in corn agroecosystems, *Ecological Application*, **7:** 1341–1349.

Bouche, M.B., 1977. Strategies lombriciens, *Ecological Bulletin*, **25:** 122–132.

Camilo, G.R., and Zou, X., 1999. Soil fauna in managed forests: lessons from the Luquillo Experimental Forest, Puerto Rico, In: *Effects of logging in the Tropics*, (A. Grajal; R. Fimble, and J. Robinson, eds.), Chapter 13, Columbia University Press, New York.

Curry, J.P., 1976. Some effects of animal manures on earthworms in grassland, *Pedobiologia*, **16:** 425–438.

Edwards, C.A., and Bohlen P.J., 1996. *Biology and ecology of earthworms*, Chapman and Hall, New York.

Edward, C.A., and Lofty, J.R., 1980. Effects of earthworm inoculation upon the root growth of direct drilled cereals, *J. of Applied Ecology*, **17:** 533–543.

Fragoso, C., and Lavelle, P., 1992. Earthworm communities of tropical rain forests, *Soil Biology and Biochemistry*, **24:** 1397–1408.

González, G., and Zou, X., 1999a. Plant and litter influences on earthworm abundance and community structure in a tropical wet forest, *Biotropica*, **31**(3): 486–493.

González, G., and Zou, X., 1999b. Earthworm influence on N availability and the growth of *Cecropia schreberiana* in tropical pasture and forest soils, *Pedobiologia*, **43:** 824–829.

González, G., Zou, X., and Borges, S., 1996. Earthworm abundance and species richness in abandoned tropical croplands: Comparisons between tree plantations and secondary forests, *Pedobiologia*, **10:** 385–391.

González, G., Zou, X., Sabat, A., and Fetcher, N., 1999. Earthworm abundance and distribution pattern in contrasting plant communities within a tropical wet forest, *Caribbean Journal of Science*, **35:** 93–100.

Grainger, A., 1988. Estimating areas of degraded tropical lands reqiring replenishment of forest cover, *International Tree Crops Journal*, **5:** 31–61.

Horn, H.S., 1974. The ecology of secondary succession, *Annual Review of Ecolology and Systematics*, **5:** 25–37.

Hendrix, P.F., 1995. *Earthworm Ecology and Biogeography in North America*, Lewis Publishers, Boca Raton, FL.

James, S.W., and Seastedt, T.R., 1986. Nitrogen mineralization by native and introduced earthworms: Effects on big bluestem growth, *Ecology*, **67:** 1094–1097.

Kalisz, P.J., and Dotson D.B., 1989. Land-use history and the occurrence of exotic earthworms in the mountains of eastern Kentucky, *American Midland Naturalist*, **122:** 288–297.

Lal, R., and Stewart, B.A., 1990. *Soil Degradation*, Springer, New York.

Lavelle, P., 1988. Earthworm activities and the soil systems, *Biology and Fertility of Soils*, **6:** 237–251.

Lee, K.E., 1985. *Earthworms, their Ecology and Relationships with Land Use*, Academic Press, New York.

Lugo, A.E., 1992. Comparison of tropical tree plantations with secondary forests of similar age, *Ecological Monograph*, **62:** 1–41.

Lugo, A.E., Sanchez, M.J., and Brown, S., 1986. Land use and organic carbon content of some subtropical soils, *Plant and Soil*, **96**: 185–196.

Makeschin, F., 1994. Effects of energy forestry on soils, *Biomass and Bioenergy*, **6:** 63–79.

Marshall, V.G., 1977. *Effects of manure and fertilizers on soil fauna: A review*, Commonwealth Bureau of Soils, Harpende.

Nyland, R.D., 1996. *Silviculture: Concepts and Applications*, McGraw-Hill.

Post, W.M., and Mann, L.K., 1990. Changes in soil organic carbon and nitrogen as a result of cultivation, In: *Soil and the Greenhouse Effect*, (A.F. Bowman, ed.), Wiley, New York, 401–407 pp.

Reiners, W.B., Bowman, A.F., Parsons, W.F.J., and Keller, M., 1994. Tropic rain forest conversion to pasture: Changes in vegetation and soil properties, *Ecological Applications*, **4:** 363–377.

Saunders, D.A., Hobbs, R.J., and Margules, C.R., 1991. Biological consequences of ecosystem fragmentation: A review, *Conservation Biology*, **5:** 18–32.

Schulze, E.D., and Mooney, H.A., 1993. *Biodiversity and Ecosystem Function*, Springer, New York.

Shukla, J., Nobre, C., and Sellers, P.J., 1990. Amazon deforestation and climate change, *Science*, **247:** 1322–1325.

Soil Conservation Service, 1988. *National Agronomy Manual*, USDA, Washington, D.C.

Standen, V., 1988. Oligochaetes in fire climax grassland and conifer plantations in Papua New Guinea, *J. of Tropical Ecology*, **4:** 39–48.

Stockdill, S.M.J., 1982. Effects of introduced earthworms on the productivity of New Zealand pastures, *Pedobiologia*, 24: 29–35.

Vitousek, P.M., 1982. Nutrient cycling and nutrient use efficiency, *American Naturalist*, **119:** 553–572.

Yeates, G.W., 1988. Earthworm and enchytraeid populations in a 13-year-old agroforestry system, *New Zealand J. of Forestry Science*, **18:** 304–310.

Zou, X., 1993. Tree species effects on earthworm density in tropical tree plantations in Hawaii, *Biology and Fertility of Soils*, **15:** 35–38.

Zou, X., and Bshkin M., 1998. Soil carbon accretion and earthworm recovery following revegetation in abandoned sugarcane fields, *Soil Biology and Biochemistry*, **30:** 825–830.

Zou, X., and González, G., 1997. Changes in earthworm density and community structure during secondary succession in abandoned tropical pastures, *Soil Biology and Biochemistry*, **29:** 627–629.

Zou, X., Zucca, C., Waide, R., and McDowell, W., 1995. Long-term influence of deforestation on tree species composition and litter dynamics of a tropical rain forest in Puerto Rico, *Forest Ecology and Management*, **78:** 147–157.

PART—III: SOIL-NUTRIENT DYNAMICS

— *Eucalyptus* plantation in Brazil: Their soil nutrient dynamics and management.
A. Spangenberg and *H. Folster* (ISSFN, Germany).

— Nitrogen fixing leguminous trees in forestry and agroforestry systems.
D.L.N. Rao (I.I.S.S., India).

— Nutrient flux control by trees for improving soil fertility in tropical agroforestry. *Johannes Lehmann* (UB and FRIFP, Germany).

— Accretion of carbon and nutrients in the forest floor and mineral soil of *Casuarina equisetifolia* plantations on the coastal sand dunes of Senegal.
Daniel Mailly (MRN, Canada) and *H.A. Margolis* (UL, Canada)

CHAPTER 12

EUCALYPTUS PLANTATIONS IN BRAZIL: THEIR SOIL-NUTRIENT DYNAMICS AND MANAGEMENT

A. Spangenberg and H. Fölster

Institute of Soil Science and Forest Nutrition,
Büsgenweg 2, 37077 Göttingen, Germany

ABSTRACT

Planted forests, especially Eucalyptus plantations, are an important branch of forestry in Brazil. The south-located industrial centres create a great demand for charcoal, pulp, industrial wood and fuelwood. Centres of supply are private plantations in south and central Brazil with short rotations of 6 to 10 years. Since the end of the seventies, a vivid scientific research has accompanied plantation forestry, mainly centred at the universities of Viçosa and Piracicaba. During a development from a preliminary focus on silvicultural aspects of yield production towards a comprehensive view of the fragile ecosystem, water, nutrient, and stand development interactions were investigated. Today, nutrient budgets are regarded as an important tool for describing dynamics of nutrient supply in plantations. They illustrate nutrient losses due to wood export (export loss), and management-dependent loss during and after harvest (transformation loss) basically caused by burning, erosion and leaching. Nutrient loss must be compensated by fertilizer input. The loss can be reduced by adequate management practices which maintain as many nutrients as possible within the system. The unavoidable loss should be assessed as closely as possible in order to develop a site, species and aim-oriented nutrient strategy. The input can be optimized by flexible adaptation of element ratios, following uptake and growth-dependent demand.

How soil-nutrient dynamics is affected by export and volatilization losses and their impact on sustainability of the plantation system are illustrated at two sites of the Amazon and one of the central Brazilian 'Cerrado' (savannah) region.

Key Words: Eucalyptus plantations, plantation forestry, soil-nutrient dynamics, nutrient management, fertilizer application.

HISTORY OF PLANTATION FORESTRY IN SOUTH AMERICA

Although the first eucalyptus to arrive in Europe was in 1774 with Cook, the earliest cultures were not planted till the middle of the 19th century. When their pharmaceutical role in combating malaria started to expand around the world. Chile introduced eucalypts to South America in 1823. Since 1856, Argentina began planting extensive shelterbelts and windbreaks with eucalypts from 1856. A small eucalypt oil industry was based on *Eucalyptus globulus*. Their seed material was introduced to Bolivia at the beginning of the 20th century. Costa Rica imported eucalypts only late in this century, in 1965.

The earliest Brazilian cultivations (*E. robusta* and *E. tereticornis*) are dated 1825 (State of Rio de Janeiro, FAO, 1981) and 1868 (Rio Grande do Sul, Lima, 1993). The first large-scale plantations were initiated by Navarro de Andrade in 1903 on behalf of the Paulista Railway Company (Lima, 1993). De Andrade is known as the "father of Brazilian plantation forestry" because he introduced more than 140 eucalypt species. Currently about 30 species are regarded economically important. Most of them are generally successful in South America, for example *E. grandis, E. tereticornis, E. saligna, E. urophylla, E. deglupta, E. citriodora, E. camaldulensis, E. pellita, E. globulus, and E. viminalis.*

Up to 1966, approximately, 400,000 ha were planted with eucalypts in Brazil. In 1966 the Brazilian government decided to subsidize afforestation programmes in order to develop an internationally competitive wood-based pulp and paper industry in the private sector. The programme continued from 1967 to 1989, and helped build up the necessary experience on growth and use of eucalypt wood for cellulose and charcoal production (Ondro *et al.*, 1995). By 1973 Brazil had established more than 1 million ha (Lima, 1993; FAO, 1981) and by 1995 about 5 million ha of eucalypt plantations (Gonçalves *et al.*, 1997). In 1990, Brazilian wood production amounted to about 263.5 million m^3, of which 24.4% accounted for eucalypt plantations. Their wood was used for cellulose, charcoal, fire wood, and chip boards. From the seventies to the nineties short fibre use in cellulose production increased fivefold (Lima, 1993), and from 1992 to 1996 wood pulp production rose by 20% (FAO, 1996). It is estimated that eucalypt production would still increase in the future.

Nearly all plantations are concentrated in south and central Brazil. In the Amazon region exist two big companies, of which the Jari Cellulose S.A. ("Jari project") is the best known (Fig. 1). It was founded in 1967, when D. K. Ludwig bought an area of more than 1,680,000 ha located at the Jari river, close to the junction with the Amazon river. To meet the world-wide growing demand for cellulose, a large-scale plantation system was established. The biggest part of the area even today is covered by natural forest, but over 100,000 ha are intensively used for plantation forestry.

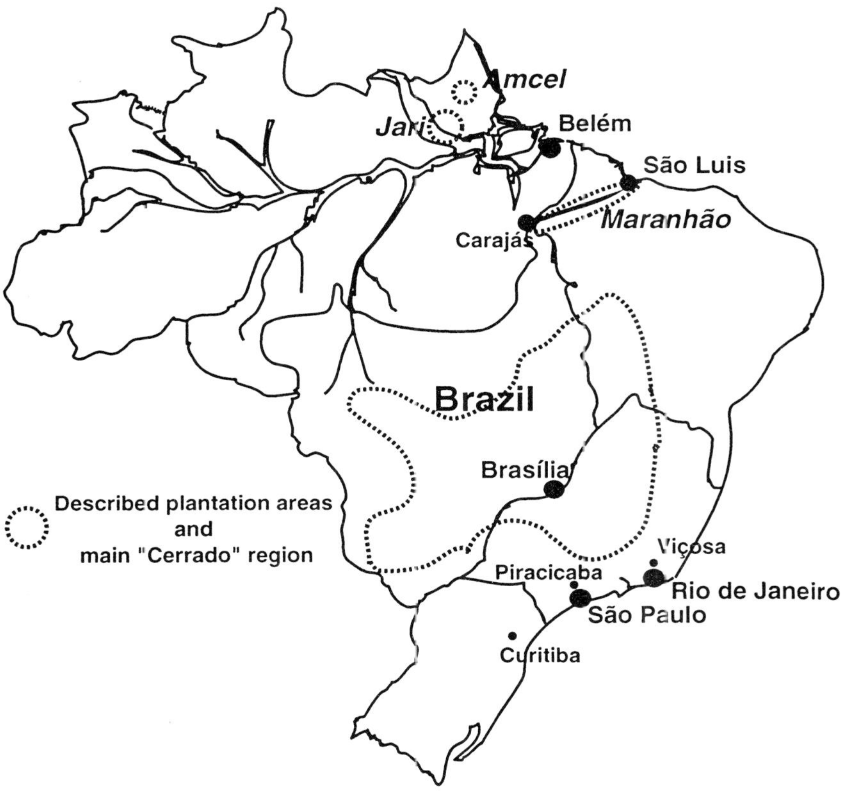

Fig. 1 Map of Brazil with described plantation areas in Amazonia and the "Cerrado" region. Besides Jari, with mainly *Eucalyptus* plantations, there exists a second company in eastern Amazonia, the "Amcel", producing mainly *Pinus* wood.

Ludwig began with *Gmelina arborea*, which reached an annual yield of 12.6 t ha^{-1} but only on the small percent of clayey soils in Jari. Ludwig expected an average productivity of 15.6 t ha^{-1} year^{-1} that was reached in Nigeria, Malaysia, and the Philippines by *Gmelina* (Jordan and Russel, 1989). After several provenance trials (Batista and Woessner, 1980) *Pinus caribaea* was successfully planted on the more sandy soil types. Creating a new hybrid, *E. urograndis* (*E. grandis X E. urophylla*), brought the final possibility of producing the necessary amount of wood for supplying Jari's own cellulose factory. Today some areas have already reached a sequence of 5 rotations.

At the project's initial phase and later, nature conservationists and economists discussed the scope of such large plantations in the humid tropics (Kalish, 1979; Fearnside and Rankin, 1980; Jordan, 1985; Jordan and Russell, 1989). In 1982, the complete project with kaolin, bauxite mining, and forestry products was sold to a consortium of 23 Brazilian industries. But in fact, the project became exemplary for other large-scale plantations in the humid tropics globally, in Brazil among others for the "Florestas Rio Doce S.A.", a subsidiary of the "Companhia Vale Rio Doce". During the eighties, this company had built the railway between the port of São Luis and the huge mining area of Carajás. The growing steel work industry increased the demand for fuel wood and charcoal, for which a network of plantations were planned along the railway.

Centres of plantation forestry were and still are the "Cerrado" (savannah) regions (Fig. 1), spread over 180 million ha in central Brazil up to the Northeast (Maranhão), and the "Mata Atlântica", stretching over 500 km along the coast from Rio Grande do Sul to Ceará. Very often, new plantations were founded to reforest degraded areas. Advantages and disadvantages of cultivating such risky areas were discussed carefully, but the general belief in helpful fertilizer application was strong (Mello *et al.*, 1970). In fact, the demand for quick supply of wood is greater in central Brazil where more than 90% of the industries are located. Therefore an absolute need of plantation forestry exists in this region (Poggiani, 1976).

BRAZILIAN PLANTATION FORESTRY RESEARCH

After establishing large-scale plantations, scientific interest in eucalypts increased and an active research began, starting from the universities of Viçosa, Curitiba and Piracicaba. Till the beginning of the eighties this interest focused on yields and biomass production of different eucalypt species and provenances. Up to today, Brazilian scientists search for optimal plant spacing to reach best growth efficiency (Rezende *et al.*, 1983, Bernardo 1995, Leles, 1995). A spacing of 3 × 2 m is often favoured (Leles, 1995).

Most soils in Brazil that are planted with plantation species, belong to the group of Oxisols, generally deeply weathered, with nutrient deficiencies and water availability as limiting factors for plantation forestry. The latter is considered to be more important in the southern regions with mean annual precipitation varying from 800 to 1500 mm, than in the north (Amazonia) with about 2000 mm. Research interests began to concentrate on soil fertility, when it became evident that wood production was often limited by soil nutrient stores (Poggiani, 1980, Poggiani *et al.*, 1983). Scientific studies investigated nutrient concentrations of tree compartments and the amount of N, P, K, Mg and Ca necessary to produce a certain amount of biomass (Barros *et al.*, 1986). A quotient of biomass and nutrient concentration leads to the question of luxury consumption, and of how many nutrients must be available in the soil for a satisfactory tree growth. It was found that element concentrations in stems of various provenances of *E. grandis* and *E. saligna* showed higher values on better sites than on poorer ones (Santana, 1994). For this trial, three even-aged (6.5 years old) provenances of *E. grandis* and of *E. saligna* were planted on each of five differently fertile sites. The variation in stem nutrient contents between the differently fertile sites was distinctly higher than the variation between the provenances (Santana, 1994). A sort of luxury consumption was obvious.

Once the nutrient status of soils was found to be the most limiting factor for continuous production in Brazil, scientific studies investigated the essential amounts of fertilizer application. The aim was no longer only an increase of growth but the compensation of nutrient losses due to export of wood and bark. This basis was then used to estimate the number of feasible rotations, assuming constant conditions (Rezende *et al.*, 1983). At this time, the high Ca concentrations in the bark of eucalypt species had already been detected, but remained without consequences for plantation management (Poggiani, 1980). Soil analyses increased to estimate the total nutrient availability. The main results manifested the low nutrient content in soil, especially of P and K. Phosphate was prognosed to be the main limiting element in the Cerrado region (Barros *et al.*, 1981; Leal *et al.*, 1988). Consequently, fertilizer experiments with raw phosphate and NPK mixtures at different plantation ages were carried out. Reis and Barros (1990) stated, that by application of one element to poor soils, the uptake of all other nutrients increased. This was supported by a study by Galo (1993) in the Cerrado region with *E. grandis.*

Presently, amounts and fertilizer applied in short-term plantations generally vary depending on the soil type, species selection, and production target (Gonçalves, 1995). As in other tropical countries, routine application of multi-nutrient compound fertilisers is widespread in Brazil. The "Jari Florestal" distributed about 200–400 kg ha^{-1} superphosphate and 100 kg ha^{-1} N:P:K 25:00:25, when some plantations had already reached the fourth rotation. For first planting after either pasture, secondary vegetation, or rainforest, no fertilizer is used except NPK mixtures directly applied to planting holes (e.g. at the Companhia

Rio Doce 100g NPK: 8-28-6). Typical ranges in plantations in the Cerrado region in Brazil are N 30–100; P 30–80; K 30–60 kg ha^{-1} (Gonçalves *et al.*, 1997). Sometimes trace elements are considered, B 0.5–1.0 and Zn 1.0–1.5 kg ha^{-1} (Gonçalves *et al.*, 1997). While N application is often described as important for eucalypts on sandy and shallow soils, for example in South Africa (Schönau, 1983), Brazilian soils especially need P and K, (Galo, 1993; Santana, 1994; Barros *et al.*, 1981) and, a more recent finding, Ca (Leal *et al.*, 1988; Spangenberg *et al.*, 1996). Especially Ca and P fertilisation are problematic because of high insolubility and no or slow mobility in the soil. Application of eucalypt litter ash showed a complete insolubility of P, and secondly a linear dependency of Ca solubility on ash dilution (Khanna *et al.*, 1994). However, the main effect of ash was to replace exchangeable Al with K, Ca and Mg added in the ash and, furthermore, a subsequent pH increase. Some recent innovative Brazilian studies investigate the application of sludge from pulp industries (Fabres *et al.*, 1994), organic waste of garbage collection in cities (Zen *et al.*, 1994), slag derived from steel industries (Oliveira *et al.*, 1994), and ash obtained from biomass burning (Moro and Gonçalves, 1995; all cited in Gonçalves *et al.*, 1997).

At the beginning of the nineties, many plantations were already in their 4th or 5th rotation. Especially poor sites showed a general decrease of nutrient availability although they received a controlled nutrient supply. In order to get an idea of the nutrient exports, the concept of quantitative assessment of nutrients was introduced by comparing exports, actual nutrient stores and inputs. Based on these nutrient balances, economic analyses showed that the necessary fertilizer application to compensate losses may cause higher costs than expected (Mackensen, 1998). The fertilizer consumption did not rise in Brazil, although their costs decreased (Louzada and Pacheco, 1992, cost-benefit analysis of a company in North Bahia) and their profit for yield production was evident. The general trend to cost reduction is evident in all plantation managements. When compensation of nutrient export will increase management costs, it becomes imperative to introduce nutrient-saving management techniques, since the more elements remain in the internal cycling, the less has to be replaced by expensive fertilizer application. This leads to an optimization of the nutrient management.

STRATEGIES OF NUTRIENT MANAGEMENT IN PLANTATION FORESTRY IN THE TROPICS

Nutrient management has to cover different aspects of nutrient dynamics in tropical plantation systems:

1. Assessment of nutrient losses and their compensation for sustained production.

2. Nutrient saving by maintaining as many nutrients as possible within the internal cycle.

3. Optimization of fertilizer application.

ASSESSMENT OF NUTRIENT LOSSES AND THEIR COMPENSATION FOR SUSTAINED PRODUCTION

Compensatory fertilizing is necessary on most tropical soils to sustain production rates, especially after the first and second rotation (Barros *et al.*, 1992). Recognizing plantation forestry with fast growing species as rather intensive land use requires to follow the old agricultural concept of replacing whatever nutrients are exported from the field in order to sustain the production level. This concept has to take into account the restricted efficiency of the applied fertilizer nutrients (leaching and fixation) but also the possible additional loss incurred during harvest by soil preparation and slash burning (volatilization and decomposition), leaching, and soil and ash erosion.

Like in agriculture, there are two possible approaches to assess fertilizer requirements, (i) foliar diagnosis and soil nutrient status and (ii) an estimate of the nutrient losses.

Diagnosis of fertilizer requirements can be based on foliar analysis and nutrient status of the topsoil as indicators. Although the nutrient status of the 0–20 cm soil layer is related to tree growth in short rotations (Gonçalves *et al.*, 1990), it may be insufficient to estimate the future increment (Cardenás, 1987). The topsoil is decisive during the first months after plantation (Melo, 1984), but subsoil root distribution of aged trees may reach >100 cm soil depth.

The concept of foliar analysis has been inherited from temperate countries though there it is less commonly applied. A direct relationship between site factors, growth, and foliar concentrations has often been observed (Herbert, 1991; Schönau and Herbert, 1982; Haag *et al.*, 1976; Lamb, 1977). Fertilizer trials gave evidence of the influence of additional nutrients on growth rates and foliar concentrations (Herbert, 1983; Herbert and Schönau, 1991; Bellote, 1990; Barros *et al.*, 1990). An international data base would be helpful for checking the validity of indicator values, and in case of satisfactory agreement this could be a practical tool for plantation managers to meet the individual demand of their stands. *E. grandis* is one of the mainly used species in tropical countries. Optimal foliar nitrogen and calcium concentrations found in it have been described as follows: N 20.0–22.2 mg g^{-1} (Haag and Mello, 1963; Lamb, 1977 (E. deglupta); Schönau and Herbert, 1982), Ca

11.5–16.0 mg g^{-1} (Haag and Mello, 1963; Herbert and Schönau, 1991). Apart from others, some Brazilian plantations fall short of the recommended Ca optimum (Spangenberg *et al.*, 1996; Haag *et al.*, 1976; Poggiani *et al.*, 1984; Bellote, 1990), which points to the real Ca depletion of many Brazilian soils (Radambrasil, 1974).

The simplest estimate of nutrient losses and the required compensation is a comparison between the expected nutrient export by wood and bark harvest and the nutrients still stored in the soil. In South America, this approach had already been applied by Chijike (1980), Hase and Fölster (1983) and Vibrans (1986). Fassbender and Bornemisza (1987) proposed to use the ratio of exported elements to its soil store as a stability indicator. Estimating export rates will require—besides the expected quantities of wood and bark—information on the element concentrations. Again, it would be useful for the plantation manager to initiate an international data base on element concentrations in wood and bark of different species, sites and age in order to facilitate this calculation.

Exported nutrients represent the main loss but not the only one. Further loss is incurred by volatilization during burning of the residual biomass, by leaching of soluble ash, by erosion of the ash, by decomposition and leaching during land preparation, as well as by continued leaching during the start of the following rotation (transformation losses). Till today, burning of the residual biomass and heavy machinery for land preparation is quite commonly applied, globally. Mackensen (1994) measured a loss of nearly 100% of the N contained in the slash, and of 25% Ca, 43% Mg, and 31% K. However, the losses vary with temperature and wind impact. Klinge (1998) studied leaching loss at the same site in eastern Amazonia. Volatilisation loss was twofold higher than leaching in case of N, but smaller (10 to 60%) in case of the base cations. But these transformation losses are site-, species- and management-dependent. As their measurement demands high investments in equipment and time, only limited data exists.

In the following, three examples are selected containing a juxtaposition of nutrient export and soil store enriched by a consideration of the loss on burning the slash. The examples are the "Jari project" in eastern Amazonia (mean annual precipitation of 2100 mm, Figs. 2 and 4), a site at Maranhão (the southeastern fringe of Amazonia, 1600–2000 mm, Figs. 3 and 5), and a site at the "Cerrado" in central Brazil (mean annual precipitation of 1500 mm, Table 2).

Figure 2 gives an overview of nutrient amounts in tree compartments (*E. urograndis*), soil and litter at Jari. All values derive from studies involving 12 plots of different soil nutrient stores and consequently different yields (data by Spangenberg *et al.* 1996). The 12 plots were grouped into 6 fields of low ("poor plots") and 6 of higher ("rich plots")

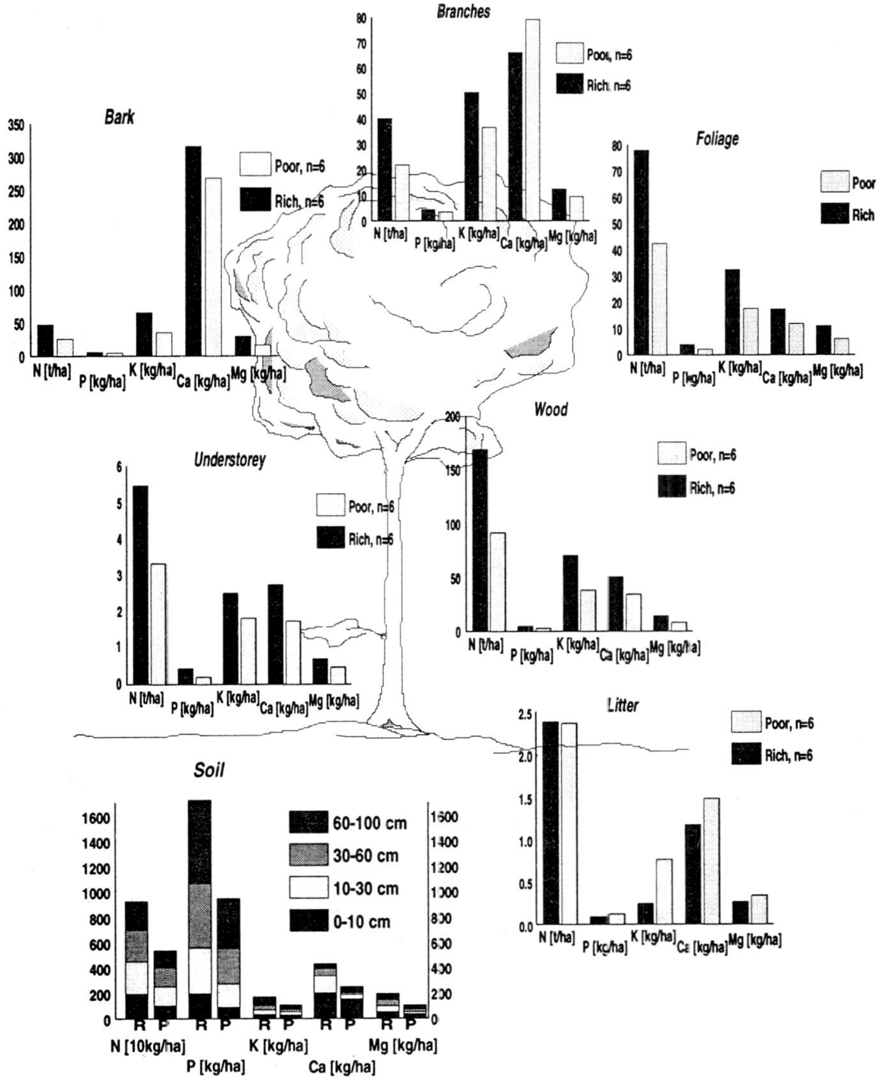

Fig. 2 Nutrient amounts in tree compartments, soil stores, understorey and litter of 4.5 years old *E. urograndis* (*E. grandis* × *E. urophylla*) plantations at the Jari project (Jari Florestal S.A.), Pará. Detailed soil description and climatic diagram are given by Spangenberg *et al.* (1996). Biomass "rich" (R) and "poor" (P) plots.

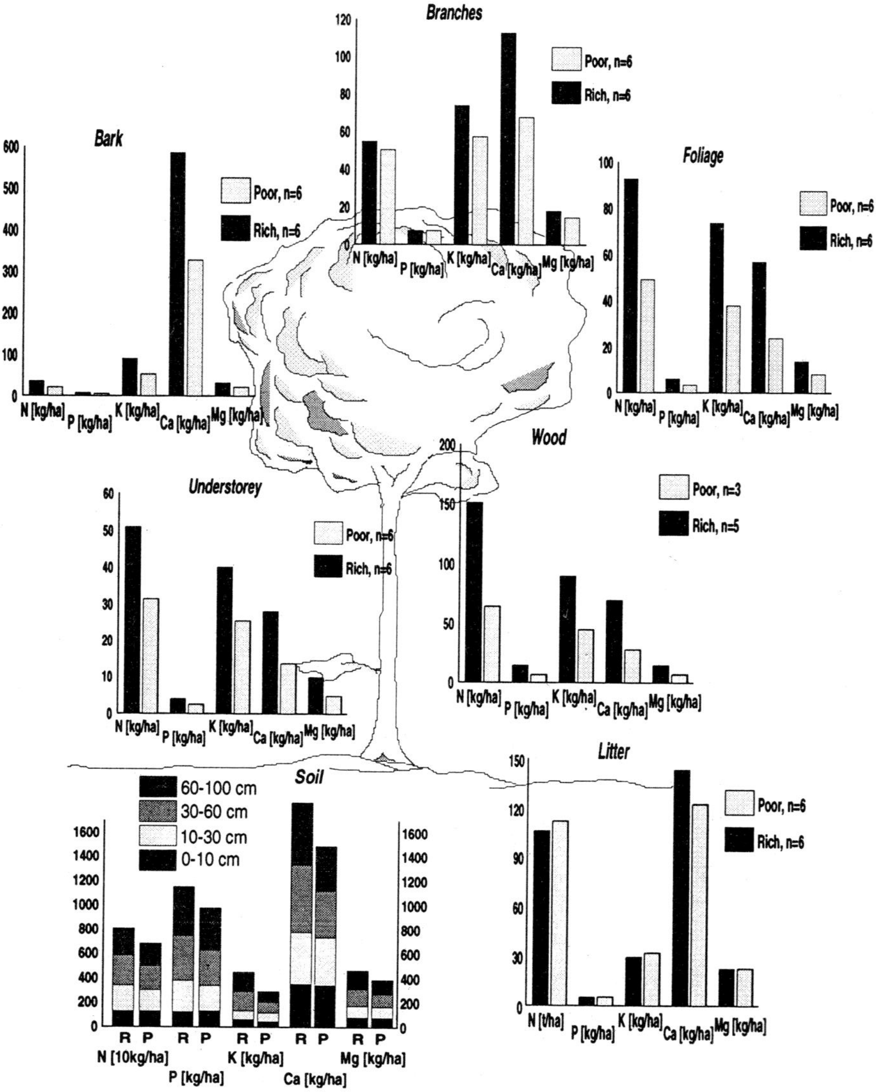

Fig. 3 Nutrient amounts in tree compartments, soil stores, understorey and litter of about 6 years old *E. tereticornis* plantations at Maranhão (Florestas Rio Doce S.A.), southeastern Amazonia. Data by Mügge, 1995. Biomass "rich" (R) and "poor" (P) plots.

biomass production, respectively. Generally, average nutrient accumulation is higher on rich sites, in soil as well as in tree compartments. Thus, the expression "rich" can be related to either the

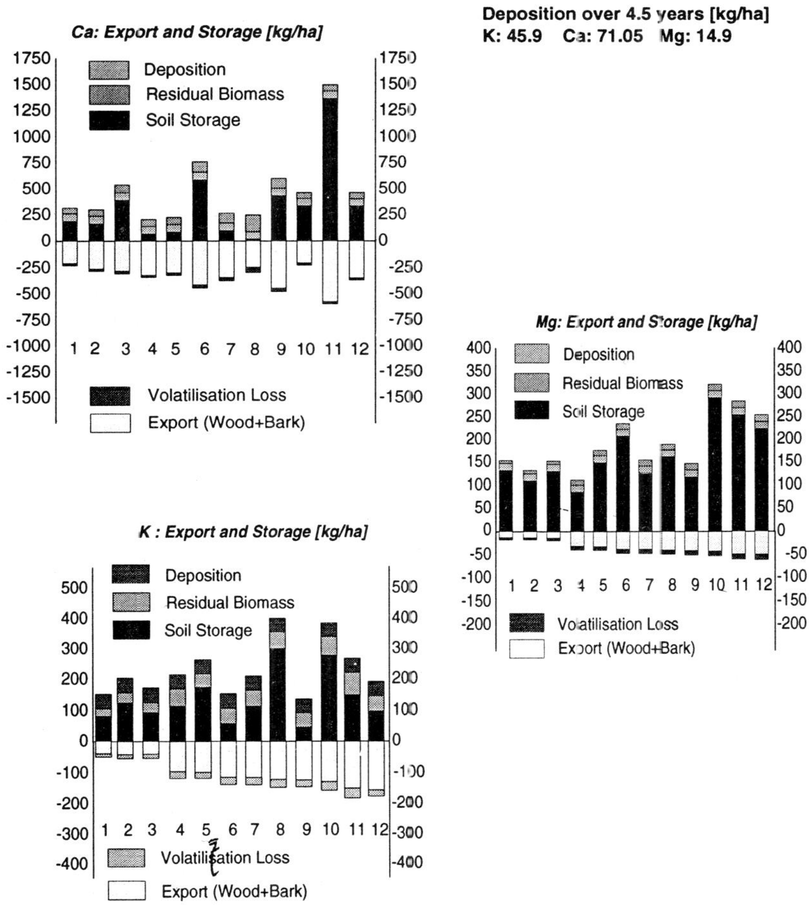

Fig. 4 Nutrient balance of inputs and outputs at the Jari Project (12 plots, sorted by increasing total biomass). Deposition values are calculated for 4.5 years according to Russell (1983). All other values are momentary pools and no fluxes. Volatilization losses are estimations according to Mackensen (1994), where 24.9% (Ca), 31.3% (K) and 42.9% (Mg) losses are pronounced. "Residual biomass" contains all tree compartments (except wood and bark), understorey and litter, which are burned together. Volatilization losses due to burning are subtracted from these values. Amounts of total N and P are not involved in budget calculations because their actual plant availability is uncertain as the fixed proportion remains unknown.

productivity or the soil fertility. Soil nutrient store under stands with high biomass production amounts to about the double of soil store under low biomass stands. Base cation stores are very low, especially in the subsoils (>30 cm). Only one plot (see plot 11, Fig. 4) represents a very clayey (80%) kaolinitic soil (first rotation).

Twelve plots of *E. tereticornis* plantations (first rotation after either pasture or young secondary forest) were investigated in southeastern Amazonia (between Marabá, Açailandia, Pindaré-Mirim and São Luis). The general distribution is similar to that at Jari (Fig. 3), except that the P stores in the soil are slightly lower and the Ca stores significantly higher. Both eucalypt species accumulate high Ca amounts in the bark, which is exported due to stem harvesting. The distinctly greater Ca soil store in most of the plots at Maranhão can be related to parent rock and lower precipitation.

The higher accumulation of Ca in the bark at the latter site reflects this difference in soil conditions, because total wood and bark biomass are very similar (mean annual increments at Jari: 5.7–21.5 t ha^{-1} y^{-1}, Maranhão: 3.4–22.75 t ha^{-1} y^{-1}), while Ca concentrations in bark are distinctly higher at the Maranhão site (mean at Jari: 24.8 mg g^{-1}, Maranhão: 41.2 mg g^{-1}). Among all macronutrients, Ca always reaches the highest export rates.

In Figs. 4 and 5, assets (soil stores, slash, in case of Jari also atmospheric deposition) are compared to the loss by wood and bark export and nutrient volatilization during the next rotation. Other inputs (weathering, fertilizer) or outputs (leaching) have not been considered. The 12 plots of each study are arranged according to increasing biomass production. While several plots of the central Amazon (Jari) indicate insufficient Ca amounts in soil to supply the next one or two rotations, this applies only to few plots of the southeastern Amazon. Next to Ca, K seems to be another limiting element.

In an *E. grandis* plantation at Espírito Santo, south of the Maranhão site, Cardenás (1987) determined mean export rates [kg ha^{-1}] of 7 (P), 49 (Mg) to 102 (K) and 404 (Ca) kg ha^{-1}. Calculating the possible number of future rotations by comparing soil stores and export rates, Cardenás (1987) concluded that the next generations will be mostly endangered by lack of P and K.

Table 2 allows a comparison with data of the "Cerrado" region (Riocell S/A, Rio Grande do Sul) by Melo (1984). The 8 plots of 7 year old *E. saligna* stands vary considerably in regard to base cation soil stores and export (wood only). The ratio is tightest in case of K. With the bark remaining at the site, the slash contains more base cations than the exported wood. Burning would reduce this asset of slowly released nutrients considerably in case of K and Mg.

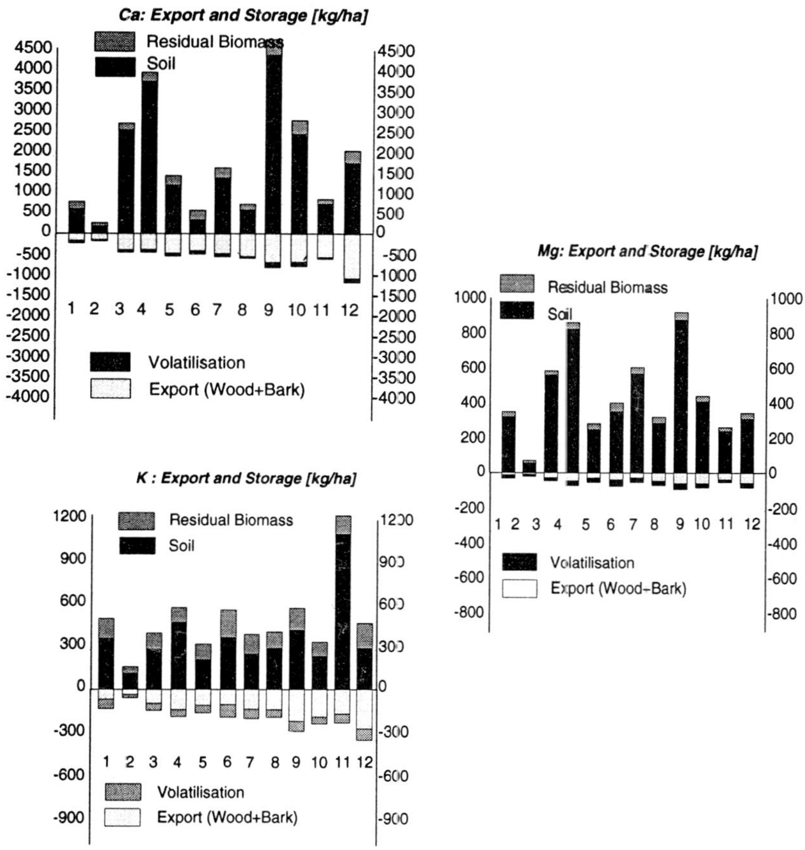

Fig. 5 Nutrient balance of inputs and outputs at Maranhão (Florestas Rio Doce S.A., 12 plots, sorted by increasing total biomass). Volatilization losses are estimations according to Mackensen (1994). "Residual biomass" contains all tree compartments (except wood and bark), understorey and litter, which are burned together. Volatilization losses due to burning are subtracted from these values. Amounts of total N and P are not involved in budget calculations because their actual plant availability is uncertain as the fixed proportion remains unknown.

Generally, the comparisons show that soil stores should not be reckoned with for sustained production. The minimum compensation rates should follow the export loss taking into account fertilizer efficiency.

Saving Nutrients within the Internal Cycle

A study by Silva *et al.* (1983) on poor soils of the "Cerrado" region describes Ca contents in the bark of five ten years old eucalypt species. They varied from 30 kg ha^{-1} (*E. propinqua*) to 53 kg ha^{-1} (*E. robusta*), 70 kg ha^{-1} (*E. dunnii*) and 71 kg ha^{-1} (*E. grandis*) to 102 kg ha^{-1} (*E. saligna*). Among these species *E. grandis* and *E. saligna* showed highest biomass production (mean annual increment of 18 t ha^{-1}) and *E. grandis* the

Table 1 Mean litter production and stand productivity of 4 Brazilian sites (variance). Data by [1]Spangenberg *et al.* (1996), [2]Mügge (1995), [3]Cardenás (1987) and [4]Ferreira (1984).

Location	*Species, age [years]*	*Stand productivity [t ha^{-1} a^{-1}] (var.)*	*Litter production [t ha^{-1} a^{-1}] (var.)*
Jari[1] n=12	*E. urograndis*, 4.5	16 (6–22)	0.064 (0.04–0.12)
Maranhão[2] n=12	*E. tereticornis*, 5.9	17 (3–23)	1.9 (1.1–3.0)
Espirito Santo[3] n=17	*E. grandis*, 7.0	15 (0.5–28)	2.2 (1.4–2.8)
Cerrado[4] n=2	*E. grandis*, about 3.0	12 (7–16)	1.5 (1.5–1.6) (litterfall) 2.1 (1.6–2.6) (humus)

Table 2 Mean values and variance of a nutrient balance of the Cerrado region by Melo (1994). Seven years old *E. saligna* stand at the Riocell S/A, Rio Grande do Sul. Volatilization loss = litter, canopy and bark-Vol. according to Mackensen (1994).

Nutrient	*Plots (n=4) [kg ha^{-1}]*		
	K	*Mg*	*Ca*
Soil store (0–100 cm)	234 (40–580)	670 (250–1400)	738 (200–2050)
Litter, Bark and Canopy	150 (95–215)	108 (90–140)	568 (420–710)
Wood (direct export)	116 (65–165)	26 (18–35)	115 (65–150)
Volatilization losses	47 (30–67)	46 (39–60)	142 (105–177)

highest nutrient accumulation. Compared to this, even the lowest Ca contents in bark of *E. urograndis* at Jari ranged between 160 and 190 kg ha^{-1}, at an age of 4.5 years and a mean annual increment of 7–21 t ha^{-1}. The mean Ca contents were 270 kg ha^{-1} (poor) and 320 kg ha^{-1} (rich), at the Maranhão site even 330 kg ha^{-1} and 590 kg ha^{-1}, respectively.

These examples show clearly that in the selection of species, not only silvicultural but also nutritional aspects play a role. Species selection may help to reduce luxury consumption (Leles, 1995). In a more general way the examples emphasize the importance of leaving the bark and crown at the site for a slow release of nutrients. If unavoidable, using the bark for energy supply should also be considered under this aspect, but ash should at least be returned to the field.

Burning the crown slash results in a loss of nutrients to the atmosphere. The quantities can be considerable (N) but generally reach a level, which is, calculated as fertilizer cost, economically relevant. The loss will be enhanced, if the remaining ash is easily eroded (Gonçalves *et al.*, 1997) or leached (Hölscher *et al.*, 1997; Klinge, 1998) at a time when the root system of the new rotation is still weakly developed. Unburned slash, on the other hand, besides only slowly releasing the nutrients, serves as a conservator of soil (erosion) and soil humus (Poggiani and Couto, 1983; Fölster and Khanna, 1997), and forces plantation management to employ a more conservative mode of land preparation (minimal soil disturbance). As a result, slash burning has already been reduced on many plantation sites in Brazil.

Further possibilities to save or gain nutrients could be the use of N-fixing plants as undergrowth (Couto *et al.*, 1982), by species mixtures (Binkley *et al.*, 1992; Khanna, 1995) or by alternating species in successive rotations, as well as the use of spontaneous ground vegetation in newly established rotations as temporary nutrient stores.

The internal cycling involves several transformation processes. Their products constitute nutrient fluxes on a solution and gaseous basis. Total amount and composition of litter represent only the small visible part of the internal cycling system and can thus function as indication of the system's status. If for example the rate of decomposition is lower than that of litter production, organic matter will accumulate resulting in an immobilization of nutrients which may lead to nutrient deficiencies on originally poor soils. The rate of decomposition depends on several factors, such as biological (soil fauna, fungi, bacteria), physical (temperature, moisture content) and chemical (nutrient status of soil and litter).

Litter collections relatively are rare in Brazil. The few available data are hardly comparable because of method incompatibilities. Taking this into account some data of the Amazon and southern savannah region may be discussed. The litter amount of the central Amazon site (Fig. 1) is extremely low, which may also be a consequence of young age. Litterfall is generally low during the first year after establishment of a plantation because of the canopy developing phase. With increasing age, woody litterfall becomes more important, although the withdrawal of nutrients rises. In contrast to the Jari site, a momentary litter inventory at the Maranhão site already shows a 10 fold higher value (Fig. 2), and at Espírito Santo (middle of Brazil) 17 plots of 7 year old *E. grandis* had a mean litter production of 2.2 [t ha^{-1} $year^{-1}$] (Table1). In the south ("Cerrado"), a maximum of annual litterfall of fresh leaves and branches of *E. grandis* was observed at an age of about 3 years, showing a mean amount of 4.6 t ha^{-1} (Ferreira, 1984). The additionally measured humus accumulation varied between 7.8 (rich site) and 6.8 (poor site) t ha^{-1} (Table 1). Although data from the Amazon regions are momentary total inventories and represent no annual flux, such high values as found in the south of Brazil cannot be accumulated over the year. The amount of litter seems to rise with increasing distance from the equator. In contrast to the humid and always warm Amazon region, the South has a seasonal climate with partly lower turnover rates.

Optimization of Fertilizer Application

This section is of course related to previous considerations, because there is no clear difference between optimization and compensation as described earlier in this chapter. But one should differentiate between maintaining sustained production and optimizing production. The first can ideally be calculated, the latter must be found out by experimentation.

In Brazil, amounts and types of fertilizers applied to a plantation vary depending on soil, species and production target (Gonçalves, 1995). Nevertheless, a standard recipe of compound fertilizer is often applied (compare fertilizer amounts at Jari, section 3.0). The indispensable specific adaptation can probably be improved in most cases by increased knowledge about nutrient fluxes and fertilizer efficiency (Mackensen, 1998). However, such adaptation is not facilitated by compound fertilizers with fixed element ratios. Moreover, nutrient supply could be reduced by following the growth and demand characteristics of a plantation (Gonçalves *et al.*, 1997). Fertilizer use is often reduced to an initial application at planting time, because of a vigorous effect on the development of the rooting system. But as the effect is short-lived (Novais *et al.*, 1983), the dose should be split into another application before canopy closure (Gonçalves *et al.*, 1997). Research on this aspect of timing will have to continue.

One should also take into account that optimization dose not necessarily mean the highest production level but could also focus on optimal production/cost ratios.

SYNTHESIS

Nutrient budgets provide an important tool to get information on nutrient fluxes and necessary changes in nutrient supply. The extent of these changes depends on the species and site, individual production target, the available nutrient stores of the site and the production implied nutrient losses. Soils used in Brazilian plantation forestry are often deeply weathered and show very low fertility. The sustainability of the system is often endangered by lack of Ca, K, P and sometimes even Mg due to transformation and export (harvest) losses. Especially the base cations have to be compensated by fertilizer application. Two case studies at eastern Amazonia showed that harvest losses are generally higher than transformation losses due to volatilization and leaching. Volatilization occurred at the study sites during burning of the residual biomass (crown slash and litter), and leaching is caused by soluble particle transport of the ash and after the clear cutting. Thus, recommendations for reducing nutrient losses and introducing nutrient-saving techniques are minimization of bark, slash and litter export, burning, and clear cutting at the plantation age of minimal nutrient concentrations in tree compartments. Minimal site preparation for avoiding wind or water erosion is also recommended (for example bulldozing is known to remove 58% of K, 80% of Ca and 67% of the Mg storage in topsoil at Jari (Russell, 1987)). Additionally, the internal nutrient cycle should be enriched by calculated use of ground vegetation as intermediate nutrient storage and by use of leguminous trees or understorey for N-fixation (Couto *et al.*, 1982; Fölster and Khanna, 1997; Khanna, 1995). These points include suitable guidelines for optimisng nutrient management in plantations.

High costs for nutrient compensations may then be avoidable. The additional ecological use of such "soft" plantation forestry should also count if deforestation of other forest systems continues and wood production is reduced to a certain extent.

ACKNOWLEDGEMENT

Thanks are due to Mrs. Celsa Zanuncio Protil for making part of the Brazilian literature accessible.

REFERENCES

Barros, N.F. de, Braga, J.M., Brandi, R.M., and Defelipo, B.V., 1981. Produção de eucalypto em solos de cerrados em resposta à aplicação de NPK e de Zn, *Revista Árvore,* **5**(1): 90–103.

Barros, N.F. de, Novais, R.F. de, Carmo, D.N. do and Neves, J.C.L., 1986. Classificação Nutricional de Sítios Florestais–Descrição de uma Metodologia, *Revista Árvore,* **10**(1): 112–120.

Barros, N.F., Novais, R.F., and Neves, J.C.L., 1990. Fertilização e correção do solo para o plantio de eucalypto, In: Relação Solo-Eucalypto. Folha de Viçosa (Barros, N.F., and Novais, R.F., eds), Viçosa, 127–186.

Barros, N.F., Novais, R.F., Neves, J.C.L., and Leal, P.G.L., 1992. Fertilising eucalypt plantations on the Brazilian savannah soils, *South African Forestry Journal,* **160:** 7–12.

Batista, M.P., and Woessner, R.A., 1980. Comparação do Incremento em Altura e Diâmetro de Espécies Novas e Exóticas no Nordeste do Pará, Brasil, *Floresta,* **11**(1): 24–32.

Bellote, A.F.J., 1990. Nährelementversorgung und Wuchsleistung von gedüngten Eucalyptus grandis-Plantagen im Cerrado von São Paulo (Brasilien). *Freiburger Bodenkundliche Abhandlungen,* vol. 26, Albert-Ludwigs-Universität Freiburgi.Breisgau, Germany.

Bernardo A.L., 1995. *Crescimento e Eficiência Nutricional de Eucalyptus spp. Sob Diferentes Espaçamentos na Região de Cerrado de Minas Gerais,* Viçosa, Minas Gerais, Brasil.

Binkley, D., Dunkin, K.A., DeBell, D., and Ryan, M.G., 1992. Production and Nutrient Cycling in Mixed Plantations of *Eucalyptus* and *Albizia* in Hawaii, *Forest Science,* **38**(2): 393–408.

Cardenás, A.C., 1987. *Exportação de Nutrientes e Produtividade de Povoamentos de Eucalypto no Litoral Norte do Espírito Santo,* Viçosa, Minas Gerais, Brasil.

Chijike, E.O., 1980. Impact on soils of fast growing species in lowland humid tropics, Vol. 21, *FAO Forestry Paper,* 111 pp.

Couto, L., Barros, N.F. de, and Rezende, G.C., 1982. Interplanting Soybean with Eucalypt as a 2-tier Agroforestry Venture in southeastern Brazil, *Aust. For. Res.,* **12:** 329–323.

Fabres, A.S., Couto, C., and Conceição, D.A., 1994. Uso de risíduo industrial de celulose em florestas, In: *Uso de Risíduos Industriais e Urbanos em Florestas,* (I.A., Guerrini, A.F.J. Bellote, and L.T. Bül, eds.), FCA/UNESP, Botucatu, 121–140 pp.

Fassbender, H.W., and Bornemisza, E., 1987. *Quimica de suelos con enfasis en suelos de America Latina,* Instituto Interamericano de Cooperacion para la Agricultura (IICA), San José, 420 pp.

FAO, 1981. Eucalypts for planting, *FAO Forestry Series no.* 11, Food and Agriculture Organization of the United Nations, Rome.

FAO, 1996. Forest products, 1992–1996, *FAO Forestry Series no.* 31, Food and Agriculture Organization of the United Nations, Rome.

Fearnside, P.M., and Rankin, J.M., 1980. Jari and Development in the Brazilian Amazon, *Interciencia,* 5(3): 146–156.

Ferreira, Maria das Graças M., 1984. An analysis of the future productivity of *Eucalyptus grandis* plantations in the "Cerrado" region in Brazil: a nutrient cycling approach, Ph.D. thesis, University of British Columbia.

Fölster, H., and Khanna, P.K., 1997. Dynamics of Nutrient Supply in Plantation Soils, In: *Management of Soil, Nutrients and Water in Tropical Plantation Forests,* (Nambiar, E.K.S., and Brown, A.G., eds.), ACIAR and CSIRO Australia, Monograph No. 43, 339–378.

Galo, M.V., 1993. *Resposta do Eucalypto à Aplicação de Potássio em Solo de Cerrado,* Viçosa, Minas Gerais, Brasil.

Gonçalves, J.L.M., Demattê, J.L.I., and Couto, H.T.Z., 1990. Relações entre a produtividade de sítios florestais de *Eucalyptus grandis* e *E. saligna* com as propriedades de alguns solos de textura arenosa e média no Estado de São Paulo, IPEF 43/44, 24–39.

Gonçalves, J.L.M., 1995. Recomendações de adubação para *Eucalyptus, Pinus* e espécies típicas da Mata Atlantica, *Documentos Florestais,* **15:** 1–23.

Gonçalves, J.L.M., Barros, N.F., Nambiar, E.K.S., and Novais, R.F., 1997. Soil and Stand Management for Short-rotation Plantations, In: *Management of Soil, Nutrients and Water in Tropical Plantation Forests*, (E.K.S. Nambiar, and A.G. Brown, eds.), ACIAR and CSIRO Australia, Monograph No. 43, 379–417.

Haag, H.P., and Mello, H.A., 1963. Composição química de *E. alba* and *E. grandis*: resultados preliminares, *Fertilité*, **18:** 9–14.

Haag, H.P., Sarruge, J.R., and Oliveira, G.D. de, Poggiani, F., and Ferreira, C.A., 1976. Análise foliar em cinco espécies de eucalyptos, IPEF, *Piracicaba*, **13:** 99–116.

Hase, H., and Fölster, H., 1983. Impact of plantation forestry with teak (*Tectona grandis*) on the nutrient status of young alluvial soils in West Venezuela, *Forest Ecology and Management*, **6:** 33–57.

Herbert, M.A., 1983. The Response of *Eucalyptus grandis* to Fertilising with Nitrogen, Phosphorus, Potassium and Dolomitic Lime on a Mispah Soil Series, *South Afr. For. J.*, **124**(3): 4–12.

Herbert, M.A., 1991. The influence of site factors on the foliar nutrient content of *Eucalyptus grandis* in Natal, *South African Forestry Journal*, **156:** 28–34.

Herbert, M.A., and Schönau, A.P.G., 1991. Fertilizing commercial forest species in southern Africa: research progress and problems, In: *Mineralstoffversorgung tropischer Waldbäume*, (W. Zech, ed.), Bayreuther Bodenkundliche Berichte., **21:** 7–43.

Hölscher, D., Möller, M.R., Denich, M., and Fölster, H., 1997. Nutrient input-output budget of shifting agriculture in eastern Amazonia, Nutrient Cycling in Agroecosystems, **47:** 49–57.

Jordan, C.F., 1985. Jari: A Development Project for Pulp in the Brazilian Amazon, *The Environmental Professional*, **7:** 135–142.

Jordan, C.F., and Russell, C.E., 1989. Jari: Pulp and Plantation in the Brazilian Amazon, *Geo. Journal*, **19**(4): 429–435.

Kalish, J., 1979. The Jari (project in Northern Brazil), *Pulp and Paper International, London*, **21**(1): 37–52.

Khanna, P.K., Raison, R.J., and Falkiner, R.A., 1994. Chemical properties of ash derived from Eucalyptus litter and its effect on forest soils, *Forest Ecology and Management*, **66:** 107–125.

Khanna, P.K., 1995. Comparison of growth and nutrition of monocultures and mixed plantings of *Eucalyptus globulus* and *Acacia mearnsii*, In: *Proceedings Intern. Congress on Soils of Tropical Forest Ecosystems*, (A. Schulte, and D. Ruhiyat, eds.), 3rd Conference on Forest Soils, vol. 4, Soils and Plantation Forestry.

Klinge, R., 1998. *Wasser- und Nährstoffdynamik im Boden und Bestand beim Aufbau einer Holzplantage im östlichen Amazonasgebiet*, Ph.D thesis. Göttinger Beiträge zur Land- und Forstwirtschaft in den Tropen und Subtropen, Vol. 122.

Lamb, D., 1977. Relationships between Growth and Foliar Nutrient Concentrations in *Eucalyptus deglupta. Plant Soil*, **47:** 495–508.

Leal, P.G.L., de Barros, N.F., de Novais, R.F., Neves, J.C.L., and Teixeira, J.L., 1988. Produção de biomassa e absorção de nutrientes em *Eucalyptus grandis* influenciadas pela aplicação de fosfato natural em solos de cerrado, *Rev. Árvore.*, **12**(2): 165–182.

Leles, P.S. dos Santos, 1995. *Crescimento, Alocação de Biomassa e Distribuição de Nutrientes e Uso de Água em E. camaldulensis* e *E. pellita Sob Diferentes Espaçamentos*, Viçosa, Minas Gerais, Brasil.

Lima, Walter de Paula, 1993. *Impacto Ambiental do Eucalypto*, Editora da Universidade de São Paulo, São Paulo, Brasil.

Louzada, P.T.C., and Pacheco, R.M. 1992. Use of fertilizer in Eucalypt Plantations: Response to Applications and Consumption Evolution. IPEF International, *Piracicaba*, **2:** 25–30.

Mackensen, J., 1994. *Nährstoffverluste in die Atmosphäre beim Verbrennen von Schlagabraum in Ost-Amazonien*, Diplomarbeit der Forstl. Fakultät Göttingen, 88 pp.

Mackensen, J., 1998. *Untersuchung zur nachhaltigen Nährstoffversorgung in schnellwachsenden Plantagensystemen in Ost-Kalimantan, Indonesien*, Göttinger Beiträge zur Land- und Forstwirtschaft in den Tropen und Subtropen, 127.

Mello, H.A., Sobrinho, J.M., Simões, J.W., and Couto, H.T.Z., 1970. Resultados da Aplicação de Fertilizantes Minerais na Produção de Madeira de *Eucalyptus saligna* Sm. Em Solos de Cerrado do Estado de S. Paulo, *IPEF*, **1:** 7–28.

Melo, V. de Freitas, 1984. *Produção de Biomassa e Acumulação de Nutrientes em Florestas de Ciclos-Curtos*, Relatório Annual, SIF-IBDF, Viçosa, Minas Gerais, Brasil.

Moro, L., and Gonçalves, J.L.M., 1995. Effect of forest biomass "ash" on the productivity of *Eucalyptus grandis* plantation and financial evaluation, *IPEF.*,48/49, 18–24.

Mügge, D., 1995. *Nährstoffvorräte und –exportverluste von Eucalyptus tereticornis-Plantagen in den Bundesstaaten Maranhão und Pará, Brasilien*, Diplomarbeit, University of Göttingen.

Novais, R.F., Barros, N.F., and Neves, J.C.L., 1983. Nutrição do eucalypto e fertilisação e correção do solo para seu cultivo, In: *IUFRO/MAB/UFV. Florestas plantadas nos neotrópicos como fonte de energia*, Viçosa, UFV, 133–169.

Oliveira, A.C., Hahne, H., Barros, N.F., and Morais, E.J., 1994. Uso de escoria de alto forno como fonte de nutrientes na adubação florestal, In: *Uso de Risíduos Industriais e Urbanos em Florestas*, (I.A. Guerrini, A.F.J. Bellote, and L.T. Bül, eds.), FCA/UNESP, Botucatu, 77–96 pp.

Ondro, W.J., Couts, L., and Betters, D.R., 1995. The status and practice of forestry in Brazil in the early 1990s, *Forestry Chronicle*, **71:** 106–119.

Poggiani, F., 1976. *Ciclo de nutrientes e produtividade de florestas implantada*, Silvicultura Nov/Dec.

Poggiani, F., 1980. Florestas para fins energéticos e ciclagem de nutrientes, *IPEF, Piracicaba, Série técnica*, **1**(2): D1–D11.

Poggiani F., Couto, H.T.Z., and Suiter-Filho, W., 1983. Biomass and Nutrient Estimates in Short Rotation Intensively Cultured Plantation of *Eucalyptus grandis*, IPEF, *Piracicaba*, **23:** 37–42.

Poggiani F., Zen, S., Mendes, F.S., and França, F.S., 1984. Ciclagem e exportação de nutrientes em florestas para fins energéticos, *IPEF, Piracicaba*, **27:** 17–30.

Radambrasil, 1974. *Levantamento de Recoursos Naturais 5. Departamento Nacional da Produção Mineral*, Projeto RADAM. Folha SA-22 Belém; Geologia, geomorfologia, solos, vegetação, uso potencial da terra. Rio de Janeiro, 271 pp.

Reis, M.G.F., and Barros, N.F., 1990. Ciclagem de nutrientes em plantios de eucalypto, In: Vegetation Management Practices in Plantation Forests of Australia and New Zealand, (Barros, N.F., and Richardson, B., ed.), *Canadian Journal of Forest Research*, **23:** 1989–2005.

Rezende, G.C. de, Barros, N.F. de, Moraes, T.S.A., and Mendes, C.J., 1983. Produção e macronutrientes em florestas de eucalypto sob duas densidades de plantio, *Revista Árvore*, **7**(2): 165–176.

Russell, C.E., 1983. *Nutrient cycling and productivity of native and plantation forests at Jari Florestal, Pará, Brazil*, Ph.D thesis, University of Georgia, Athens, 133 pp.

Russell, C.E., 1987. Plantation Forestry, case study no. 9: The Jari Project, Pará, Brazil, In: *Amazonian Rain Forests: Ecosystem Disturbance and Recovery*, (C.F. Jordan, eds.), Ecological Studies, **60:** 76–89.

Santana, R.C., 1994. *Crescimento de Eficiência Nutricional de Procedências de Eucalyptus grandis e Eucalyptus saligna em Sítios de Estado de São Paulo*, Viçosa, Minas Gerais, Brasil.

Schönau A.P.G., and Herbert, M.A., 1982. Relationship between Growth Rate and Foliar Concentrations of Nitrogen, Phosphorus and Potassium for *Eucalyptus grandis*. *South Afr. For. J.*, **120**(3): 19–23.

Silva, H.D. da, Poggiani, F., and Coelho, L.C., 1983. Biomass, nutrient content and concentration in five *Eucalyptus* species cultivated on low soil fertility, *Bol. Pesq. Flor.*, Curitiba, 6/7: 9–25.

Spangenberg, A., Grimm, U., da Silva, J.R.S., and Fölster, H., 1996. Nutrient store and export rates of *Eucalyptus urograndis* plantations in eastern Amazonia (Jari), *For. Ecol. Manage.*, **80:** 225–234.

Vibrans, A., 1986. *Untersuchungen zum Nährstoffexport aus Kiefernbeständen in Südbrasilien*, Graduate thesis, University of Göttingen, Germany, 82 pp.

Zen, S., Bellote, A.F.J., Silva, E.D., and Ferreira, C.A., 1994. Resíduos urbanos como fonte de nutrientes em povoamentos de eucalypto, In: *Uso de Risíduos Industriais e Urbanos em Florestes*, (I.A. Guerrini, A.F.J. Bellote, and L.T. Bül, eds.), FCA/UNESP, Botucatu, 25–39 pp.

CHAPTER 13

NITROGEN-FIXING LEGUMINOUS TREES IN FORESTRY AND AGROFORESTRY SYSTEMS

D.L.N. Rao

Indian Institute of Soil Science, Nabi Bagh, Bhopal-462 038, India

ABSTRACT

The increasing concern over deforestation and land degradation in the tropics has given rise to intensive researches on plantation forestry and agroforestry during the last three decades. Nitrogen is the single most limiting factor for increasing productivity. Use of nitrogen fixing trees and shrubs (NFTS) offers good opportunities for optimizing the input of nitrogen but is limited by lack of knowelege on various aspects of silviculture and biological N_2 fixation (BNF) process in nitrogen fixing trees. Traditional knowledge on beneficial exploitation of NFTS is available in various cropping and plantation systems in Asia and Africa for soil fertility improvement, forage, shade, timber, fuel wood and other benefits like wind breaks, dune stabilization etc. Many scientific studies on quantification of the actual benefits/disadvantages of trees and shrubs in traditional as well as newer forms of agroforestry like alley farming have emerged in the last three decades allowing researchers to optimize the right mix of trees required for various crops in various agro-ecological zones in India and numerous developing world countries.

Most of the caesalpinoid trees do not nodulate (~70%): most of the mimosoid trees nodulate (~70%) and papilinoid trees are almost universally capable of nodulation (~90%). Nodulation surveys have been conducted worldwide. Benefits of inoculating trees nurseries with *Rhizobium* and vesicular-arbuscular mycorrhizae (VAM) have been shown by many workers. In India most of the rhizobia of arid tree legumes are thought to belong to the 'cowpea' miscellany group and have high limits of tolerance to salinity. Soil moisture is the single most important constraint in restricting nodule function and longevity. Nodulation occurs mostly in surface soils (0–15 cm) and younger trees exhibit better nodulation than older ones. Nodulation peaks in monsoon season and declines till winter. Leaf nitrogen content of ~3% distinguishes fixing from non-fixing trees in semi-arid tropics. ^{15}N isotope dilution and traditional nitrogen balance methods have been used to estimate nitrogen fixation, reported to range from 60 to more than 300 kg N/ha/year. Benefits of nutrient recycling through litter fall, direct transfer of nitrogen to companion crops as well as rotational benefits to crops grown in sequence has been demonstrated unequivocally in the field showing the benefits of culitvating nitrogen fixing trees. A noticeable lack of research on economics of tree plantations and assigning of monetary benefits for the ecological benefits accruing from planting trees is an urgent concern that needs to be addressed. The past progresses of 30 years on various aspects listed above are discussed in this article and an outlook on future research needs is presented.

Key Words: Nitrogen fixing trees & shrubs, forestry, agro forestry, nodulation, nutrient recycling.

Deforestation and general land degradation throughout much of the tropics has resulted in shortage of food, fuel wood and fodder. In India, of the 70 million ha of forests, 24.9 m ha are either completely (less than 10% forest cover) or partially (10–40% cover) degraded. During 1993, the country faced a deficit of 570 mt of green fodder, 276 mt of dry fodder, 9 mt of timber, 6 mt of pulp wood and 80 mt of fuel wood. There is an urgent need for all round efforts to increase biomass production by protecting existing forests, increasing forest productivity, afforesting marginal lands, creating energy plantations by encouraging farm forestry etc. The National Commission on Agriculture made a strong plea nearly 25 years ago for utilizing all available wastelands for growing multi-purpose trees and for intensive cultivation of woodlots so as to help create renewable firewood resources in the farm itself. Besides alleviating firewood shortages, additional ecological benefits of afforestation include soil improvement, sustenance of soil fertility, restoring traditional role of forests for providing clean air and water from watershed area, conservation of plant, microbe and wildlife diversity etc. Benefits accruing from reforestation of erodable soils for sand dune stabilization or as wind breaks, rehabilitation of mine spoils are among several other well known practical benefits of afforestation. As a result, the recent past has witnessed an explosion of interest in designing and use of forestry and agroforestry systems all over the tropics.

Nitrogen is most often the limiting factor for increasing productivity, nitrogen-fixing trees and shrubs (NFTS) offer a good opportunity to increase yield of subsequent crops in agroforestry systems. The rising cost of nitrogen fertilizers and the emissions of greenhouse gases resulting from their manufacture and use are compelling reasons for using trees as a means for producing biomass and nitrogen in an environmentally sustainable way. On a global basis, about 170 million tons of nitrogen are fixed biologically each year, about 80% of it is fixed by nodulated legumes. It has been estimated that 4100 × 10^6 ha of the earth's forest and woodlands account for 10 kg/ha per year of biological N_2 fixation or 40 million metric tons/yr (Burns and Hardy, 1975). This is 29% of the total estimated to occur in terrestrial ecosystems, compared with 32% estimated to occur in permanent meadows and grasslands, 25% in land supporting leguminous crops, and 14% in other ecosystems. However, data on N_2 fixation are available for only a limited number of forest communities (Boring *et al.*, 1988). Data from tropical forests are noticeably lacking, although recent studies on the biology and cultivation of N_2-fixing trees suggest that N_2 fixation may be relatively more important in tropical forests than in temperate forests (Dommergues *et al.*, 1984; Halliday, 1984). Legume/*Rhizobium* associations are undoubtedly the most important N_2-fixing symbiosis, and play a role of varying importance in pastures, arable cropping systems, plantations and agroforestry. Agro-forestry systems are increasingly spreading to newer areas rapidly and when one of the components (tree or crop) is a legume, considerable benefit from nitrogen fixation may accrue. BNF is now increasingly being recognized as a potential silvicultural tool and there has recently been a notable increase in the use of leguminous, N_2-fixing trees and shrubs. Cultivated genera include *Acacia, Albizzia, Bauhinia, Calliandra, Dalbergia, Erythrina, Gliricidia, Inga, Leucaena, Prosopis, Robinia* and *Sesbania* among others. Some of the advantages of a N-fixing tree are that it ensures a regular supply of available N on a sustained basis; being organic in nature, the N is less susceptible to be lost via leaching or volatilization. It increases soil organic matter through litter production, fine roots turnover etc.; improves soil aggregation etc. Some of the disadvantages may be that in comparison with application of industrially fixed nitrogen, enhancement of biological nitrogen fixation is currently limited by lack of knowledge of silvicultural aspects. Exploiting them in agroforestry systems, is likely to be more expensive on a unit applied nitrogen basis, slower in producing fertility and may be a source of vegetation competition for other species (Gordon and Wheeler, 1983). Till about 20 years ago we were ignorant about those N fixing trees which are likely to be most useful on a large scale. Intensive researches since then have led to information in most areas (Singh *et al.*, 1994).

TRADITIONAL FORESTRY/AGROFORESTRY SYSTEMS INVOLVING NFTS

Legume trees are used in agroforestry as wind shields, shade trees in plantations, to provide wood for fuel and for browsing by animals in addition to providing prunings which can be used as fodder or a green manures. Nitrogen fixing trees have been maintained or grown in various situations e.g., as trees in crop lands (spatial-mixed), as pure tree stands, as rotational or spatial-zoned systems with crops. There are several examples where the preponderance of leguminous trees in specific ecological niches has been for a large part due to their ability to fix nitrogen. Nitrogen limits plant productivity of arid soils more severely than water (Felker and Clark, 1980), because of the virtual absence of organic matter in the sandy sites. Perhaps the capability of some tree legumes to deal with nitrogen deficiency aspects of deserts has led them to co-evolve complementary drought resistant mechanisms. According to Felker (1981) there is no other theory available to explain why leguminous trees should have evolved to dominate the arid and semi-arid ecosystems. In arid parts of northern India, *Prosopis cineraria* are maintained in fields used for arable crops and *Faidherbia albida* is widely distributed in cropping fields throughout the Sahel and parts of Southern Africa. Crop yields often double beneath the tree than outside the canopy and this is largely due to the improved organic matter content of the soil and the associated improvements in moisture retention and the supply of N and other nutrients. Similar effect has been seen beneath *Acacia* trees in Australia, East Africa and the Sahel. The beneficial effects of the building up of soil fertility have been attributed to the N_2-fixing ability of the trees. *Acacia* (*Faidherbia*) *albida* is used in the more mesic (400–600 mm/yr) regions of Sahelian West Africa where yields of millet, sorghum and peanuts are increased from approximately 500 to approximately 900 kg/ha (Dancette and Poulian, 1969) due to several fold increase in soil N and organic matter content and water holding capacity under the tree. Recognition of the benefits of nitrogen-fixing (non-leguminous) tree associations to agriculture predates recognition of their N_2-fixation capabilities. Since ancient times, alder (*Alnus*) has been known as a soil-improving tree in the Aomori district of northern Japan, where cut-over alder sites are usually used as crop land (Yamaya, 1968). Indigenous farmers of Central America have also traditionally used trees in their agricultural practices.

The use of *Inga* sp. (formerly *Pithecellobium*) in coffee and *Albizzia* sp. in tea plantations as shade trees is well known as is the role of *A. tortilis, A. lebbek, Dichrostachys natans* and *Prosopis juliflora* as shelter belts and wind breaks in the Thar desert in Rajasthan. The cultivation of *Sesbania grandiflora* as support for betelvine cultivation in south India; use of *A. falcataria, L. leucocephala* for home-stead multistorey farming in high rainfall areas of coastal Karnataka and Kerala; planting of the non-leguminous nitrogen-fixing *Casuarina equisetifolia* around rice bunds in south India, and in the coastal sands of peninsular India, and the interplanting of *Sesbania sesban* in sugarcane fields in western and south India are some of the indigenous and popular practices involving N fixing trees. In the highlands of Tamilnadu: *A. melanoxylon, A. dealbata, A. decurrens* and *A.mearnsii; A. amara* at lower elevations of 1500–200 feet; *Albizzia, Calliandra, Inga, Robinia* in the Himalayas are some of the important N-fixing trees in cooler climes.

In India soils under forage crops are very low in nitrogen and fertilizer application is not a common practice because of several socio-economic constraints. In this situation inclusion of nitrogen fixing trees has been found very promising. With substantial recent emphasis on reforestation in our country, pure stands of trees have been raised by individual large farmers, voluntary organizations, forestry departments and by industries. Among other species, the NFTS most widely planted are *L. leucocephala* in many parts of India and *A. nilotica* and *P. juliflora* in arid and semi-arid areas, especially in Haryana, Punjab, Rajasthan and Gujarat for the rehabilitation of salt affected soils. The latter two are also the most widely studied species in agroforestry systems. Even commercial plantations of slow growing leguminous trees like *Dalbergia, Intsia, Pericopsis* and *Pterocarpus* which are valuable timber trees in their own right, can be inter-planted with other fast growing species like *Acacia, Acrocarpus, Albizia, Leucaena, Mimosa, Sesbania,* which can fix high amounts of nitrogen in short periods and benefit the timber trees.

Substantial emphasis has been placed in humid and sub-humid and sub-humid tropical regions of Africa on a form of spatial-zoned agroforestry known as 'alley farming' or hedge-row intercropping in which trees are grown within arable fields in rows and are pruned periodically and maintained as hedges. The prunings are used as green manure for the crops grown in alleys between the hedge rows. It is an attractive system but has not been widely adopted as yet, because it maximizes the contact between crops and trees (tree-crop interface) and the competition for both light (above ground) and for water and nutrients (below ground). Two of the most popular species used in alley farming are *Senna* (*Cassia*) *siamea* (although it is a non-nodulator) and *Gliricidia sepium.* Work at CRIDA, Hyderabad showed that *Leucaena* hedge rows (providing mulch/fodder) with sorghum/pearlmillet grown in 7 m wide alleys was a compatible system for the dry zones of Deccan. In the authors studies on alley farming in an alkali soil at Karnal in North India, only *L. leucocephala* and *Sesbania sesban* performed well whereas, *Pongamia pinnata* and *A. lebbek* were medium and *Cassia siamea* and *Dalbergia sissoo* performed poorly (Rao *et al.,* 1998).

Trees compete with crops for nutrients and moisture, the latter especially in dry areas, such that there can be negatives effects on crop growth. For example *G. sepium* widely grown on field bunds has a drying effect on the soil. In an alkali soil, Batra and Kumar (1994) showed that Rhodes grass yields were similar whether grown alone or in combination with *Acacia, Dalbergia or Casuarina* in case the grass was planted at the same time as the trees. Introducing the grass after one or two years of tree establishment reduced grass yield significantly. Vigorous growth of grass also reduced the tree growth. Total pruned biomass from trees was reduced significantly when the grass was introduced at the same time as the tree compared to introduction after one or two years of tree establishment obviously due to competition for nutrients and moisture. But the use of nodulated nitrogen-fixing trees can at least reduce root competition with crops for nitrogen. Nitrogen is a relatively mobile nutrient. If the tree obtains its supplies partly by fixation this reduces soil depletion around roots, so allowing more nitrogen to be taken up by inter-planted non-nitrogen-fixing crops.

NODULATION AND NITROGEN-FIXATION

Nodulation Potential and Rhizobia of Tree Legumes

The leguminosae forms one of the largest families of flowering plants, ranking third in terms of world-wide occurrence with about 600 genera and 18,000 species. Leguminosae is generally sub-divided into four sub-families, Caesalpinoidae, Mimosoideae, Swartzoideae and Papilionoideae. Less than 15% of the legume species have been examined for presence of nodules (Lim and Burton, 1982). Not all leguminous plants form nodules. The caesalpinoids which are mainly trees are characterized largely by non-nodulating species, about 70% lack nodules. The Mimosoideae are largely trees and shrubs and a majority of them (60–70%) are nodulated. The papilionoids which contain few trees, being mainly shrubs and herbs are almost universally capable of nodulation (90–95%). There are reportedly at least 72 genera of the Leguminosae with species indigenous to India (see Table 1 for a partial list). Many of these include woody perennial species.

Basak and Goyal (1975, 1980a) reported nodulation in 19 species indigenous to the Indian desert. Felker and Clark (1980) reported nodulation in 12 *Prosopis* species. In a nodulation survey in Western ghats in Nilgiris and Kodaikanal areas, Rangarajan *et al.*, (1984) found nodulation in 24 genera out of 30 genera surveyed. *Bauhinia, Cassia, Delonix, Hardwickia, Parkia, Parkinsonia and Peltophorum* were found to be non-nodulating. In the hilly areas of western Maharashtra nodulation was reported for the first time (Nimbalkar *et al.*, 1986) in 19 wild plants-13 shrubs, one climber (*Mucuna monosperma*), and 5 trees (*Butea monosperma, D. lanceolaria, Erythrina stricta, A. tomentosa*).

Basak and Goyal (1980a) assigned the rhizobia of tree legumes to the 'cowpea miscellany group'. However other studies showed that tree legume rhizobia do not confirm to familiar pattern in crop species. Growth rate (fast, intermediate, slow, very slow) varies both within and between host genera. Many are rather promiscuous, nodulating a variety of species, so there is likely to be a high range in nitrogen-fixing efficiency. Nodulation by fast and slow growing strains was reported in *Acacia spp.* by Dreyfus and Dommergues (1981). *A. mangium* is a specific host whereas *A. auriculaeformis* is a promiscuous host (Galiana *et al.*, 1990). Attempts to isolate rhizobia for tree legumes from soils have usually involved raising seedlings from surface sterilized seeds on agar slopes followed by inoculation of soil suspensions to induce nodulation and then isolation of rhizobia from these nodules. In such cases it is not possible to know the effectivity or competitiveness of a given strain under actual field situations. Also it is not possible to know if the strains can induce a successful symbiosis under stress conditions. Therefore, we adopted an approach to isolate rhizobia directly from effective nodules in well nodulated trees (representing 6 genera and 12 species) in specific ecological situations followed by their purification and testing (Rao *et al.*, 1998). Contrary to the general belief that rhizobia of tropical legumes are slow growing, alkali producers, it was found that majority (49 out of 70 strains tested) were fast growing and in some tree species both fast and slow growers were found to nodulate. A majority of them (61 out of 70) nodulated with cowpea supporting the contention that rhizobia of tree legumes could be placed in cowpea miscellany group.

Rhizobia of tree legumes have been reported to tolerate 0.8–1.8% NaCl and some upto 3% NaCl (Basak and Goyal, 1980b). Rao *et al.*, 1998 found that rhizobia of tree legumes were present and surviving even at high pH of 9.6 and salinity (ECe) of nearly 16–20 dS m^{-1} and in one case even upto 90 dS m^{-1}. Laboratory studies showed that they were highly tolerant to alkalinity of pH 10.5–12.0 in artificial culture media and salinity of 3–4% NaCl corresponding to electrical conductivity (EC) of 44–57 dS m^{-1}. So it appeared that like the rhizobia of annual legumes, the salt tolerance limits of rhizobia of tree legumes is very high and much above the tolerance limits of the host tree legume.

It is particularly important to introduce only the best microbial strains when inoculating in areas devoid of appropriate symbionts for the introduced host plant. The use of pure cultures of selected bacterial symbionts, rather than crushed nodule or soil inocula, can avoid the problems of introducing undesirable strains, pathogens and pests to virgin sites (Diem and Dommergues, 1990). Some tree legumes are more responsive than others, to inoculation with their specific rhizobia.

Table 1 Important nitrogen-fixing tree legumes of India (modified from Chamberlain and Brewbaker, 1988).

Sub-family	*Indigenous*	*Naturalized*	*Possibly indigenous but of uncertain species and/or N-fixing status*
CAESALPINOIDEAE	*Acrocarpus fraxinifolius*	*Tamarindus indica*	*Intsia palembanica*
	Intsia bijuga		
MIMOSOIDEAE	*Acacia leucophloea*	*Acacia mearnsii*	*Acacia ferruginea*
	Acacia nilotica	*Acacia farnesiana*	*Acacia jacquemonti*
	Acacia polycantha	*L. leucocephala*	*Acacia lenticularis*
	Acacia senegal	*Pithecellobium dulce*	*Acacia modesta*
	Albizia amara	*Prosopis juliflora*	*Acacia suma*
	Albizia chinensis	*Prosopis pallida*	*Prosopis spicigera*
	Albizia lebbek		
	Albizia lucidior		
	Albizia odoratissima		
	Albizia procera		
	Dichrostachys cinerea		
	Parkia speciosa		
	Prosopis cineraria		
PAPILIONOIDEAE	*Dalbergia latifolia*	*Gliricidia sepium*	*Caragana brevispinia*
	Dalbergia paniculata	*Robinia pseudoacacia*	*Dalbergia lanceolaria*
	Dalbergia sissoo	*Tephrosia candida*	*Dalbergia sericea*
	Desmodium latifolium		*Dalbergia sissoides*
	Erythrina arborescens		*Erythrina stricta*
	Erythrina suberosa		*Hardwickia binata*
	Hardwickia pinnata		*Inga umbrosa*
	Ougenia oojeinensis		*Pterocarpus dalbergioides*
	Piptanthus nepalensis		*Pterocarpus santalinus*
	Pongamia pinnata		
	Pterocarpus marsupium		
	Sesbania sesban & S. grandiflora		

When tree legumes were inoculated with their own rhizobial isolates under sterilized conditions (Basak and Goyal, 1980a) the response in terms of number of nodules formed was more pronounced especially in *Acacia catechu, A. nilotica, A. salicina, A. tortilis, A. victoriae, Albizzia lebbek, Dichrostachys natans.* In a pot study on red soil with *L. leucocephala* K-8, greater beneficial effect was obtained by combined inoculation of microbial inoculants (*Glomus fasciculatum, R. loti*-TAL 582 and *Pseudomonas striata*) in respect of nodulation, root length, herbage yield as well as N and P uptake by the plant (Pahwa, 1994). Dual inoculation of *Leucaena* and *Prosopis* seedlings with *Glomus* and *Rhizobium* led to their better establishment in saline soils (Dixon *et al.*, 1993). The mycorrhizal influence is probably related to improvement of P nutrition rather than directly influencing the salt tolerance of the plants. The potential for benefiting by inoculation with 'hardy' rhizobia was demonstrated by Shoustari and Pepper (1985) in mesquite. The desert strain AZ-MI was more competitive and survived better in a saline soil at pH 9.5, ESP 27 and ECe 11.2 dS m^{-1} as compared to commercial strain 31A5, although neither strain was indigenous to any of the soils tested. The above findings indicated that there is a possibility for evolving suitable rhizobial inoculants for tree legumes to ensure their better establishment in tree plantations in all kinds of soil and climatic situations.

It is pertinent to mention that besides symbiotic nitrogen fixation by rhizobia there is also a significant input of fixation due to non-symbiotic microorganisms in forest soils. This issue is exhaustively dealt with in Dawson (1992). There is evidence that N_2-fixation is enhanced in the rhizosphere of trees and forest plants (Richards and Voight, 1964). Many forest soils are acidic; hence N_2-fixing rhizosphere bacteria in forests are likely to be in genera with acid-tolerant species such as *Beijerinckia, Clostridium,* and *Pseudomonas* (Richards, 1973). Nitrogen-fixation in tree rhizosphere is enhanced by waterlogged conditions (Balandreau and Burgos-Leonn, 1972) and in wet mangrove forests (Zuberer and Silver, 1978).

Field Studies on Nodulation and Nitrogen Fixation

Is there a direct evidence for a substantial role of N_2 fixation in most of these cases and especially in semi-arid and arid areas? While several studies showed that NFTS are capable of potentially nodulating under defined laboratory or greenhouse experiments, do they actually nodulate adequately and fix significant amounts of N in actual field situations? Vandenbeldt (1991) found little evidence for abundant nodulation in the field. Most studies on nodulation of tree legumes have been confined to small time scales in laboratory and green-house screenings. An assessment of whether or not to inoculate tree species with rhizobia requires reliable information on nodulation status of trees at actual field sites. There is lack of reliable information on nodulation of mature trees in field conditions, extent of longevity of nodules in mature trees, annual/perennial nature of nodules, influence of the roots of non-legumes growing in close vicinity to legumes on the nodulation of tree legumes, factors limiting nodulation (salinity, alkalinity and moisture stress) etc. Rao *et al.*, (1998) made extensive nodulation surveys for leguminous trees (69 species representing 16 genera) for three years in actual field sites (1050 no.) in normal/reclaimed alkaline soils, alkaline, saline-waterlogged soils and in soils of the arid zone to study the above aspects. Nodulation wherever present was evident in the surface 15 cm soil itself. In normal (pH 8.0 ± 0.5) as well as alkaline soils (pH 9.0 ± 0.7) nodulation occurred in two-thirds (63 and 64%), in saline-waterlogged soils (ECe 19 ± 10 dS m^{-1}) in 80% and in arid soils in only 14% of the samples. A greater number of younger trees (79%) exhibited nodulation compared to older ones (45%) indicative of the plant's need to fix N in early stages. Trees may thus benefit from inoculation only from nursery establishment to young (1–3 yr) plantations.

Low water or high soil N both inhibit the nodulation of tree legumes. Miettinen (1988) demonstrated that moisture was needed for the nodulation of mesquite. Shoustari and Pepper (1985) also showed that nodulation of mesquite trees was abundant when moisture was plentiful and also that nodulation was inhibited by high soil N concentrations. Inhibition of nodulation of *Acacia* sp. by N was reported by Umali-Garcia *et al.*, (1988). In a semi-arid alfisol Venkateswarlu *et al.*, (1990) reported in *L. leucocephala* that nodule weight was very low in May and started increasing after the onset of the rainy season in July and reached a peak in August. The acetylene reduction activity of nodules was low in summer and markedly high during July to September. Most of the nodules were present on fine roots (<4 mm dia) and >85% in under surface 25 cm of soil layer. We found (Rao *et al.*, 1998) that in normal soils, the extent of nodulation was highest in the monsoon season (August–September) and declined till winter (November–February). In alkaline soils it was poor in monsoon due to high pH and surface water stagnation, but improved later in post-monsoon (October). In arid zone soils nodulation was very poor due to low moisture, but even in monsoon season it was only 14%. In saline-waterlogged soils extent of nodulation was highest, because of the presence of shallow water-table (~1m) due to which surface soils which were in the capillary fringe, were in a good moist state (~15 %) and nodules remained healthy and functional despite high salinity. Adequate moisture not only promoted the survival and proliferation of rhizobia and helped nodule functioning but also helped in overcoming the adverse effect of salinity stress on the plant. Therefore longevity of nodules was highest in saline-waterlogged soils where they were healthy and functional even in summer (May–June), ten months after onset of monsoon while it had reduced by spring (March–April) in all other sites. Soil salinity and alkalinity had little influence on the extent of nodulation in various tree species except an adverse influence of alkalinity on nodulation in *A. nilotica* and *S. sesban*. Once the nodules are formed under non-saline conditions in monsoon season their further functioning is affected to a lesser extent by salinity. In the post-monsoon season in the arid-zone, nodulation was found only where there was surface proliferation of roots and good moisture. Even in monsoon, majority of species did not nodulate, showing that besides the principal constraint-soil moisture, there were other factors operating in the sandy desert soils. Nodulation of trees growing in crop lands was better than those on bare soils which also may be due to higher moisture status of irrigated crop lands. Our results point to the sustained benefits that can accrue from planting NFTS on waterlogged sites, canal sides subject to seepage etc., where besides producing biomass and controlling water-tables, they can add significant amounts of biologically fixed nitrogen to crops.

A rough guide to the nitrogen fixing nature of a tree is the leaf N content. As a broad generalization, most diazotrophs have over 2% N in their foliage, while forest trees growing in similar situations normally have much lower N levels (Silvester, 1983). In our studies (Rao *et al.*, 1998) the leaf N of nodulated trees (3.2–3.5%) was higher than that of non-nodulated (2.8–2.9%); 3% leaf N was the critical limit to distinguish N_2 fixing trees. In the arid zone and in other sites where nodulation was not found, nodulation in deeper layers was suspected, especially in *Prosopis* spp. since the leaf N content was more than 3% and indicated N fixation in lower layers. In some areas *F. albida* tree roots have been reported to be as deep as 40 m where they are probably reaching a permanent water table; compatible bradyrhizobia have been found at 30–35 m (Dupuy and Dreyfus, 1992) so nodulation may be occurring. Where a deep water-table is reached by the deep roots of trees then nodulation may occur at depth. Nodulation at depths of 4-6 m has been shown with *Prosopis* growing in the Sonoran desert and large populations of rhizobia have been found at the water-table (Virginia *et al.*, 1986). ^{15}N natural abundance methods have confirmed N-fixation under such conditions. Studies of Virginia *et al.* (1981) showed that *Prosopis* plants growing in Sonoran desert in U.S.A had a δ15N of +3.0 (soil + 6.0) even though nodules were not found on surface, but it is confirmatory of nitrogen fixation, so nodulation must have occurred in deeper layers.

Abundant nodulation of *Casuarina* roots was reported under *Eucalyptus* roots in a mixed plantation (Dommergues, 1987) in southeastern China. Nodules and fine roots of *I. jinicuil* were concentrated around the trunks of coffee trees within or just below the litter layer (Roskoski, 1981). Presence of nodulated roots close to maize plants which belonged to *Gliricidia* hedgerows 3.5 metres away, was noted by M.van Noordwijk in a field experiment in southern Sumatra. There are also reports of presence of large (5–8 cm in dia) nodules of *A. lebbek* intertwined with sorghum roots 15 metres away from the tree in Senegal. Based on the above observations Noordwijk and Dommergues (1990) hypothesized that roots of nitrogen fixing trees have more nodules, where nitrogen fixation takes place, when they are in close contact with roots of non-nitrogen fixing plants. This increased nodulation may lead to the direct transfer of nitrogen to the non-nodulating plant in inter-cropping, avoiding the pathway of uptake by the tree and return to the soil via prunings just like in grass/legume mixtures. The uptake of the fixed N creates areas of N depletion thus stimulating nodulation. The fact that all roots leak nutrients mean that closely neighboring roots containing little nitrogen could utilize nitrogen exuded from roots of high nitrogen content. And because a large part of the fine roots, including nodules can be expected to die after the tree's shoot/root balance is disturbed by pruning, transfer is even more likely in a hedge-row inter-cropping system.

Estimates of Nitrogen Fixation

Estimates of maximal annual N_2 fixation potentials of fully stocked legume stands range from several hundred kilograms per hectare a year in temperate regions to 1000 kg/ha per year in the humid tropics, where nutrients and environmental conditions are optimal (Sprent, 1983). The measurement of nitrogen fixation in the field for trees needs lot of research work. Unfortunately we cannot adequately assess the contribution of N_2 fixing trees to the N cycle in agroforestry systems. "Traditional" methods used in glasshouse studies—^{15}N isotope dilution method and acetylene reduction assay of nitrogenase activity have serious shortcomings in the field because of logistic and sampling problems associated with trees and partly because such tests as acetylene reduction give only one point in time (even if one could overcome problem of acetylene introduction into soil). ^{15}N natural abundance methods have been used with success wherever spatial variability is not a problem, and using it Shearer *et al.* (1983) reported that 70% of mesquite N was derived from fixation. A further complication in measurements is the remobilization of nitrogen from the roots at different phases of growth, such that current production (or amounts of fixed N) in prunings may not reflect current N fixation activity. Even with its

Table 2 Nitrogen fixation by trees and shrubs*.

Species	*Nitrogen fixation*	*Reference kg/N/ha/year*
Acacia albida	20	Nair, 1984
Acacia mearnsii	200	Dommergues,1987
Casuarina equisetifolia	58–110	Dommergues,1987
Coffee+Inga	35	Roskoski & van Kessel, 1985
Erythrina poeppigiana	60	Dommergues, 1987
Inga jinicuil	50	Roskoski, 1982
Leucaena leucocephala	100–500	Dommergues, 1987
L.leucocephala (in alley farming)	75–120	Mulongoy, 1986
L.leucocephala	100–130 (6 months)	Sanginga *et al.*, 1987
Prosopis glandulosa	40–50	Virginia *et al.*, 1986
Prosopis tamarugo	200	Nair, 1984
Sesbania sesban	260–332	Rao *et al.*, 1990

*Nair (1984) and Dommergues (1987) are compilations from primary sources.

limitations the best method so far has been the increase of N in soil and plant biomass (biomass analysis) at a field site. For perennial plants the nitrogen returned in litter constitutes a high proportion of that taken up in fixation and soil uptake. Leaf litter fall has also been used as an indication of nitrogen fixation by nodulated plants but it neglects the N stored in the plant growth and N turnover underground.

The number of estimates of nitrogen fixation in trees is limited and in many cases we only have estimates of proportion of N from N_2 fixation which are relatively easy to make by sampling prunings of trees and comparing their natural ^{15}N abundance with that of non-fixing plants growing nearby. Few studies have been made where destructive harvesting has been used to give estimates of the total amount of N fixed. Exceptions are the studies of Sanginga *et al.*, 1986 in *L. leucocephala* and Rao *et al.* (1990) for *S. sesban* discussed later. Estimates of nitrogen fixed on an annual per ha basis vary from as low as 58 kg N for *C. equisetifolia* on a sandy site, 80–120 kg for *Leucaena*, 300 kg for densely planted acacias and casuarinas. For *S. sesban* grown in various spacings nitrogen fixation ranged from 260–332 kg/ha/yr; for a high density plantation it was 358 kg/ha/yr (Rao *et al.*, 1990, Table 3). Most data refer to the tree in a pure stand, but that for coffee with *Inga* and alley cropping with *Leucaena* are for the cultivation in spatial-mixed and spatial-zoned agroforestry systems respectively. The range is largely 20–200 kg/ha/yr with the exception of *Leucaena* and *Sesbania* alone which are capable of higher values under favorable climatic and soil conditions.

Table 3 Nitrogen balance (kg ha^{-1}) in *Sesbania sesban* grown at selected spacings in an alkaline soil (1985–87)

N in kg/ha	*Inter-row spacing* (m)			
	0.5	*1.0*	*2.0*	*3.0*
Initial soil N	2986	3006	3026	3132
Tree biomass N	444	336	237	168
Litter N	247	228	243	220
Final soil N	3221	3487	3421	3574
Nitrogen balance	+876	+995	+825	+780
Nitrogen fixation (kg/ha/yr)	292	332	275	260

(Rao *et al.*, 1990)

Among woody annuals particular attention has been paid to *Sesbania* and *Cajanus.* At early stages in 'dhaincha' (*S. cannabina*), peak nitrogenase (acetylene reduction) activity of root nodules occurred at 33 days resulted in maximum N accumulation in shoots at 40 days (Rao and Ghai, 1995a). Green matter production of 17 annual accessions of *Sesbania* at 54 days, averaged 26.0 mg/ha (5.9 mg DM), N uptake was 158 kg/ha. N fixation was of the order of 105–150 kg/ha, or a range of 2.0 to 2.8 kg N fixation/ha/day (Rao and Gill, 1993). At 100 DAS, green stem and leaf production in semi-reclaimed alkaline soils was 21.5 and 9.4 mg/ha. Biofertilizer value in leaf plus upper tender stems was 125 N, 5 P, 8.1 K and 12 S (kg/ha) respectively (Rao and Gill, 1995b). *Sesbania* is also useful as a source of woody biomass; at 200 DAS woody stem yield was 20 mg (dry)/ha with a calorific value of 4730 Kcal/kg. Pigeonpea (*Cajanus cajan*) yielded 9.1 mg/ha of dry woody stem; litter fall was 1.9 mg/ha, thus recycling 39.5 kg N, 2.1P, 7.3K and 2.1S (kg/ha) of nutrients, benefiting the next crop in rotation. N-fixation in the growing season was nearly 115 kg/ha (Rao and Gill, 1995c).

Sesbania sesban can tolerate alkalinity upto pH 9.5; of the various species of perennial *Sesbania, S. sesban* had highest biomass production and nitrogen fixing ability. In a 2 × 2m plantation in alkaline soil, N accumulation in aerial and root parts was 180 and 41 g/tree amounting to 449 and 102 kg/ha (Rao *et al.*, 1990). In high density plantations managed by coppicing for six years, dry biomass production ranged between 25–35 mg/ha/yr and N fixation was nearly 358 kg/ha/yr in first three years. In next three years it ranged from 170–240 kg N/ha/yr. In first three years buildup of soil N in the 0–60 cm layer was 540 kg/ha. In highly alkaline soils of pH 10.5, nutrient recycling in *A. nilotica* was about 40 kg/ha/yr (Gill *et al.* 1987). With 40 trees/ha of *A. albida* in the Sahel in W. Africa, leaf litter was 2.7 (DM) mg/ha/yr contributing 45 kg N/ha. Nitrogen recycled through litter was nearly 65.6 kg/ha/yr in a *P. juliflora* plantation (Singh, 1996). These figures represent the approximate amount of N fixed.

NUTRIENT RECYCLING

Nutrients Recycled through Litter Production

It is important to distinguish between nitrogen fixation, an input to the plant-soil system, and nitrogen addition to the soil in litter or prunings, which is an internal transfer. Much of the nitrogen in litter in young plantations has been taken up from the soil, originating either as fertilizer input or stored soil reserves. But in mature stands as the nutrient cycles become more closed, litter fall represents net N added and is thus a good indicator of N fixation. Scientific data on litter production and amount of nutrients recycled through litter-fall in plantations of multipurpose tree species under tropical and sub-tropical regions are limited. Such information is of special significance when working out *in situ* applications of specific tree species to agroforestry systems, for integration and optimization of productivity without harming sustainability of the soil productivity under diverse situations. Investigations were, therefore conducted at an alkaline site in Karnal, Haryana in Northern India to estimate litter production and the amounts of important nutrients recycled in plantations of *L. leucocephala* and *S. sesban* by Gill *et al.* (1990) as both of these are potential candidates and most widely advocated multipurpose tree species for agroforestry intervention under tropical and sub-tropical regions. In *Leucaena* the plantation geometry was 6 × 3 m; in *Sesbania* it was a high density plantation, 3 × 1 m, the canopy of this species tended to dense and closed up just after a year of planting whereas it took almost more than two years in case of *L. leucocephala.*

Annual yields of litter (oven-dry) of *Leucaena* were 3285, 4976, 6198 and 7168 kg/ha during 1985 to 1988 (mean, 5406 kg/ha), in *Sesbania* it was 6428, 7486, 5982 and 3866 kg/ha during 1984 to 1987 (mean, 5940 kg/ha). Differences in litter production behaviour of *L. leucocephala* and *S. sesban* plantations owe to their widely different stocking rates and the growth characteristics that determine nature and extent of the structure and density of canopy. Periodic litter-fall in *L. leucocephala* and *S. sesban* showed a unimodal pattern with its peak in the winter season (November–February). Average litter fall that occurred during the winter periods in *L. leucocephala* and *S. sesban* plantations accounted for 59 and 61 per cent of the annual litter yields of the four years. The foliage component accounted for 79 to 85 per cent of the total fall in *L. leucocephala*. This was followed by the reproductive component (11 to 16 per cent) represented collectively by flowers, pods and seeds etc. and the woody matter component (3 to 5 per cent) during various study years. In case of *S. sesban* again the foliage component constituted 87 to 90 per cent of the total fall whereas reproductive and woody matter components accounted for 4–6 per cent and 4–9 per cent of the total yearly litter yields respectively.

Average composition of the litter for important elements showed a general decline with the successive growth years in both of the plantations. Both of the species demonstrated their tendency to avoid excessive accumulation of Na and hence checked its recycling. Average concentration of N, P, K, Ca, Mg, S and Na in litter of *L. leucocephala* and *S. sesban* are given in Table 4. As an indicator of nitrogen fixation it may be seen that the average annual amount of nitrogen recycled through litter was 127 kg/ha in *Sesbania* and 95 kg/ha in *Leucaena* respectively. Seasonal fluctuations in concentrations of all the nutrients did not follow a regular trend during different years of these investigations in both of the plantations. These irregularities are generally attributable to seasonal physiological adjustments in growth vis-à-vis absorption of nutrients from the soil. Decline in concentration of nutrients with successive growth years is expected to occur upon dilution effect and cessation of root activity when growth stagnates after completion of active growth period in short duration species like *S. sesban*. Similar trend in concentration of these nutrients was observed (Gill *et al.*, 1987) earlier in plantations of *A. nilotica* and *Eucalyptus tereticornis*.

The comparative amounts of nutrients recycled through the litter of *S. sesban* was greater than the corresponding quantities through the litter of *L. leucocephala* (Table 4) and owes to its much densely built canopy with much higher stocking rate (3333 plants/ha) than that (555 plants/ha) in plantation of the latter species. The former is also highly tolerant to saline and alkaline conditions (Rao *et al.*, 1986) and hence is suitable for planting in programmes aiming at both production and reclamation purposes. Seasonal trend of nutrients recycling reflected similarly with that of the litter-fall pattern notwithstanding minor seasonal fluctuations in concentration of nutrients in the litter materials. From this study it was concluded that both *L. leucocephala* and *S. sesban* produce significant amounts of litter rich in nutrients like N, Ca, P, K and S. However, their litter production behavour varies a lot. The latter attains peak levels within 2 to 3 years after planting whereas the former indicated progressive increase with the successive growth years. Thus,

Table 4 Concentration of nutrients in litter (4 yr average) of *Leucaena leucocephala* and *Sesbania sesban* and total amount of nutrients recycled through litter (kg/ha) in four years.

Element	*L. Leucocephala*		*S sesban*	
	Conc.	*Amt (kg/ha)*	*Conc.*	*Amt (kg/ha)*
Nitrogen, g/kg	21.70	379.20	22.30	508.80
Phosphorus, g/kg	1.32	27.70	1.90	45.40
Potassium, g/kg	16.80	352.70	17.90	428.50
Calcium, g/kg	18.20	385.80	18.60	445.50
Magnesium, g/kg	8.70	182.20	9.20	221.10
Sulphur, g/kg	2.90	60.90	3.55	85.40
Sodium, g/kg	0.25	5.26	0.53	12.60
Iron, mg/kg	540.60	11.47	477.30	11.44
Manganese, mg/kg	94.60	1.92	130.20	3.12
Zinc, mg/kg	21.80	0.43	45.80	1.08
Copper, mg/kg	12.70	0.26	14.60	0.33

their management for different agroforestry systems need to be synchronized accordingly. *L. leucocephala* appears an ideal choice for alley cropping to grow trees and the crops concurrently whereas *S. sesban* due to its relatively shorter growth period may be one of the best multipurpose tree species for sequential type of agroforestry systems. However, results imply the need to study further the extent and efficiency of nutrients cycling under the different set of different agroforestry interventions.

Transfer of Tree N to Crops

As compared with agricultural systems, agroforestry systems contribute towards improved nutrient cycling and sustainability through-greater mineralization of nutrients from unavailable reserves, addition of nutrients in plant litter/tree residues (returned directly or via animals), more closed nutrient cycles as a result of greater uptake by plant roots and less leaching losses, and achieving a balanced supply of nutrients including micronutrients because of the organic nature of nutrient supply. Studies in *Leucaena*-maize alley cropping in Kenya showed that *Leucaena* could add 173 kg N per ha as green-leaf manure (Jama *et al.*, 1986). Mulch prunings of *Leucaena* maintained a maize yield of 2 mg/ha/yr for 4 years whereas in control soils where prunings were removed yields decreased from 1 mg/ha in 1st year to 0.3 mg/ha by the 4th year. Nitrogen gains of 120–130 kg/ha/yr have been reported for *S. sesban*-guinea grass silvipastural system, 90 kg for *A. lebbek*-guinea grass at IGFRI, Jhansi in Central India (Hazra, 1989). Mixed cropping of *S. sesban*

(*Shevri*) with fodder Sorghum recorded maximum green fodder (53 mg/ha) which was higher than that recorded with sole crops of Sorghum (38.2 mg/ha) or *Sesbania* (19.6 mg/ha) grown alone (Patil and Gill, 1986). Sorghum performed much better in association with *Sesbania* than alone obviously due to benefits derived from N-fixation. Further, *Leucaena*-Maize and *Leucaena*-oats intercropping have been found very promising at Jhansi, and nitrogen contribution by *Leucaena* in such systems was upto the order of 300 kg/ha/yr (Pathak *et al.*, 1990).

Decomposition of leaf litter adds considerable amounts of organic matter and nitrogen and has an ameliorative effect on soil. Rao and Ghai (1985) showed that in soil under *P. juliflora* in alkali soil there was considerable enrichment of organic matter (increased from 0.1% to 1.5% in 12 yrs) and nitrogen (from 0.03% to 0.15%) as well as increased biological (dehydrogenase) activity. Singh (1996) reported that in a *P. juliflora* plantation raised in a highly deteriorated alkali soil by auger-hole method of planting, there was considerable soil amelioration, pH of surface soil (0–15 cm) decreased from 10.3 to 9.5 in six years, organic C increased from 0.17 to 0.36% and available N from 82 to 118 kg/ha.

The nitrogen fixed by these N fixing trees can also be utilized once the trees are harvested by subjecting these lands to cropping with cereals. Although there are data on soil changes under shifting cultivation, there are no data under rotational agroforestry systems. In the authors studies on 'sequential agroforestry' a 4-yr-old plantation of a nitrogen-fixing tree, *S. sesban* raised in a reclaimed alkali soil in Karnal in northwest India was harvested, biomass was removed and rice and wheat were grown for next 6 years. N fixation by *S. sesban* in the legume phase ranged from 260–330 kg/ha/yr (table 3). Soil N enrichment in the 0–60 cm soil layer was 388 kg /ha. In cereal phase, rice and wheat were grown in the same plots, but without application of nitrogen. In control plots (which lay fallow for 4 years), rice and wheat were grown similarly, without applying N as well as by addition of graded doses of urea. Maximum residual benefit of *Sesbania* was observed in the first two years, after which yields of rice stabilized for the next four years at 5.3 mg ha^{-1} compared to 4.1 mg ha^{-1} in control. Wheat yields stabilized at 1.9 mg ha^{-1} in *Sesbania* plots in comparison to 1.5 mg ha^{-1} in control. Average yields of rice and wheat with 120 kg urea-N ha^{-1} application were 6.4 and 4.6 mg ha^{-1} respectively. The residual effects of *Sesbania* resulted in additional grain production of 1.2 mg ha^{-1} of rice (+26.4%) and 0.5 mg ha^{-1} of wheat (+24.1%) each year. Apparent N uptake efficiency from applied urea-N was 33.4% in rice and 54.5% in wheat. Additional crop N uptake each year in *Sesbania* plots was 30.8 kg N ha^{-1} (20.6 kg in rice and 10.2 kg in wheat). The fertilizer-N equivalence of this residual effect amounted to 80.4 kg of urea-N ha^{-1}. Soil organic C, total N, available N and microbial activity in *Sesbania* plots were at par or were higher as compared to 120 kg N ha^{-1} plots. This system demonstrated a way to ensure efficient availability of the fixed nitrogen to crops grown after tree harvest and the sustained benefits which can be drawn through planting nitrogen fixing trees.

FUTURE RESEARCH NEEDS

The study of biological N_2 fixation in forest and agroforestry systems has been neglected. So far BNF research has been driven in an agricultural context because the availability and price of wood relative to food crops has created less demand for research to improve and increase wood supplies. Unlike the situation with food crops, there are alternatives to wood for fuel and building materials. However, the pressures of population growth and economic development, especially in tropical and arid regions, have reduced the availability of wood for fuel and shelter in some regions in which affordable alternatives do not exist. It is imperative therefore to accelerate research on nitrogen fixing trees and shrubs. Unfortunately however there are too many uncertainties and gaps in our awareness to fully exploit nitrogen fixing tree systems for their multi-purpose nature and many questions need answering.

There is increasing concern about loss of biodiversity of legume species and hence there is a need to preserve and conserve the germplasm of N fixing trees and shrubs. For biomass production in field conditions to approach the theoretical maximum, many climatic and edaphic factors have to be met. Besides choice of species and correct planting techniques, appropriate rotation and spacing, direction orientation, irrigation, weed control and fertilization are also important. Inoculation of tree saplings at nursery stage with specific *Rhizobium* strains is an area which has been almost neglected and requires urgent attention. Although the isolation of rhizobia from woody legumes and their use in research trials and demonstration plots are now commonplace, comprehensive isolations, screenings, and trials of strains for specific host plants and locales are uncommon. Systematic evaluation of host and microbial partners for optimal growth and survival at specific sites and locales is required. Role of VAM in helping plants acquire P and micronutrients, and scavenge small amounts of N as well as in improving survival and establishment in adverse sites is well known. Although inoculation is practiced by mixing soil from beneath older plantations in potting mixtures, it is not always possible to do so with many new tree species, is not always the best practice and better response can be obtained by inoculation with specific strains. Awareness of these facts by nurserymen is a prime consideration and to achieve targets, training of nurserymen in the art of rhizosphere biology of nitrogen fixing tree legume must be undertaken in India.

When trees are grown in short rotations of two to three years the objective of improving a degraded site may not be met. Attempts to intensify production based on the use of non-legume trees may founder as the area available for nutrient capture for each tree is reduced, whereas if legume trees are fixing their own N then a degree of intensification of tree planting may have long term sustainability. If we are to assess the long term sustainability of agroforestry systems we must know how inputs from N_2-fixation vary with age of tree and with age of the agroforestry system. Well nodulated trees will fix considerable N especially if soil N is low as in degraded soils. As trees mature, and litter from the trees falls to the soil, or is recycled close to the trees in prunings, then the N status of the trees will improve which in turn will lead to an inhibition of nodulation and N fixation in the tree. Besides nitrogen fixation we also need to know what is the contribution of the tree component to a more efficient supply to the crop, of the nitrogen stored in the soil or added as fertilizer? Such questions can be answered by nitrogen balance studies, taking into consideration all inputs and outputs to the plant-soil system, and internal flows. Data for atmospheric input, non-symbiotic fixation, gaseous losses and leaching will need to be estimated by comparison with specialized studies from similar environmental conditions. One of the major benefits cited in favour of agroforestry systems is that of making nutrient cycles less open by increasing the ratio between internal cycling and external losses (Nair, 1984). This needs to be proven by setting up long-term experiments. Numerous studies have shown the substantial negative nutrient balances, unless compensated by fertilizers, in systems of permanent and semi-permanent annual cropping. There is a need for thorough studies of all components in nutrient cycling to be conducted for agroforestry systems in tandem with agricultural plots for comparison.

There are very few studies on roots, their spread, depth of rooting, in order to predict moisture competition, allelopathic effects etc. Roots of non-legumes tend to have much more substantial root systems which may extend upto 50 m from trees. Shade effect of trees on crops, soil recouping factor (R-factor) for various plantations, N addition by leaves and roots, quality of wood produced in short-rotations etc. are the other areas requiring attention. Progeny testing in NFTS is a very slow process and caution has been advocated against collecting seeds from plantations raised in last few years. Even the debates on exotic vs native NFTS, and about indiscriminate advocation of alley farming for all situations has not yet been conclusive.

Since most of the trees in Caesalpinoideae cannot nodulate, non-nodulating trees of this group should be evaluated against non-legumes since they have important potential. Although most mimosoid legumes can nodulate they may not actually do so in arid and saline soils, since nodulation requires available water. However, mimosoid trees are very important and have deep roots (phraeotophytes). In nutrient poor soils, investment of plant resources in root systems is of more potential value than investment in nodules which only alleviates N deficiency. While nitrogen fixation is a desirable trait there has not been any comprehensive study of the economics of wood/fodder production in plantations fixing their own nitrogen versus non-fixing trees fertilized with inorganic nitrogen. There is an aura of biological uncertainty in data currently used for economic analysis of nitrogen management in forestry. These uncertainties pertain to rates of N fixation, transfer into crop trees and the response in tree growth to N fixed.

Overall, the outlook for exploitation of nitrogen fixing systems in forestry and agroforestry is bright in tropical environments like ours due to low soil nitrogen and holds a special promise in rehabilitation of degraded waste lands. It is hoped that researches along the lines proposed above will throw light and we will find answer to most of the questions posed above in the next 20 years.

ACKNOWLEDGEMENTS

The research work presented in this review was done at the Central Soil Salinity Research Institute, Karnal over a course of 15 years (1983-99). I would like to thank successive Directors' of CSSRI, Drs. I.P. Abrol, N.T. Singh and N.K. Tyagi for their support and encouragement of the agroforestry programmes. My gratitude to all my former colleagues Drs. H.S. Gill, Lalita Batra, Parbodh Sharma for participation in various studies and unstinted cooperation, and to Sh. E.S. Tomar and Sahib Singh for technical assistance. I am grateful to Indian Council of Agricultural Research and Department of Biotechnology, New Delhi for funding support.

REFERENCES

Balandreau, J., and Burges-Leon,W., 1972. Molecular-nitrogen fixation in the rhizosphere of some plants of temperate regions, *Bull. Ecol. Natl. Super. Agron. Ind.*, Aliment., Nancy, **14:** 65–67.

Basak, M., and Goyal, S.K., 1975. Studies on tree legumes I. Nodulation pattern and characterization of the symbiont, *Ann. Arid Zone*, **14:** 367–370.

Basak, M., and Goyal, S.K., 1980a. Studies on tree legumes II. Further additions to the list of nodulating tree legumes, *Plant and Soil*, **56:** 33–37.

Basak, M., and Goyal, S.K., 1980b. Studies on tree legumes III. Characterization of the symbionts and direct and reciprocal cross inoculation studies with tree legumes and cultivated legumes, *Plant and Soil*, **56:** 39–51.

Batra, L., and Kumar, A., 1994. Biomass production and nitrogen accumulation by Rhodes grass and three tree species in an agroforestry system under alkali soil conditions, In: *Agroforestry systems for degraded lands, Vol. 2*, (Punjab Singh, P.S. Pathak, and M.M. Roy, eds.), Oxford & IBH, New Delhi, 713–719 pp.

Boring, L.R., Swank, W.T., Waide, J.B., and Henderson, G.S., 1988. Sources, fates and impacts of nitrogen inputs to terrestrial ecosystems: Review and synthesis, *Biogeochemistry*, **6:** 119–159.

Burns, R.C., and Hardy, R.W.F., 1975. *Nitrogen-fixation in Bacteria and Higher Plants*, Springer-Verlag, New York.

Chamberlain, J.L., and Brewbaker, J.L., 1988. *Species profiles of important nitrogen fixing trees of India*, Paper presented at Workshop on NFTS for India, 12–13 November 1988, New Delhi.

Dancette, C., and Poulain, J.F., 1969. Influence of *Acacia albida* on pedoclimatic factors and crop yields, *Afr. Soils*, **14:** 143–184.

Dawson, J.O., 1992. Nitrogen fixation in forests and Agroforestry, In: *Soil Microbial Ecology: Applications in Agriculture and Environmental Management*, (F.B. Metting, ed.), Marcel Dekker Inc., New York, 227–253 pp.

Diem, H.G., and Dommergues, Y.R., 1990. Current and potential uses and management of Casuarinaceae in the tropics and subtropics, In: *The Biology of Frankia and Actinorhizal Plants*, (C.R. Schwintzer, and J.D. Tjepkema, eds.), Academic Press, New York, 317–342 pp.

Dixon, R.K., Garg, V.K., and Rao, M.V., 1993. Inoculation of *Leucaena* and *Prosopis* seedlings with *Glomus* and *Rhizobium* species in saline soil: Rhizosphere relations and seedling growth, *Arid. Soil. Res. Rehab.*, **7:** 133–144.

Dommergues, Y.R., 1987. The role of biological nitrogen fixation in agroforestry, In: *Agroforestry: a decade of development*, ICRAF, Nairobi (H.A. Steppler, and P.K.R. Nair eds.), 245–271 pp.

Dommergues, Y.R., Diem, H.G., Gauthier, D.L., Dreyfus, B.L., and Cornet, F., 1984. Nitrogen-fixing trees in the tropics: Potentialities and limitations, In: *Advances in Nitrogen Fixation Research*, (C. Veeger, and W.E. Newton, eds.), Nijhoff Junk Publ., The Hague, 7–15 pp.

Dreyfus, B.L., and Dommergues, Y.R., 1981. Nodulation of *Acacia* species by fast and slow growing tropical strains of *Rhizobium*, *Appl. Environ. Microbiol.*, **41:** 97–99.

Dupuy, N., and Dreyfus, B., 1992. Presence of bradyrhizobia under *Acacia albida*, In: *Faidherbia albida* in the West African Semi-arid Tropics, (R.J. Vandenbeldt, ed.), ICRISAT/ICRAF, Andhra Pradesh/Nairobi, 145–148 pp.

Felker, P., 1981. Uses of tree legumes in arid regions, *Econ. Bot.*, **35:** 174–186.

Felker, P., and Clark, P.R., 1980. Nitrogen fixation (acetylene reduction) and cross-inoculation in 12 *Prosopis* (mesquite) species, *Plant and Soil*, **57:** 177–186.

Galiana, A., Chaumont, J., Diem, H.G., and Dommergues, Y.R., 1990. Nitrogen fixing potential of *Acacia mangium* and *Acacia auriculiformis* seedlings inoculated with *Bradyrhizobium* and *Rhizobium* spp. *Biol. Fert. Soils*, **9:** 261–267.

Gill, H.S., Abrol, I.P., and Samra, J.S., 1987. Nutrient recycling through litter production in young plantations of *Acacia nilotica* and *Eucalyptus tereticornis* in a highly alkaline soil, *For Ecol. Manage.*, **22:** 57–69.

Gill, H.S., Rao, D.L.N., and Abrol, I.P., 1990. Litter yield and its quality in plantations of *Leucaena leucocephala* and *Sesbania sesban* in alkaline soils, In: *Multipurpose tree species for agroforestry systems*, (P.S. Pathak, R. Deb Roy, and Panjab Singh, eds.), IGFRI, Jhansi, 187–196 pp.

Gordon, J.C., and Wheeler, C.T., 1983. *Biological nitrogen fixation in forest ecosystems: foundations and applications*, (ed.) Nijhoff Junk, The Hague, 342 pp.

Halliday, J., 1984. Integrated approach to nitrogen fixing tree germplasm development, *Pequisa Agropec. Bras.*, **19** (Spec. issue): 91–117.

Hazra, C.R., 1989. Forage and soil productivity under agroforestry and silvi-pasture systems in Bundelkhand region, In: *Promotion of fodder and fuelwood trees*. (N.G. Hegde, L.L. Relwani, and V.D. Kelkar, eds.), BAIF Dev. Res. Foundation, Pune, 83–88 pp.

Jama, B., Getahun, A., Ngugi, D., and Macklin, B., 1986. *Leucaena* alley cropping for the Kenyan coast, In: *Amelioration of soil by trees*, (R.T. Prinsley, and M.J. Swift, eds.), Commonwealth Science Council, London, 155–165 pp.

Lim, G., and Burton, J.C., 1982. Nodulation status of the leguminosae, In: *Nitrogen fixation vol. 2, Rhizobium*, (W.J. Broughton, ed.). Clarendon Press, Oxford, 1–34 pp.

Miettinen, P., Luukkanen, O., Johansson, S., Eklund, E., and Mulatya, J., 1988. *Rhizobium* nodulation in *Prosopis juliflora* seedlings at different irrigation levels in Eastern Kenya, *Plant and Soil*, **112:** 233–238.

Mulongoy, K.,1986. Nitrogen cycling in alley cropping systems. IITA, *Res. Briefs*, **7:** 3–5.

Nair, P.K.R., 1984. *Soil Productivity aspects of Agroforestry*, ICRAF, Nairobi, 85 pp.

Nimbalkar, S.S., Godbole, S.H., and Kulkarni, Y.S., 1986. New additions to the nodulated species of wild legumes, *Ind. J. For.*, **9:** 132–136.

Noordwijk, M.V., and Dommergues, Y.R., 1990. Root nodulation: the twelfth hypothesis, *Agrofor. Today*, **2**(2): 9–10.

Pahwa, M.R., 1994. Potential of nitrogen fixers in fodder tree species and range grasses for ameliorating productivity, In: *Agroforestry systems for degraded lands, vol. 2*, (Punjab Singh, P.S. Pathak, and M.M. Roy, eds.), Oxford & IBH, New Delhi, 739–744 pp.

Pathak, P.S., Deb Roy, R. and Singh, Panjab (eds.), 1990. Multipurpose tree species for agroforestry systems, *Proc. Natnl. Wkshp.*, Range Management Society of India, Jhansi, 244 pp.

Patil, B.D., and Gill, A.S., 1986. *Shevri*-A multipurpose shrub. *Ind. Far.*, **35**(10): 24–25.

Rao, D.L.N., and Ghai, S.K., 1985. Urease and dehydrogenase activity of alkali and reclaimed soils, *Aust. J. Soil. Res.*, **23:** 661–65.

Rao, D.L.N., and Gill, H.S., 1993. Nitrogen fixation, biomass production and nutrient uptake by annual *Sesbania* species in an alkaline soil, *Biol. Fert. Soils*, **15:** 73–78.

Rao, D.L.N., and Ghai, S.K., 1995a. Predicting nitrogen fixation and N accumulation in field grown annual *Sesbania* spp. *Proc. Ind. Natnl. Sci. Acad.* (B), **61:** 57–62.

Rao, D.L.N., and Gill, H.S., 1995b. Biomass and biofertilizer production by *Sesbania cannabina* in alkaline soils, *Biores. Tech.*, **53:** 169–172.

Rao, D.L.N, and Gill, H.S., 1995c. Biomass production and nutrient recycling through litter by pigeonpea (*Cajanus cajan* L. Millsp), *Biores. Tech.*, **54:** 123–128.

Rao, D.L.N., Gill, H.S., and Abrol, I.P., 1986. Nitrogen fixation and biomass production by *Sesbania sesban* in alkali soils, In: *Amelioration of Soil by Trees—A Review of Current Concepts and Practices*, (R.T. Prinsley, and M.J. Swift, eds.), Commonwealth Science Council, London, 74–79 pp.

Rao, D.L.N., Gill, H.S., and Abrol, I.P., 1990. Regional experience with perennial *Sesbania* in India, In: Perennial *Sesbania* species in Agroforestry Systems, (B. Macklin, and D.O. Evans, eds.), *Proc. of Wkshp.* ICRAF, Nairobi, NFTA Spl. Publ. (90–91), Hawaii, USA, 189–198 pp.

Rao, D.L.N., Batra, L., Sharma, P.C., Chauhan, K.K.S., and Gautam, K., 1998. *Rhizobium-tree* legume associations in Agroforestry systems. Project completion report, Department of Biotechnology, Govt. of India, 50 pp

Rangarajan, M., Kandasamy, D., Udayashanker, K., Narayanan, R., Ponnaiah, C., Govindarajan, K., and Oblisami, G. 1984. *Advances in the studies on biofertilizers technology and its application for ecodevelopment*, Project Report, TNAU, Coimbatore, 36 pp.

Richards, B.N., 1973. Nitrogen fixation in the rhizosphere of conifers, *Soil Biol. Biochem,* 5: 149–152.

Richards, B.N., and Voight, G.K., 1964. Role of mycorrhiza in nitrogen fixation, *Nature,* **201**: 310–311.

Roskoski, J.P., 1981. Nodulation and N_2 fixation by Inga jinicuil, a woody legume in coffee plantations. 1. Measurements of nodule biomass and field C_2H_2 reduction rates, *Plant Soil,* **59:** 201–206.

Roskoski, J.P., 1982. Nitrogen fixation in a mexican coffee plantation, In: *Nitrogen cycling in ecosystems of Latin America and the Caribbean,* (G.P. Robertson, R. Herrera, and T. Rosswall, eds.), Nijhoff, The Hague, 283–292 pp.

Roskoski, J.P., and van Kessel, C., 1985. Annual, seasonal and diel variation in nitrogen fixing activity by *Inga jinicuil,* a tropical leguminous tree, *Oikos,* **44:** 306–312.

Sanginga, N., Mulongoy, K., and Ayanaba, A.,1986. Nitrogen fixation by *Leucaena leucocephala* (Lam.) de wit under Nigerian field conditions estimated by N-15 technique, *Biol. Agr. Hort.,* 3: 347–352.

Shearer, G., Kohl, D.H., Virginia, R.A., Bryan, B.A., Skeeters, J.L., Nilsen, E.T., Sharifi, M.R., and Rundel, P.W., 1983. Estimation of N_2-fixation from variation in the natural abundance of ^{15}N in Sonoran desert ecosystems, *Oecologia (Berl.),* **56:** 365–373.

Shoushtari, N.H., and Pepper I.L., 1985. Mesquite rhizobia isolated from the Sonoran desert, Competitiveness and survival in soil, *Soil Biol. Biochem.,* **17:** 803–806.

Silvester, W.B., 1993. Analysis of nitrogen fixation, In: *Biological Nitrogen Fixation in Forest Ecosystems: Foundations and Applications,* (J.C. Gordon, and C.T. Wheeler, eds.), Nijhoff Junk, The Hague, 173–212 pp.

Singh, G., 1996. Effect of site preparation techniques on *Prosopis juliflora* in an alkali soil, *For. Ecol. Manage.,* **80:** 267–278.

Singh, Punjab, Pathak, P.S., and Roy, M.M. (eds.), 1994. *Agroforestry systems for degraded lands,* Vol. 1 and 2, Oxford & IBH, New Delhi.

Sprent, J.I., 1983. Agricultural and horticultural systems: Implications for forestry, In: *Biological Nitrogen fixation in Forest Ecosystems: Foundations and Applications,* (J.C. Gordon, and C.T. Wheeler, eds.), Nijhoff Junk Publ., The Hague, 213–232 pp.

Umali-Garcia, M., Lisbuit, J.S., and Baggayan, R.L., 1988. Effects of Rhizobium inoculation on growth and nodulation of *Albizzia falcataria* (L.) Fosh. and *Acacia mangium* Willd. in the nursery, *Plant Soil,* **108:** 71–78.

Venkateswarlu, B., Korwar, G.R., and Singh, R.P., 1990. Studies on nitrogen fixation and nutrient addition by *Leucaena leucocephala* in a semi-arid alfisol, *Leucaena Res. Rep.,* **11:** 65–67.

Vandenbeldt, R.J., 1981. Rooting systems of western and southern African *Faidherbia albida* (Del.) A. Chev. (syn. *Acacia albida Del.*)—a comparative analysis with biogeographic implications, *Agrofor. Syst.,* **14:** 233–244.

Virginia, R.A., Jarrell, W.M., Kohl, D.H., and Shearer, G.B., 1981. Symbiotic nitrogen fixation in a *Prosopis* (leguminosae) dominated desert ecosystem, In: *Current Perspectives in Nitrogen fixation,* (A.H. Gibson, and W.E. Newton, eds.), Aust. Acad. Sci., Canberra, 483 pp.

Virginia, R.A., Jenkins, M.B., and Jarrell, W.M., 1986. Depth of root symbiont occurrence in soil, *Biol. Fertil. Soils,* **2:** 127–130.

Yamaya, K., 1968. On the influence of alder (*Alnus inokumae*) on soil properties in northern Japan, In: *Biology of Alder,* (J.M. Trappe, J.F. Franklin, R.F. Tarrant, and G.M. Hansen, eds.), UDSA Forest Service, Pac. NW For. Ran. Exptl. Stn., Portland, USA, 197–207 pp.

Zuberer, D.A., and Silver, W.S., 1978. Biological dinitrogen fixation (acetylene reduction) associated with Florida mangroves, *Appl. Environ. Microbiol.,* **35:** 567–575.

CHAPTER 14

NUTRIENT FLUX CONTROL BY TREES FOR IMPROVING SOIL FERTILITY IN TROPICAL AGROFORESTRY

Johannes Lehmann*

Institute of Soil Science and Soil Geography, University of Bayreuth, 95440 Bayreuth, and Federal Research Institute for Forestry and Forest Products, 21027 Hamburg, Germany

**Present address: Department of Crop and Soil Sciences, Cornell University, Ithaca, NY 14853, USA.*

ABSTRACT

Maintenance of soil fertility is one of the central aims of agroforestry management especially in the humid tropics, where rapid mineralization, strong weathering and high precipitation pose large risks for nutrient leaching. The spatial distribution of soil fertility is not homogenous in agroforestry systems, but a distinct vertical and lateral heterogeneity can be observed due to a heterogeneity of nutrient fluxes. These nutrient fluxes can be controlled by system design, management and species selection. In the present contribution, special emphasis is given to management options that control nutrient fluxes, since their importance has been less recognized up to now. Fertilization, pruning, spatial design and soil preparation can be used to control the nutrient fluxes such as biological N_2 fixation, uptake into aboveground biomass, location of nutrient uptake and nutrient return to soil. Shoot pruning was identified as an important tool which influences nutrient fluxes. Higher nutrient leaching losses into the subsoil and lower nutrient uptake were frequently observed after pruning. The intensity of pruning should become a focus for nutrient flux management. The presented examples illustrate that not only agroforestry system design and species selection govern nutrient fluxes, but also that system management can be equally effective. The described heterogeneity in agroforests offers a wide possibility of controlling nutrient fluxes thereby enlarging the options for soil fertility management in tropical landuse.

Key Words: Nitrogen flux, trees, soil fertility, tropical agroforestry management, biological N-fixation, species selection.

BASIC CONSIDERATIONS

In many agroforestry systems increasing soil fertility is an important goal (Young, 1997). Integrating trees into cropland can improve soil fertility, and much information is already available on how trees affect soil fertility in agroforestry systems (Sanchez, 1995; Buresh and Tian, 1998; Szott *et al.*, 1999). The pure service tree for fertility improvement in simultaneous agroforestry systems, e.g. in hedgerow intercropping, was not successful on a large scale due to low fertility effects, difficult management and restricted site adaptation (Sanchez, 1995). Trees which have other values apart from soil fertility improvement alone, such as producing subsistence or marketable products, may be preferred by landusers (Sanchez *et al.*, 1997a). However, even those systems may depend on soil fertility improvement by trees for an ecologically and economically sustainable production.

It is a myth that agroforestry systems can be sustainable without nutrient sources from outside, since nutrients are exported out of the soil with harvested products. N is a possible exception to this rule due to biological N_2 fixation. However, soil fertility management will always entail application of fertilizer nutrients, especially for P (Buresh and Tian, 1998). Nutrient additions through mulch or even animal manure are in most cases not a nutrient input, but a redistribution of nutrients already present in the system (Palm *et al.*, 1997). The amount of nutrients provided by atmospheric deposition, mineral weathering or in the case of N the biological fixation of atmospheric N_2, rarely match natural outputs in tropical forests (Bruijnzeel, 1991) and will not compensate harvest exports in agro-ecosystems in the long run. However, agroforestry techniques may help in utilizing available nutrient resources, both inherent and applied, more efficiently than monocultures.

The central requirement for a higher resource use efficiency of agroforestry over monocultures is that a tree provides a service to the farmer or the cropping environment that is not provided by another crop (e.g., water and nutrient use, harvestable product, wind-brake, shade, fence, pest control) and which increases the productivity of the whole cropping system in the specific socioeconomic context. If nutrient availability is an important issue addressed by agroforestry techniques at a particular site, it requires that trees take up nutrients which are not available for another crop (compare Cannell *et al.*, 1995), either while enriching the soil with available nutrients or sparing soil nutrients which are then available for associated or subsequent crops.

These synergistic effects in agroforestry will result in spatial and temporal heterogeneity of soil fertility, which is a fundamental difference to monoculture plantations. In order to manage soil fertility through agroforestry techniques, it is not only important to assess the heterogeneity of soil fertility, which is a static property, but also the heterogeneity of nutrient fluxes, which is a dynamic property. Nutrient flux control determines the soil fertility level (quantity factor) and resilience to nutrient exports (quality factor). In this contribution, I will demonstrate the heterogeneity of soil fertility (related to nutrients) and nutrient fluxes, and discuss the possibilities for controlling nutrient fluxes through agroforestry techniques.

HETEROGENEITY IN AGROFORESTRY SYSTEMS

Heterogeneity of Soil Fertility

There are several sources for heterogeneity of soil fertility in agroforestry systems. First, spatial inhomogeneity of soil genesis, previous vegetation and landuse history can affect soil fertility distribution on a field or farm level. These sources are not related to agroforestry but have to be dealt with when evaluating the effects of agroforestry on soil fertility. Since agroforestry studies and experimentation increasingly take place on farmers' fields, this challenge has become an important issue (Buresh, 1999). Rabits were found to create fertile microsites in Western Texas, which increased seedling survival, growth and nutrient uptake of a perennial bunchgrass (Dhillon, 1999). For a savanna ecosystem with *Faidherbia albida* in Niger, Geiger *et al.* (1994) hypothesized that a large part of the frequently found "albida effect" on soil fertility was rather caused by pre-existing heterogeneity of soil conditions than by trees. But also on experimental stations, soil fertility often varies considerably especially because of the large area which is needed for experimentation with trees. Large plot sizes are especially needed for multi-strata agroforestry systems such as cocoa or coffee with shade trees (Beer *et al.*, 1998). The number of samples required for an acceptable certainty for soil fertility assessment was determined for a Xanthic Ferralsol on a 19 ha field cleared from a fertilized rubber plantation (Table 1; Teixeira and Villani, 1996). Whereas only very few samples were necessary for pH and organic C characterization, the number increased dramatically for P, Mg and especially Ca. Similar relationships were found for smaller scale heterogeneity at the same site in 2500 m^2 plots using 20 measurements (Teixeira and Villani, 1996).

Table 1 Number of samples required for estimation of soil chemical parameters of a Xanthic Ferralsol on a previously cleared rubber plantation of 19 ha, determined with 90 samples taken in a grid at 0–0.15 m depth (from Teixeira and Villani, 1996).

Parameters	*Mean values*	*Coefficient of variation* [%]	*Number of required samples probability level* p > 95 %		p > 80%	
			Allowed variation			
			20%	*5%*	*20%*	*5%*
pH (H_2O)	4.7	4	1	2	1	1
Corg [mg Kg^{-1}]	3.9	19	4	57	1	24
P [kg m^{-3}]	5.8	39	15	242	6	102
K [kg m^{-3}]	31	27	7	112	3	47
Ca [mmol cm^{-3}]	2.4	90	80	1277	34	539
Mg [mmol cm^{-3}]	1.3	53	27	440	12	185

The second source of heterogeneity is inherent to agroforestry systems. Tree and annual crop species have different effects on soil fertility and may receive different amounts and types of fertilizers (mineral fertilizer, compost, manure, mulch) in any one agroforestry system. Most of these effects of trees on soil nutrient contents, availability and distribution have been extensively reviewed (Palm, 1995; Sanchez, 1995; Rhoades, 1997; Young, 1997; Buresh and Tian, 1998; Rao *et al.*, 1998; with regional focus: Drechsel and Reck, 1998; Jama *et al.*, 2000) and they will not be repeated here. These effects on soil fertility, however, are not equally distributed in mixed cropping but they are patchy under trees or annual crops similar to savanna ecosystems (Rhoades, 1997). In a multi-strata agroforest with fruit trees, the soil inorganic P was strongly influenced by fertilizer applications, which were concentrated under the trees (Fig. 1). However, even the tree effect on organic P was not homogeneous under the canopy of one species (Fig. 1). Similarly, increases of soil organic P were inhomogeneously distributed under unfertilized *Ziziphus joazeiro* in Yucatan and larger at one side of the canopy (83 μg g^{-1}) than on the other (26 μg g^{-1}) (Tiessen *et al.*, 1999). The fact that total P contents did not increase in the latter example from Yucatan indicated that this was a tree effect and not an effect of P inputs. Similar heterogeneity was also found in several other studies from savanna ecosystems as reviewed by Rhoades (1997).

The patchiness of soil fertility is caused by uneven nutrient applications (e.g. mineral fertilizer, mulch) as shown above, by nutrient

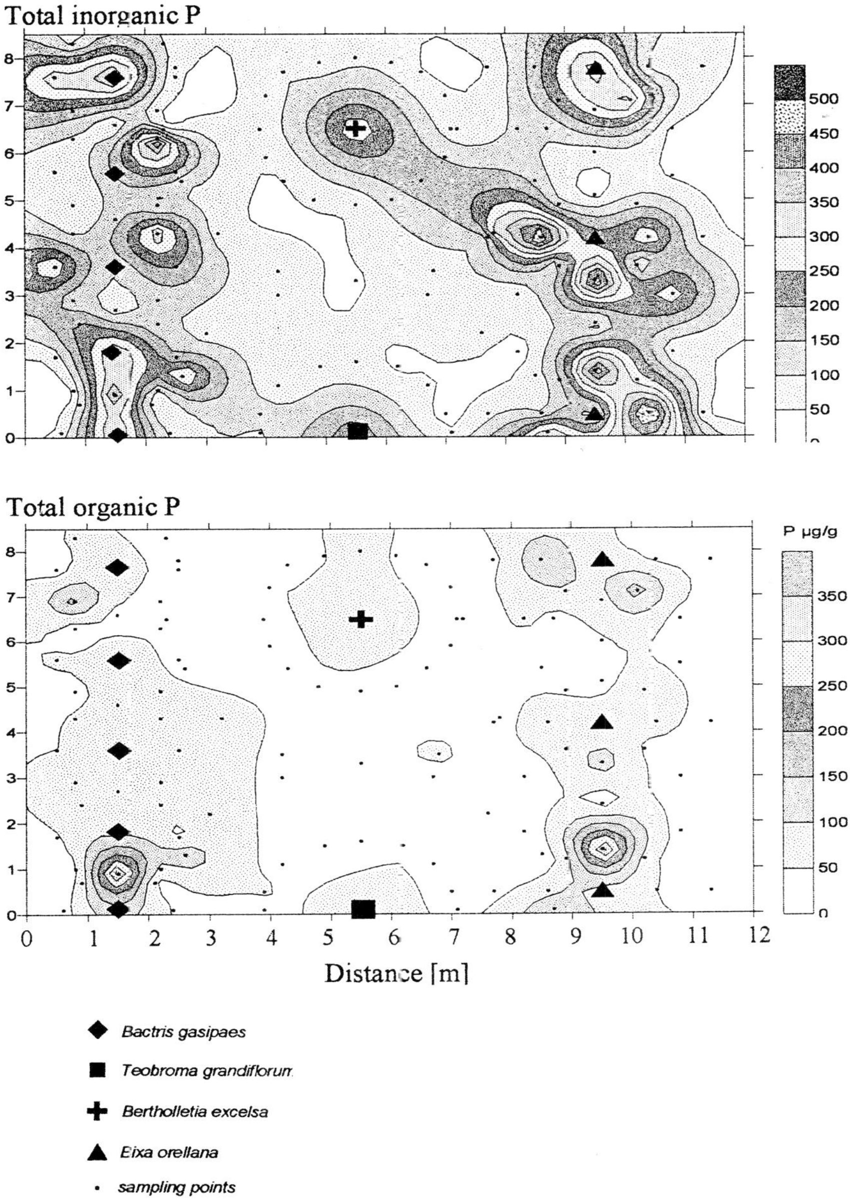

Fig. 1 Spatial distribution of soil inorganic and organic P in an agroforest with four different fruit and timber trees on a Xanthic Ferralsol in the central Amazon (D. Günther, and J. Lehmann, unpubl. data). Soil inorganic P enrichment was not homogeneously distributed under the canopies probably due to uneven application.

uptake of trees either from different soil areas (e.g. nitrate adsorbed in the subsoil), at different times (e.g. available nutrients between cropping cycles of annual crops) or from different soil pools (e.g. fertilizer phosphorus immobilized in recalcitrant soil pool). The nutrients taken up are subsequently returned to soil by litterfall, throughfall/stemflow, mulching, root turnover or exudation. This enrichment will not be homogeneously distributed throughout the agroforestry system.

Not only are tree and annual crop species and their management in agroforestry systems responsible for spatial or temporal heterogeneity of soil fertility but also preferences of farmers for certain crops (Woomer *et al.*, 1998). Individual farmers may prefer to fertilize a particular crop within the system due to its market value, medicinal value or because of dietary reasons.

Heterogeneity of soil fertility is also observed at a micro-scale, e.g. in micro-sites, at the aggregate scale or in different soil pools. The occurrence of the micro-scale spatial heterogeneity is not specific to agroforestry but is frequently observed in natural ecosystems as shown for N and P in a dry tropical forest in India (Roy and Singh, 1995). A combination of plant species, such as seen in agroforestry systems, may utilize this heterogeneous source of nutrients more efficiently than a single crop, because root systems of different plant species possess different plasticity to respond to micro-site fertility as summarized by Schroth (1999). Whether different plants can use different nutrient sources at an aggregate scale is less clear. This depends on the ability of the tree to form very fine roots or associations with mycorrhiza. The varying accessibility of soil nutrient pools by different plant species, however, is better documented (e.g., by root exudation etc.; see sections below).

Heterogeneity of Nutrient Fluxes

The fluxes and transformations which create the described heterogeneity of soil fertility in the field are not the same throughout the whole system. A single tree or even a hedgerow will most likely take up more nutrients from the subsoil underneath the canopy than from the subsoil further away from the stem. This will lead to an uneven nutrient depletion, and explains the heterogeneity encountered when measuring nitrate profiles even in relatively homogenous agroforestry systems such as intensive fallows with *Sesbania sesban* in Kenya (Mekonnen *et al.*, 1999). In alley cropping with tagasaste (*Chamaecytisus proliferus*) in Australia, soil water (and therefore probably nutrients) was taken up from at least the same lateral distance from the tree as depth, resulting in a highly heterogeneous depletion pattern (Lefroy and Stirzaker, 1999).

In mixed tree stands as in an Amazonian seasonal forest, root distribution was highly heterogeneous, and pockets of roots from understorey plants were found as deep as 4 m, while growth of roots from emergent trees was as shallow as 1 m (Sternberg *et al.*, 1998). Also from the same soil volume associated plant species may access different soil pools in crop combination systems. For example, Ae *et al.* (1990) demonstrated that pigeon pea (*Cajanus cajan*) was able to mobilize and take up P occluded in iron oxides in contrast to intercropped maize in Nigeria.

In simultaneous systems not only the nutrient uptake but also the nutrient return is usually very heterogeneous. In an Amazonian agroforest the annual P return by litterfall was 30 times higher under annatto (*Bixa orellana*) than under cupuaçu (*Theobroma grandiflorum*) (K. Uguen, unpubl. data), being one of the reasons for the heterogeneous soil P contents shown in Fig. 1. In an arid environment, the below-ground N input by root turnover was found to be 2–4 times higher under the hedgerow of *Acacia saligna* than in the alley and 3–8 times higher in the topsoil than in the subsoil (Lehmann and Zech, 1998). These examples illustrate that the nutrient uptake from soil as well as the nutrient return to soil is highly heterogeneous in agroforestry systems. This heterogeneity requires specific conceptual approaches for their evaluation especially in highly complex multi-strata agroforests, i.e. monitoring single-tree effects, which was developed from natural ecosystems and applied to agroforestry by several authors (see review by Schroth *et al.*, 2001).

CONTROLLING NUTRIENT FLUXES IN AGROFORESTRY

Nutrient fluxes can be controlled by agroforestry through three factors: the type of agroforestry system, the management of the system, and the species selection. All three factors influence each other. A certain type of agroforestry system can also determine the species to be used, if we need to choose a deep rooting tree species that captures subsoil nutrients and exerts little competition in a hedgerow intercropping system. The species selection may in turn determine the management, e.g. how to prune it to achieve maximum nutrient recycling. The management in turn may affect the type of agroforestry system: intercropping may not be possible, if pruning cannot be done in the intensity required for reducing nutrient competition between tree and crop, e.g. when fruit production is the main objective of a particular small-holder as shown for *Parkia biglobosa* in West Africa (Timmer *et al.*, 1996). In this contribution, the type of agroforestry systems and species selection will be only briefly discussed and more emphasis will be given to the possibilities of controlling nutrient fluxes by agroforestry management.

Type of Agroforestry System

The type of agroforestry system used very strongly influences nutrient fluxes. Intensive fallows can have large N inputs by biological N_2-fixation, large nutrient retrieval from the subsoil and nutrient return by litter and by root turnover (Szott *et al.*, 1999). In western Kenya, subsoil N could be efficiently retrieved from as deep as 4 m by *S. sesban* fallows and recycled with litterfall within a few years (Hartemink *et al.*, 1996; Mekonnen *et al.*, 1997). For an enrichment with Ca, however, fallow periods of at least 2–6 years are needed even on high base status soils (Szott *et al.*, 1999). Simultaneous agroforestry systems are usually less effective in their ability to capture subsoil nutrients and return significant amounts of nutrients to the topsoil than sequential systems. However, recycling of nutrients between tree biomass and topsoil can be significant on a small-scale even in simultaneous systems. In a multistrata agroforest with perennial crops in the central Amazon, annatto (*B. orellana*) returned 30 g P per tree and year with litterfall (K. Uguen, unpubl. data), which almost matched the annual P applications by fertilizer.

In many cases, however, the type of agroforestry system may not primarily be chosen according to processes linked to nutrient fluxes, but to preferences of harvested products or be determined by environmental constraints. If maize production is envisaged on an infertile soil in a subhumid climate with little access to mineral fertilizer, soil fertility replenishment is a major objective and intensive fallows will offer feasible solutions. However, where nutrient availability is not a main constraint as on high base status soils, where land availability is limited, and where timber or fruit production (exception: semi-perennial fruit trees) is a major aim of production, sequential systems may not be used at all.

Management

Fertilizer Application

The incorporation of nutrients in the biomass and their subsequent return to soil is controlled by the amount of available soil nutrients. If biomass production is limited by specific soil nutrients, a recapitalization of the soil nutrient stocks of the whole cropping system (Sanchez *et al.*, 1997b) can lead to a larger nutrient cycling. Thus, the replenishment of a specific nutrient deficiency can serve as a start-up for the whole system. This may be the case in P-fixing soils of the central Amazon, when initial P fertilization can establish a legume cover crop that enriches N levels by biological N_2-fixation and keeps much of the applied P in available form by a rapid nutrient cycling. The legume cover crop *Pueraria phaseoloides* was able to utilize fertilizer K applied to fruit trees in a central Amazonian agroforest and to increase the cycling of other nutrients (Lehmann *et al.*, 2000b). A "jump-start" with organic fertilizer in the form of *Tithonia diversifolia* mulch was suggested by Jama *et al.* (2000). A replenishment of soil fertility by mulch applications, however, is only possible on a small scale, as mulching is rather a redistribution of nutrients than an addition at the farm or landscape scale (Palm *et al.*, 1997; Buresh and Tian, 1998).

The amount of nutrient leaching losses and nutrient uptake is directly affected by the amount of fertilizer application. Additionally, different nutrients are not equally susceptible to leaching and may be taken up at different amounts by plants. While N is rapidly nitrified in tropical aerobic soils and leached as nitrate, P is often fixed to oxides and not transported in large amounts in the profile, especially in finely textured and highly weathered tropical soils with high oxide contents (van Wambeke, 1992). Also type, placement and timing of application greatly determines losses by leaching and uptake. These are commonplace results and not specific to agroforestry systems. However, if tree mulch applications or nutrient redistribution by trees in cropland affect leaching and uptake, these may be controlled by fertilizer management or vice versa. Temporary immobilization of applied fertilizer N to reduce leaching risks should conceptually be possible by joint application with low-quality mulch, although Palm *et al.* (1997) reported low nutrient use efficiency and high N losses especially by denitrification from several experiments. Also from high quality litter, large N losses were reported with 25–41% of applied N with *Leucaena leucocephala* in semi-arid Australia (Xu *et al.*, 1993). However, negligible amounts of applied mulch N from *A. saligna* to sorghum were lost in comparison to 37% from applied ammonium sulfate in a soil with high pH (Lehmann *et al.*, 1999a).

Biological N-fixation is usually reduced by N fertilization (Marschner, 1995). Nitrogen-fixation of two *Sesbania* species decreased from 64 to 2 % with increasing N additions (Becker and George, 1995). However, if the plant mineral N demands are not met by the soil N supply, legume N fixation may even continue at high soil mineral N levels as shown for soybean (George and Singleton, 1992). Thus, N fertilization of 20 mg kg^{-1} soil at tree planting was beneficial for growth and nodule formation of *L. leucocephala* (Sanginga *et al.*, 1988). Higher soil P, Ca, Mo, Fe and Co contents as well as lower soil acidity generally improve biological

N fixation (Marschner, 1995). However, Reddell *et al.* (1997) reported a lower P demand for maximum N_2 fixation than for maximum growth of *Casuarina cunnighamiana* seedlings in a laboratory study. Therefore, the often stated high P demand for maximum N_2-fixation capacity seemed not to hold for all legumes (Sprent, 1999). Rhizobial strains differ markedly in their P demand and response (Smart *et al.*, 1984). A better insight into nutrient acquisition of legumes is needed to evaluate the responses of N_2-fixation to fertilization.

The amount and type of fertilization also affect the nutrient flux from recalcitrant soil pools (i.e., recalcitrant organic nutrients, nutrients fixed to mineral surfaces, nutrients in primary minerals). First, there is no need for the plant to explore recalcitrant soil pools if there is a sufficient amount of available nutrients in the soil solution, from where plants take up their nutrients. This applies to plant uptake as well as to secondary uptake mechanisms such as associations with mycorrhiza, P solubilizing bacteria, or root exudation. With an addition of soluble P, the activity of bacteria to solubilize recalcitrant P pools is greatly diminished, as shown by Illmer and Schinner (1995). Similarly, P additions decreased ecto-mycorrhizal infection of loblolly pine (Marx *et al.*, 1977), and Fe deficiency was reported to increase root exudation and subsequently Fe uptake by soybean (Jolly and Brown, 1987). Therefore, both the benefits by N fixation and by nutrient acquisition from recalcitrant soil pools are often larger when soil nutrients (or a specific nutrient) are limiting plant growth.

Nutrient capture from the subsoil or from a lateral distance may be controlled by the amount and placement of fertilizer applications. A high fertilization around the stem of *T. grandiflorum* was probably responsible for the low N utilization from the area underneath an associated legume cover crop (Lehmann *et al.*, 2000a). In the same agroforestry system, the low N acquisition from the alley by the tree also led to a low utilization of biologically fixed N from the legume intercrop. Wahid *et al.* (1993) concluded from several field trials that the fertilizer application at 0.25 m depth adopted in India leads to a deeper root growth of coconut (*Cocos nucifera*) than a surface application. A deeper root growth may increase the uptake of leached nutrients or nutrients released by weathering from parent material. The amount of leached nitrate adsorbed to the subsoil of acid tropical soils can be very high with reported values of 70 to 2200 kg N ha^{-1} for different sites in Kenya (Buresh and Tian, 1998). In some locations also Ca or K from mineral weathering can be found in appreciable amounts at greater depths even under tropical forests as reviewed by Bruijnzeel (1991). No direct quantification of the utilization of this nutrient source by trees is currently available.

With fertilization, foliar nutrient concentrations increase and therefore litter and mulch decomposition is enhanced. Decomposition and N release were found to be positively related to the N content and inversely to the lignin and/or polyphenol content of the original plant material (Fox *et al.*, 1990; Palm and Sanchez, 1991; Oglesby and Fownes, 1992; Handayanto *et al.*, 1995; Tian *et al.*, 1995; Lehmann *et al.*, 1995). Handayanto *et al.*, (1995) could control the foliar N as well as polyphenol contents with legume N nutrition and therefore also the N release patterns. Generally, the polyphenol contents of leaves increase with N deficiency and decrease with N fertilization as reviewed by Mafongoya *et al.* (1998). Therefore, fertilizer applications not only increase the amount of nutrients taken up by a higher biomass production and higher foliar nutrient contents, but also increase the speed of nutrient return by accelerating nutrient release during decomposition. Increasing the biomass nutrient contents and uptake may improve the overall benefit of mulching systems as shown by the generally higher soil fertility effects of hedgerow intercropping on fertile than infertile soils (Rao *et al.*, 1998).

Shoot Pruning

The most extensively studied effect of shoot pruning on nutrient fluxes in agroforestry is the direct transfer of the pruned biomass from tree to associated crops. This nutrient flux can be controlled by the leaf litter quality, the time of application, the pre-treatment (e.g. drying), the type of application (incorporation or surface application) and the mixture of prunings from different species. These direct effects have been extensively reviewed (Palm, 1995; Mafongoya *et al.*, 1998) and will not be discussed here. Secondary effects of mulch applications include depressions of biological N_2-fixation due to elevated amounts of soil mineral N as reported by Kadiata *et al.*, (1998b). This decline was only found if the prunings were incorporated into the soil. Surface applied biomass even increased total amounts of fixed N (but not the proportion), probably due to a better supply with other nutrients or organic carbon.

Shoot pruning caused a complete disappearance of nodules of the N_2 fixing *Erythrina poeppigiana* (Nygren and Ramirez, 1995) and halved nodule N of *Albizia lebbek* and *L. leucocephala* (Kadiata *et al.*, 1998a). It can therefore be assumed that N_2-fixation is greatly reduced after pruning as shown by a 35 % lower N_2-fixation after pruning of *Gliricidia sepium* in Nigeria (Sanginga *et al.*, 1994). However, Kadiata *et al.* (1998b, 1997) did not find a lower N_2-fixation after pruning of *G. sepium* and *L. leucocephala*. Similarly, N_2-fixation did not decrease with time in frequently pruned *Calliandra calothyrsus* or *G. sepium* in subhumid

Australia (Peoples *et al.*, 1996). This leads us to assume that the ability to fix atmospheric N_2 can be rapidly restored after pruning as shown from the rapid increase of nodule biomass by *S. sesban* and *L. leucocephala* within two weeks (Fownes and Anderson, 1991), but only to the extent of regrowth of the above- and below-ground biomass. Consequently, the total amount of biological N_2-fixation decreased while the proportion of fixed N remained the same (Kadiata *et al.*, 1998b), but only for low pruning intensities (Kadiata *et al.*, 1997). With such a flexible adaptation, N contribution to the cropping system by biological N_2-fixation is then determined by the impact of pruning on biomass production.

Pruning decreased total biomass production of *L. leucocephala* in Hawaii (Guevarra *et al.*, 1978), of tea in Sri Lanka (Kandiah *et al.*, 1984), of *Acacia saligna* in Kenya (Droppelmann *et al.*, 2000) and of *A. mangium* in Indonesia (Tuomela *et al.*, 1996). After six prunings in four month intervals 83% of planted *S. sesban* died in semi-arid Rwanda (Balasubramanian and Sekayange, 1991). Therefore, total N fluxes into aboveground biomass were reduced as demonstrated for *G. sepium, S. grandiflora* and *L. leucocephala* (Guevarra *et al.*, 1978; Duguma *et al.*, 1988; Kadiata *et al.*, 1997, 1998b). The same was shown for increasing pruning intensities and for lower heights of pruning (Guevarra *et al.*, 1978). However, partial pruning of only lower branches increased the foliage efficiency (biomass increment per unit foliage area) for *Pinus radiata* by 78 and 23% in monoculture and with an understorey of lucerne in New Zealand (Bandara *et al.*, 1999).

The new formation of the tree top reduces the carbohydrate reserves in stem and belowground storage tissues (Kandiah *et al.*, 1984; Eissenstatt and Duncan, 1992; Erdmann *et al.*, 1993) and regular pruning leads to a transfer of assimilates from root to shoot biomass (Schroth and Zech, 1995). The regrowth and recovery after pruning depends on the extent of allocation of starch reserves to the stem and the roots (Nygren *et al.*, 2000), which is highly species dependent. *L. leucocephala* is very resistant to pruning and had a 30% higher fraction of belowground biomass than *S. sesban* which was observed to have lower biomass production after cutting (Fownes and Anderson, 1991). Also the pruning height determines the lag-time of regrowth. Nutrient uptake decreased with increased pruning height for two *L. leucocephala* cultivars due to a larger carbohydrate transfer into pods instead of into roots (Guevarra *et al.*, 1978).

The ability of trees to retrieve nutrients from neighboring areas (e.g., the alley between tree rows) or the subsoil is greatly influenced by pruning. The relative P uptake by peachpalm (*Bactris gasipaes*) determined with the double labeling approach using P radioisotopes was relatively higher in the subsoil and lower in the topsoil without pruning (for fruit production) than with regular pruning (for heart of palm production) (Lehmann *et al.*, 2001c). Lower root length density in the subsoil accompanied by a higher water uptake was reported from *A. nilotica* and *Prosopis juliflora* as a result of pruning in north-east Nigeria (Jones *et al.*, 1998). Also in the alley between tree rows of *A. saligna* live root length density was significantly reduced by pruning in Northern Kenya (Fig. 2; Peter and Lehmann, 2000). Lower uptake from the alley in hedgerow intercropping systems decreases competition, but at the same time also decreases the nutrient use efficiency especially during times of the year when annual crops do not take up nutrients (Schroth, 1999). Additionally, a low nutrient uptake can lead to increased losses through leaching as indicated by the higher mineral N and soil water contents in the alley between

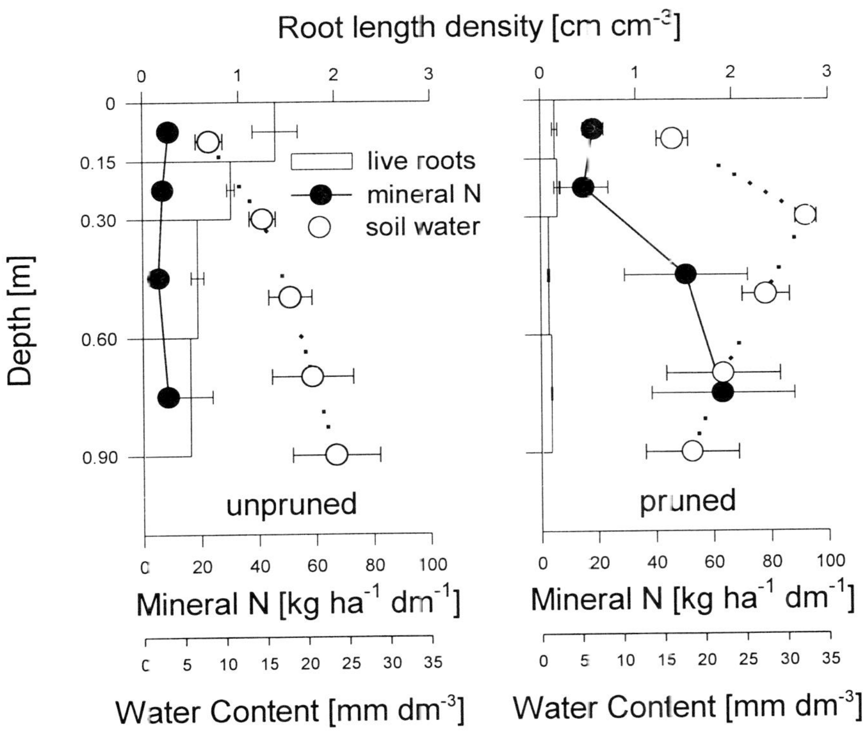

Fig. 2 Root length distribution of *Acacia saligna* (Labill.) H.L. Wend., soil mineral N (Peter and Lehmann, 2000) and soil water contents (recalculated after Droppelmann *et al.*, 2000) as affected by pruning in Northern Kenya during the dry season in December 1996 (means and standard errors; n=3).

pruned than unpruned hedges of *A. saligna* (Fig. 2; Peter and Lehmann, 2000). Water and presumably also nutrient fluxes into the subsoil were higher after heavy pruning of *Grevillea robusta* intercropped with maize in Kenya (Jackson *et al.*, 2000). This was mainly an effect of reduced water uptake, but partly also of lower rainfall interception in the canopy.

Higher nutrient contents and amounts in the aboveground biomass usually translate into higher uptake and return, as well as more rapid decomposition (see section above). However, as a result of pruning, *Erythrina poeppigiana* had an almost three times higher N return to soil and a higher relation of nutrient turnover to nutrients in the biomass than unpruned *Cordia alliodora* planted as shade trees in coffee plantations. At the same time, *C. alliodora* had higher amounts of N in the aboveground biomass than *E. poeppigiana* (Fassbender *et al.*, 1991). Therefore, pruning increased nutrient fluxes despite lower aboveground nutrient contents.

Shoot pruning may also directly affect foliar nutrient contents and control nutrient fluxes to and from the biomass. Thus, pruning significantly reduced the foliar P contents of *G. sepium, Senna siamea,* and *S. spectabilis* (Sanginga *et al.*, 1994) and *A. saligna* (Peter and Lehmann, 2000). This can be explained by the lower root length density after shoot pruning which was shown by several authors (Schroth and Zech, 1995; Lehmann *et al.*, 1998; Peter and Lehmann, 2000) and consequently by the difficulty of replacing the nutrients lost with the removed biomass. Leaf regrowth can therefore not utilize nutrients from old leaves. The uptake of nutrients which are relatively immobile in the soil (such as P) decrease, whereas relatively mobile nutrients (such as Ca) are less affected.

In contrast to the above-ground nutrient input by pruning, little is known about the magnitude and dynamics of nutrient inputs by roots after pruning. Root death increases after shoot pruning (see above) and therefore also the nutrient input by roots. The complete turnover of nodules through pruning comprised 10% of the aboveground N input by litterfall and prunings of the N_2 fixing *E. poeppigiana* (Nygren and Ramirez, 1995). Other authors also report a drastic reduction of nodules after shoot pruning (Fownes and Anderson, 1991; Sanginga *et al.*, 1994; Kadiata *et al.*, 1998a). However, total N release in the root zone of *Albizia lebbek* and *L. leucocephala* was lower with than without shoot pruning in a pot experiment (Kadiata *et al.*, 1998a). Shoot pruning increases the below-ground nutrient input over a short term as seen from the listed examples, but the constant turnover of a larger belowground biomass from unpruned trees than pruned trees may yield a higher nutrient input over a longer period.

Few studies have been published using different pruning intensities (e.g., only removal of a percentage of the above-ground biomass; Kandiah *et al.*, 1984; Jackson *et al.*, 2000; Nygren and Cruz, 1998). With a partial pruning it may be possible to reduce root death and nodule shedding in comparison to a complete removal of the aboveground foliar biomass as shown for tea in Sri Lanka (Kandiah *et al.*, 1984) or *G. sepium* in Guadeloupe (Nygren and Cruz, 1998). However, nitrogenase activity of *G. sepium* seedlings was lower at 28 days after partial than complete defoliation in a greenhouse study (Nygren *et al.*, 2000). A partial pruning may also alleviate other negative effects of pruning on nutrient fluxes such as lower foliar nutrient contents or higher nutrient leaching, but little information is available up to now.

Spatial Layout

The number of trees per unit area determines the total nutrient uptake of the vegetation cover. However, a higher planting density does not always increase biomass production and nutrient uptake as seen from an acacia intercropping system in Northern Kenya, which had the same biomass production irrespective of planting density (833 vs. 2500 trees ha^{-1}; Droppelmann *et al.*, 2000). Tree growth was limited by available soil resources (in this case soil water), which was efficiently used by fewer trees. However, Guevarra *et al.* (1978) reported 10 and 20% higher N yields of *L. leucocephala* with 66 000 and 133 000 in comparison to 45 000 plants ha^{-1}.

Nutrient fluxes from the subsoil or from neighboring areas can be controlled by changing the planting density and distance between associated crops. Increasing the planting density of *Eucalyptus grandis* from 82 to 2150 trees per ha led to a 3.6 times higher root length density at 1–1.2 m depth (Eastham and Rose, 1990) and a shift of water uptake from 0–1 m (39%) to below 5 m (35%; 1.2 m distance from the trunk) in silvipastoral agroforestry in Australia (Eastham *et al.*, 1990). The same was observed for apple trees in England (Atkinson, 1978). This may indicate that subsoil nutrient capture is enhanced by a higher planting density. Also with pruned *A. saligna* hedgerows in Northern Kenya, live root length density in the subsoil and water uptake increased with higher planting density (Fig. 3). Resource competition at the topsoil is the driving force for this root displacement (Schroth, 1999). How strongly allelopathy affects root growth of associated plants is still rather speculative (Rao *et al.*, 1998). A close contact between intercropped plants is needed in simultaneous systems if nutrient inputs by roots (particularly biologically fixed N from a legume) should improve soil nutrient contents which are to be utilized by an associated crop, such as in hedgerow intercropping systems with a tree legume (Haggar *et al.*, 1993; Lehmann *et al.*, 1999b), in coffee with legume shade trees (Snoeck, 1995) or in mixed fruit tree systems with a legume cover crop (Lehmann *et al.*, 2001a). However, a N benefit of an annual non-legume by intercropping with a tree legume without mulch applications or litter return was difficult to prove due to methodological difficulties (Lehmann *et al.*, 2001a). Also in annual intercropping systems, transfer of biologically fixed N_2 from a legume to a non-legume is difficult to determine (Danso *et al.*, 1987; Viera-Vargas *et al.*, 1995).

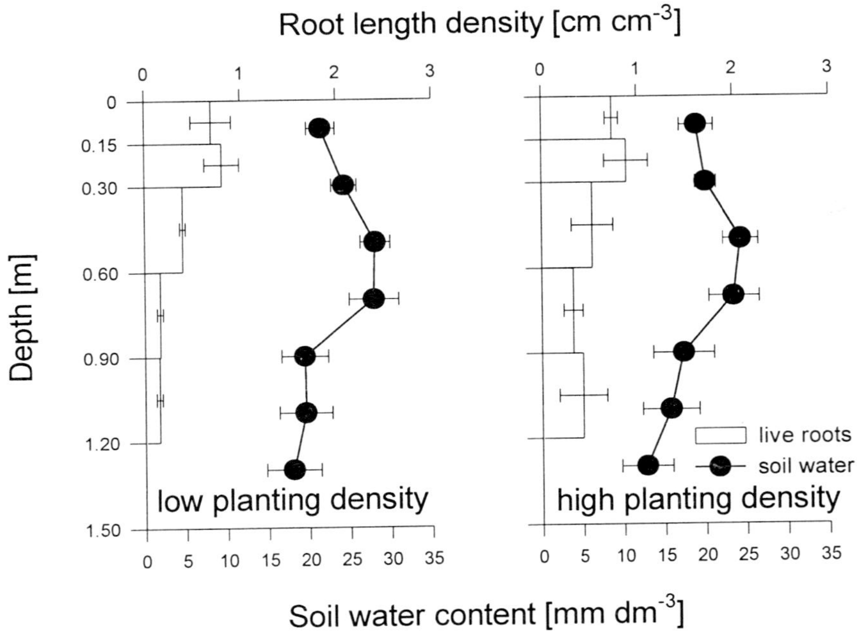

Fig. 3 Root length distribution of *Acacia saligna* (Labill.) H.L. Wend. (Lehmann, unpublished data) and soil water contents (recalculated after Droppelmann *et al.*, 2000) as affected by planting density at 833 and 2500 trees ha^{-1} in Northern Kenya during the dry season December/ January 1995/96 (means and standard errors; root distribution n=3; soil water n=12 and 11 for low and high planting density, respectively).

Dinitrogen fixation itself was found to increase from 65 to 96% in faba bean when intercropped with barley due to competition for soil N (Danso *et al.*, 1987). Therefore, N input into a legume intercrop system can be increased by soil N depletion through closer spacing and/or higher planting density.

Also nutrients which are mobilized from recalcitrant soil pools (e.g., occluded P, weathering of primary minerals) by rhizosphere effects of one plant may be used by an associated crop, if both plants are grown with intermingling roots systems. In Nigeria, the recalcitrant Fe-bound P which was mobilized by root exudation of pigeon pea could be taken up by the intercropped maize (Ae *et al.*, 1990). Similarly, Braum and Helmke (1995) observed a benefit for the Cu, Fe and Zn nutrition of soybean when intercropped with white lupine on an acidic soil due to nutrient mobilization by root exudates. The same authors also found a mobilization of recalcitrant soil P by lupine, but could not find a benefit for the interplanted soybean. Possibly, the lower mobility of P than of the other nutrients in the soil precluded transport to the root surface of the soybean.

A better nutrient utilization of associated plants by closer spacing has to be weighed against a higher nutrient competition and is a trade-off between positive and negative effects of a close rhizosphere contact (Schroth, 1999).

Soil Preparation

Most techniques of soil preparation for planting annual crops will cut roots e.g. of tree hedges or boundary plantings before crop sowing. Root death in the topsoil will cause an input of the nutrients contained in it, which can be a substantial addition depending on root mass (Schroth and Zech, 1995; Lehmann and Zech, 1998), its nutrient content and decomposition rate (Sandhu *et al.*, 1990; Lehmann *et al.*, 1995; Arunachalam *et al.*, 1996). This nutrient input will probably not be perceived as an incentive for soil preparation, but should be considered in nutrient flux calculations. A more important management aspect is how root distribution and therefore lateral nutrient retrieval and nutrient capture from the subsoil can be influenced by soil preparation. With deep plowing to a depth of 0.8 m the roots of black locust hedges (*Robinia pseudoacacia*) were cut and total root length density decreased in the topsoil under alley cropped lucerne near the trees (Fig. 4). It can therefore be assumed that the lateral nutrient uptake by the trees from the alley was also reduced by plowing. On the other hand, zero-tillage can increase nutrient fluxes from the alley to the hedges as seen from the higher belowground competition than with conventional soil preparation (Ssekabembe, 1985). Hoe-ridging or plowing were not different from each other in terms of reducing tree competition by adjacent *S. siamea* (Schroth *et al.*, 1995). The complex effects of soil preparation on nutrient fluxes make an interpretation extremely difficult, because they are a mixture of the

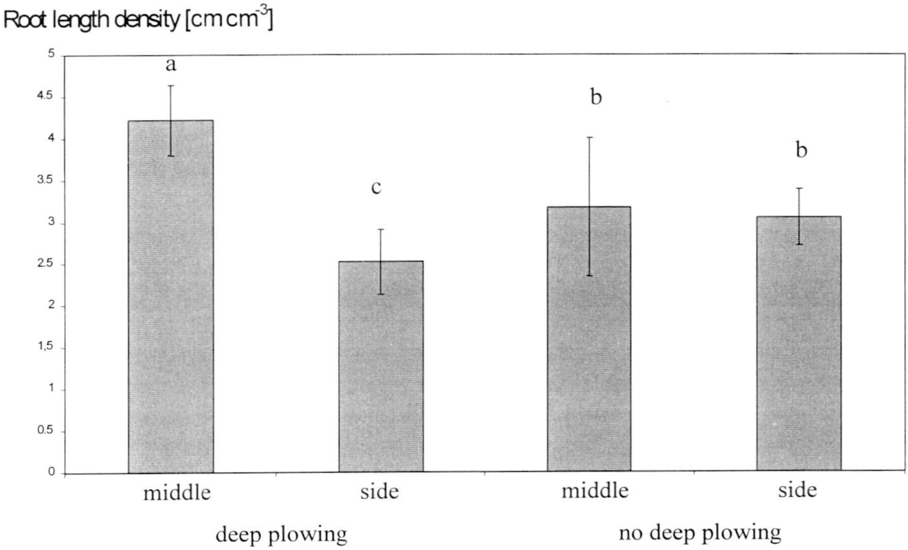

Fig. 4 Root length density of alley cropped lucerne (*Medicago sativa*) with black locust (*Robinia pseudoacacia*) at 0–0.45 m depth near the hedge and in the middle of the alley (18 m row spacing) as affected by subsoiling (deep plowing at 0.8 m) in Eastern Germany in July 13, 1999; bars with the same letters are not significantly different at $p<0.05$ ($n=5$; H. Grünewald, unpublished data; with permission).

effects of cutting the tree roots, of changing soil physical properties, nutrient mineralization and erosion.

An extreme version of soil preparation for controlling root distribution is root pruning, where a trench is dug by hand or with conventional soil preparation methods (plough) along the trees. This root management effectively decreases lateral resource uptake by trees as seen from the lower soil water depletion in the alley and higher biomass production of sorghum between root-pruned (0.3 m) *L. leucocephala* hedges in India (Korwar and Radder, 1994). In central Togo, soil water uptake and presumably also nutrient uptake by *S. siamea* from the alley was reduced by root pruning to a depth of 0.5 m (Schroth *et al.*, 1995). Trenching with inserted plastic sheets was frequently shown to reduce resource use by trees from the alley (Ong *et al.*, 1991) but is not practical for farmer's use. Whether root pruning and therefore the prevention of lateral root growth can increase vertical root growth is less known and warrants further research.

Root trenching not only cuts the tree but also the crop roots. Therefore, crop roots will not be able to use the soil underneath the trees. Lehmann *et al.* (1998) could show that sorghum root length density was higher under associated *Acacia* hedges than in the alley or in monoculture. Similarly, maize yields decreased near the hedgerow of *G. sepium*, if a plastic sheet separated the root systems of tree and crop (Schroth and Lehmann, 1995). These results indicate that also nutrient retrieval by the crops from the area underneath the trees are controlled by root trenching.

Soil preparation methods after intensive fallow or after application of mulch may control the time in which the above- and belowground biomass is decomposed and therefore the nutrients released. Incorporating leaf litter into the soil generally increases nutrient release (Wilson *et al.*, 1986; Lehmann *et al.*, 1995; Mafongoya and Nair, 1997). For root litter, the available information is ambiguous. Trench plot experiments where fine roots are left in their environment and only cut around the plot showed decomposition rates that were twice as high as litterbag studies where the roots were removed from their place of growth (Vogt *et al.*, 1991). For coarse roots, the opposite was observed probably due to higher starch reserves in phloem tissues which were an experimental artefact and would not occur if roots decay after "normal" root death and carbohydrates are resorbed by the plant (Vogt *et al.*, 1991).

Several management options to control nutrient fluxes have not been discussed such as inoculation with mycorrhiza and rhizobia. Additionally not all effects of the described management techniques were listed such as the consequences of soil preparation for pore continuity and nutrient leaching. Although they are important aspects in soil management and influence nutrient fluxes to a large extent, they are not specific to agroforestry.

Species Selection

Static Criteria for Species Selection

Biological N_2-fixation has been regarded as one of the most important incentives for incorporation of trees into cropland. Indeed, the amounts fixed by some trees are impressive, e.g. with 100–500 kg N ha^{-1} yr^{-1} of *L. leucocephala* in monoculture plantations. Even in hedgerow intercropping, biological N-fixation was reported to be very high with 75–100 kg N ha^{-1} yr^{-1} (Young, 1997). However, usually less than 20% of the N transferred to the alley as mulch is utilized by a crop in hedgerow intercropping (Palm, 1995). In agroforestry systems with shade trees, the amount of N_2-fixation by *Erythrina, Inga* or *Gliricidia* was generally below 60 kg N ha^{-1} yr^{-1} (Beer *et al.*, 1998). The ability to fix large amounts of atmospheric N_2 may have been oversold in comparison to other tree properties as pointed out by Sanchez (1995) or Rao *et al.* (1998). However, long-term effects of intercropping with a legume on soil N availability (Haggar *et al.*, 1993) and incorporation of legume N into soil organic matter and subsequent use by maize were observed (Vanlauwe *et al.*, 1999).

The area of soil nutrient uptake and therefore the root activity distribution varies greatly with tree species. Typically, tree crops take up more nutrients from the topsoil than the subsoil (IAEA, 1975). However, the total uptake from the subsoil during the whole year may still be substantial as seen from an uptake of 30–100 kg N ha^{-1} yr^{-1} by improved fallows with *C. calothyrsus*, *S. sesban* and *E. grandis* (Mekonnen *et al.*, 1997; Jama *et al.*, 1998). Some trees may even have a higher root activity in the subsoil than in the topsoil, such as peachpalm (*B. gasipaes*) with 1.7 times higher ^{32}P uptake at 0.6 m than at 0.1 m (Lehmann *et al.*, 2001c).

Generally, the harvested crop should have a high soil nutrient use efficiency and should require only small amounts of nutrients. If a soil nutrient benefit by higher nutrient acquisition and nutrient return of a plant is envisaged, nutrient uptake of the crop or associated service plant (e.g., legume cover crop; hedgerow in a mulch system) should be high. Numerous studies have shown that legume hedges can accumulate large amounts of nutrients (Young, 1997). A large nutrient recycling means large amounts of nutrients in comparison to applied fertilizers or crop nutrient demands. In contrast to a large nutrient recycling also a rapid nutrient recycling may be advantageous depending on the specific nutrient and soil properties. For example, a rapid nutrient cycling may be important in P fixing soils to constantly provide labile P forms. In sandy soils a slower nutrient recycling may be wanted to decrease the danger of nitrate leaching. A rapid nutrient recycling requires rapid nutrient release from litter or prunings and high amounts of nutrients returned per unit nutrient contained in the biomass.

What are the critical levels and a minimum dataset for the evaluation of tree species properties? Certainly, a list of tree properties will fail to be complete and critical levels will vary with site conditions. However, some characteristics may help to rapidly assess if a tree is likely to be suitable for a certain system or management such as plant tissue quality (Palm and Rowlands, 1997), ratios of nutrient turnover to plant contents and of harvestable to cycled nutrients, or above- and belowground nutrient turnover.

Criteria for a rapid nutrient release from plant litter are tissue contents of 18 to 22 mg N g^{-1} and 2.5 mg P g^{-1} (Palm *et al.*, 1997). The N release, however, is not only determined by the tissue N contents alone but also by organic constituents such as lignin and polyphenols (Mafongoya *et al.*, 1998). The relationship between these organic compounds, the tissue N content and decomposition and nutrient release have been often shown using various ratios (see above). Above 150 mg g^{-1} lignin and 40 mg g^{-1} polyphenols, net N immobilization may occur depending on the N content, for which a simple decision tree was introduced by Palm *et al.* (1997).

The properties of different tree species and the relevance of specific properties have not been fully explored up to now, and more information needs to be gathered about tree species effects on nutrient fluxes especially of trees with only regional importance.

Dynamic Criteria for Species Selection

Not only the single-species characteristics but also the "system-species" and "management-species" characteristics are important criteria for the suitability of a certain tree species. A tree may utilize nutrients from the topsoil in a monoculture system, but from the subsoil in an agroforestry combination, if the associated plants are more competitive. In a hedgerow intercropping system in Kenya, the tree species had a deeper root system and the annual crop a shallower root system than when grown in monoculture (Lehmann *et al.*, 1998). To achieve complementarity of nutrient uptake and deeper nutrient capture by trees, the topsoil competition between a shallow-rooted grass and a tree may even be used to achieve a downward displacement of tree root activity. This will require a flexible tree root system (Schroth, 1999).

Furthermore, not only the system-species characteristics but also the reaction to management is important, e.g. how much N a legume species fixes under heavy pruning (see section above). These effects of management and system design on tree properties are highly dependent on the tree species.

CONCLUSIONS

Management interventions such as shoot and root pruning, fertilization, spacing and soil preparation are powerful tools for controlling nutrient fluxes. The effect of management, however, depends on species selection and system design. These interactions are quite complex and do not allow easy prediction. The heterogeneity of nutrient fluxes is the key for sustaining soil fertility through agroforestry, because it promotes a higher resource use efficiency (e.g., through diversity of nutrient acquisition) and flexibility (more management options). Agroforestry management can be equally effective in controlling nutrient flows as system design or species selection, and shoot pruning was identified as an important tool for manipulating these fluxes. However, the assessment of the heterogeneity of nutrient fluxes in a given agroforestry system is difficult and the development of methodologies and concepts warrants further research.

ACKNOWLEDGEMENTS

The comments of Götz Schroth during the preparation of the manuscript and the corrections of Aimée Murphy-Lehmann are much appreciated.

REFERENCES

Ae, N., Arihara, J., Okada, K., Yoshihara, T., and Johansen, C., 1990. Phosphorus uptake by pigeon pea and its role in cropping systems of the Indian subcontinent, *Science*, **248:** 477–480.

Arunachalam, A., Pandey, H.N., Tripathi, R.S., and Maithani, K., 1996. Fine root decomposition and nutrient mineralization patterns in a subtropical humid forest following tree cutting, *For. Ecol. Manage.*, **86:** 141–150.

Atkinson, D., 1978. The use of soil resources in high density planting systems, *Acta Horticulturae*, **65:** 79–89.

Balasubramanian, V., and Sekayange, L., 1991. Effects of tree legumes in hedgerows on soil fertility changes and crop performance in the semi-arid highlands of Rwanda, *Biol. Agric. Hort.*, **8:** 17–32.

Bandara, G.D., Whitehead, D., Mead, D.J., and Moot, D.J., 1999. Effects of pruning and understorey vegetation on crown development, biomass increment and above-ground carbon partitioning in *Pinus radiata* D. Don trees growing at a dryland agroforestry site, *For. Ecol. Manage.*, **124:** 241–254.

Becker, M., and George, T., 1995. Nitrogen fixation response of stem- and root-nodulating Sesbania species to flooding and mineral nitrogen, *Plant and Soil*, **175:** 189–196.

Beer, J., Muschler, R., Kass, D., and Somarriba, E., 1998. Shade management in coffee and cocoa plantations, *Agrofor. Syst.*, **38:** 139–164.

Braum, S.M., and Helmke, P.A., 1995. White lupin utilizes soil phosphorus that is unavailable to soybean, *Plant and Soil*, **176:** 95–100.

Bruijnzeel, L.A., 1991. Nutrient input-output budgets of tropical forest ecosystems: a review, *J. Trop. Ecol.*, **7:** 1–24.

Buresh, R.J., 1999. Phosphorus management in tropical agroforestry: current knowledge and research challenges, *Agrofor. Forum.*, **9:** 61–66.

Buresh, R.J., and Tian, G., 1998. Soil improvement by trees in sub-Saharan Africa, *Agrofor. Syst.*, **38:** 51–76.

Cannell, M.G.R., van Noordwijk, M., and Ong, C.K., 1996. The central agroforestry hypothesis: the tree must acquire resources that the crop would not otherwise get, *Agrofor. Syst.*, **34:** 27–31.

Danso, S.K.A., Zapata, F., Hardarson, G., and Fried, M., 1987. Nitrogen fixation in faba bean as affected by plant population density in sole and intercropped systems with barley, *Soil Biol. Biochem.*, **19:** 411–415.

Dhillon, S.S., 1999. Environmental heterogeneity, animal disturbances, microsite characteristics, and seedling establishment in a *Quercus havardii* community, *Rest. Ecol.*, **7:** 399–406.

Drechsel, P., and Reck, B., 1998. Composted shrub-prunings and other organic manures for smallholder farming systems in southern Rwanda, *Agrofor. Syst.*, **39:** 1–12.

Droppelmann, K.J., Lehmann, J., Ephrat, J.E., and Berliner, P.R., 2000. Water use efficiency and uptake patterns in a runoff agroforestry system in an arid environment, *Agrofor. Syst.*, **49:** 223–243.

Duguma, B., Kang, B.T., and Okali, D.U.U., 1988. Effect of pruning intensities of three woody leguminous species grown in alley cropping with maize and cowpea on an alfisol, *Agrofor. Syst.*, **6:** 19–35.

Eastham, J., and Rose, C.W., 1990. Tree/pasture interactions at a range of tree densities in an agroforestry experiment. I Rooting patterns, *Aust. J. Agric. Res.*, **41:** 682–695.

Eastham, J., Rose, C.W., Cameron, D.M., Rance, S.J., Talsma, T., and Charles-Edwards, D.A., 1990. Tree/pasture interactions at a range of tree densities in an agroforestry experiment. II Water uptake in relation to rooting patterns, *Aust. J. Agric. Res.*, **41:** 697–707.

Eissenstatt, D.M., and Duncan, L.W., 1992. Root growth and carbohydrate responses in bearing citrus trees following partial canopy removal, *Tree Physiology*, **10:** 245–257.

Erdmann, T.K., Nair, P.K.R., and Kang, B.T., 1993. Effects of cutting frequency and cutting height on reserves carbohydrates in *Gliricidia sepium* (Jacq.), *Walp. For. Ecol. Manage.*, **57:** 45–60.

Fassbender, H.W., Beer, J., Heuveldop, J., Imbach, A., Enriquez, G., and Bonnemann, A., 1991. Ten year balances of organic matter and nutrients in agroforestry systems in CATIE, Costa Rica, *For. Ecol. Manage.*, **45:** 173–183.

Fownes, J.H., and Anderson, D.G., 1991. Changes in nodule and root biomass of *Sesbania sesban* and *Leucaena leucocephala* following coppicing, *Plant and Soil*, **138:** 9–16.

Fox, R.H., Myers, R.J.K., and Vallis, I., 1990. The nitrogen mineralization rate of legume residues in soil as influenced by their polyphenolic, lignin and nitrogen contents, *Plant and Soil*, **129:** 251–259.

Geiger, S.C., Vandenbeldt, R.J., and Manu, A., 1994. Variability in the growth of *Faidherbia albida:* the soils connection, *Soil Sci. Soc. Am. J.*, **58:** 227–231.

George, T., and Singleton, P.E., 1992. Nitrogen assimilation traits and dinitrogen fixation in soybean and common bean, *Agron. J.*, **84:** 1020–1028.

Guevarra, A.B., Whitney, A.S., and Thompson, J.R., 1978. Influence of intra-row spacing and cutting regimes on the growth and yield of leucaena, *Agron. J.*, **70:** 1033–1040.

Haggar, J.P., Tanner, E.V.J, Beer, J.W., and Kass, D.C.L., 1993. Nitrogen dynamics of tropical agroforestry and annual cropping systems, *Soil Biol. Biochem.*, **25:** 1363–1378.

Handayanto, E., Cadisch, G., and Giller, K.E., 1995. Manipulation of quality and mineralization of tropical legume tree prunings by varying nitrogen supply, *Plant and Soil*, **176:** 149–160.

Hartemink, A.E., Buresh, R.J., Jama, B., and Janssen, B.H., 1996. Soil nitrate and water dynamics in sesbania fallow, weed fallow, and maize, *Soil Sci. Soc Am. J.*, **60:** 568–574.

IAEA, 1975. Root Activity Patterns of some Tree Crops, *Technical Report Series 170*, International Atomic Energy Agency, Vienna, 154 pp.

Illmer, K., and Schinner, M., 1995. Phosphate solubilizing microorganisms under non-sterile conditions, *Bodenkultur*, **46:** 197–204.

Jackson, N.A., Wallace, J.S., and Ong, C.K., 2000. Tree pruning as a means of controlling water use in an agroforestry system in Kenya, *For. Ecol. Manage.*, **126:** 133–148.

Jama, B., Buresh, R.J., and Sheperd, K.D., 1998. Vertical distribution of roots and soil nitrate: tree species and phosphorus effects, *Soil Sci. Soc. Am. J.*, **62:** 280–286.

Jama, B., Palm, C.A., Buresh, R.J., Niang, A., Gachengo, C., Nziguheba, G., and Amadalo, B., 2000. *Tithonia diversifolia* green manure for improvement of soil fertility: A review from western Kenya, *Agrofor. Syst.*, **49:** 201–221.

Jolly, V.D., and Brown, J.C., 1987. Soybean response to iron-deficiency as related to iron supply in the growth medium, *J. Plant Nutr.*, **10:** 637–651.

Jones, M., Sincair, F.L., and Grime, V.L., 1998. Effect of tree species and crown pruning on root length and soil water content in semi-arid agroforestry, *Plant and Soil*, **210:** 197–207.

Kadiata, B.D., Mulongoy, K., and Isirimah, N.O., 1997. Influence of pruning frequency of *Albizia lebbek, Gliricidia sepium* and *Leucaena leucocephala* on nodulation and potential nitrogen fixation, *Biol. Fert. Soils*, **24:** 255–260.

Kadiata, B.D., Mulongoy, K., and Mambani, B., 1998a. Pruning effect on nitrogen release in the root zone of *Albizia lebbek* and *Leucaena leucocephala, Biol. Fert. Soils*, **26:** 187–193.

Kadiata, B.D., Mulongoy, K., and Isirimah, N.O., 1998b. Effect of tree pruning and pruning application to trees on nitrogen fixation by *Leucaena* and *Gliricidia, Agrofor. Syst.*, **39:** 117–128.

Kandiah, S., Wettasinghe, D.T., and Wadasinghe, G., 1984. Root influence on shoot development in tea (*Camellia sinensis* (L.) O. Kuntze) following shoot pruning, *J. Hort. Sci.*, **59:** 581–587.

Korwar, G.R., and Radder, G.D., 1994. Influence of root pruning and cutting interval of *Leucaena* hedgerows on performance of alley cropped rabi sorghum, *Agrofor. Syst.*, **25:** 95–109.

Lefroy, E.C., and Stirzaker, R.J., 1999. Agroforestry for water management in the cropping zone of southern Australia, *Agrofor. Syst.*, **45:** 277–302.

Lehmann, J., Schroth, G., and Zech, W., 1995. Decomposition and nutrient release from leaves, twigs and roots of three alley-cropped tree legumes in central Togo, *Agrofor. Syst.*, **29:** 21–36.

Lehmann, J., and Zech, W., 1998. Fine root turnover of hedgerow intercropping in Northern Kenya, *Plant and Soil*, **198:** 19–31.

Lehmann, J., Peter, I., Steglich, C., Gebauer, G., Huwe, B., and Zech, W., 1998. Below-ground interactions in dryland agroforestry, *For. Ecol. Manage.*, **111:** 157–169.

Lehmann, J., Feilner, T., Gebauer, G., and Zech, W., 1999a. Nitrogen uptake of sorghum from tree mulch and chemical fertilizer under high leaching conditions estimated by ^{15}N enrichment, *Biol. Fert. Soils*, **30:** 90–95.

Lehmann, J., Weigl, D., Peter, I., Droppelmann, K., Gebauer, G., Goldbach, H., and Zech, W., 1999b. Nutrient interactions of alley-cropped *Sorghum bicolor* and *Acacia saligna* in a runoff irrigation system in Northern Kenya, *Plant and Soil*, **210:** 249–262.

Lehmann, J., da Silva Jr., J.P., Schroth, G., Gebauer, G., and da Silva, L.F., 2001a. Nitrogen use of mixed tree crop plantations with a legume cover crop, *Plant and Soil*, **225:** 63–72.

Lehmann, J., da Silva Jr., J.P., Trujillo, L., and Uguen, K., 2000b. Legume cover crops and nutrient cycling in tropical fruit tree production, *Acta Horticulturae*, **531:** 65–72.

Lehmann, J., Gebauer, G., and Zech, W., 2001a. Nitrogen cycling assessment in an agroforestry system using ^{15}N enrichment, *Nut. Cycl. Agroecosys*, in press.

Lehmann, J., Günther, D., Mota, M.S., Almeida, M., Zech, W., and Kaiser, K., 2000b. Single-tree effects on soil P and S pools in a multi-strata agroforestry system, *Agrofor. Syst.*, in press.

Lehmann, J., Muraoka, T., and Zech, W., 2001c. Root activity patterns in a tropical agroforest determined by ^{32}P, ^{33}P, and ^{15}N applications, *Agrofor. Syst.*, in press.

Mafongoya, P.L., and Nair, P.K.R., 1997. Multipurpose tree prunings as a source of nitrogen to maize under semiarid conditions in Zimbabwe. I. Nitrogen-recovery rates in relation to pruning quality and method of application, *Agrofor. Syst.*, **35:** 31–46.

Mafongoya, P.L., Giller, K.E., and Palm, C.A., 1998. Decomposition and nitrogen release patterns of tree prunings and litter, *Agrofor. Syst.*, **38:** 77–97.

Marschner, H., 1995. *Mineral Nutrition of Higher Plants.* Academic Press, London, 889 pp.

Marx, D.H., Hatch, A.B., and Mendicino, J.F., 1977. High fertility decreases sucrose content and susceptibility of Loblolly pine roots to ectomycorrhizal infection by *Pisolithus tinctorius*, *Can. J. Bot.*, **55:** 1569–1574.

Mekonnen, K., Buresh, R.J., and Jama, B., 1997. Root and inorganic nitrogen distributions in sesbania fallow, natural fallow and maize fields, *Plant and Soil*, **188:** 319–327.

Mekonnen, K., Buresh, R.J., Coe, R., and Kipleting, K.M., 1999. Root length and nitrate under *Sesbania sesban*: Vertical and horizontal distribution and variability, *Agrofor. Syst.*, **42:** 265–282.

Nygren, P., and Ramirez, C., 1995. Production and turnover of N_2 fixing nodules in relation to foliage development in periodically pruned *Erythrina poeppigina* (Leguminosae) trees, *For. Ecol. Manage.*, **73:** 59–73.

Nygren, P., and Cruz, P., 1998. Biomass allocation and nodulation of *Gliricidia sepium* under two cut-and-carry forage production regimes, *Agrofor. Syst.*, **41:** 277–292.

Nygren, P., and Vaillant, V., Desfontaines, L., Cruz, P., and Domenach, A.M., 2000. Effects of nitrogen source and defoliation on growth and biological dinitrogen fixation of *Gliricidia sepium* seedlings, *Tree Physiol.*, **20:** 33–40.

Oglesby, K.A., and Fownes, J.H., 1992. Effects of chemical composition on nitrogen mineralization from green manures of 7 tropical leguminous trees, *Plant and Soil*, **143:** 127–132.

Ong, C.K., Corlett,, J.E., Singh, R.P., and Black, C.R., 1991. Above and below-ground interactions in agroforestry systems, *For. Ecol. Manage.*, **45:** 45–57.

Palm, C.A., 1995. Contribution of agroforestry trees to nutrient requirements of intercropped plants, *Agrofor. Syst.*, **30:** 105–124.

Palm, C.A., and Sanchez, P.A., 1991. Nitrogen release from leaves of some tropical legumes as affected by their lignin and polyphenolic contents, *Soil Biol. Biochem.*, **23:** 83–88.

Palm, C.A., and Rowlands, A.P., 1997. A minimum dataset for characterisation of plant quality for decomposition, In: *Driven by Nature, Plant Litter Quality and Decomposition*, (G. Cadisch, and K.E. Giller, eds.), CAB International, Wallingford, 379–392 pp.

Palm, C.A., Myers, J.K., and Nandwa, S.M., 1997. Combined use of organic and inorganic nutrient sources for soil fertility maintenance and replenishment, In: *Replenishing Soil Fertility in Africa*, (Buresh, R.J., Sanchez, P.A., and Calhoun, F., eds.), Special Publication No. 51, American Society of Agronomy, Madison, 193–218 pp.

Peoples, M.B., Palmer, B., Lilley, D.M., Duc, L.M., and Herridge, D.F., 1996. Application of ^{15}N and xylem methods for assessing N_2 fixation of three shrub legumes periodically pruned for forage, *Plant and Soil*, **182:** 125–137.

Peter, I., and Lehmann, J., 2000. Pruning effects on root distribution and nutrient dynamics in an acacia hedgerow planting, *Agrofor. Syst.*, **50:** 59–75.

Rao, M.R., Nair, P.K.R., and Ong, C.K., 1998. Biophysical interactions in tropical agroforestry systems, *Agrofor. Syst.*, **38:** 3–50.

Reddell, P., Yun, Y., and Shipton, W.A., 1997. Do *Casuarina cunnighamiana* seedlings depend on symbiotic N_2 fixation have higher phosphorus requirements than those supplied with adequate fertilizer nitrogen?, *Plant and Soil*, **189:** 213–219.

Rhoades, C.C., 1997. Single-tree influences on soil properties in agroforestry: lessons from natural and savanna ecosystems, *Agrofor. Syst.*, **35:** 71–94.

Roy, S., and Singh, J.S., 1995. Seasonal and spatial dynamics of plant-available N and P pools and N-mineralization in relation to fine roots in a dry tropical forest habitat, *Soil Biol. Biochem.*, **27:** 33–40.

Sanchez, P., 1995. Science in agroforestry, *Agrofor. Syst.*, **30:** 5–55.

Sanchez, P.A., Buresh, R.J., and Leakey, R.R.B., 1997a. Trees, soils, and food security. Phil., *Trans. R. Soc. Lond.*, **352:** 949–961.

Sanchez, P.A., Shepherd, K.D., Soule, M.J., Place, F.M., Buresh, R.J., Izak, A.M., Mokwunye, A.U., Kwesiga, F.R., Ndiritu, C.G., and Woomer, P.L., 1997b. Soil fertility replenishment in Africa: an investment in natural resource capital, In: *Replenishing Soil Fertility in Africa*, (Buresh, R.J., Sanchez, P.A., and Calhoun, F., eds.), Special Publication No. 51, American Society of Agronomy, Madison, 1–46 pp.

Sandhu, J., Sinha, M., and Ambasht, R.S., 1990. Nitrogen release from decomposing litter of *Leucaena leucocephala* in the dry tropics, *Soil Biol. Biochem.*, **22:** 859–863.

Sanginga, N. Mulongoy, K., and Ayanaba, A., 1988. Nodulation and growth of *Leucaena leucocephala* (Lam.) de Wit as affected by inocculation and N fertilizer, *Plant and Soil*, **112:** 129–135.

Sanginga, N., Danso, S.K.A., Zapata, F., and Bowen, G.D., 1994. Influence of pruning management on P and N distribution and use efficiency by N_2 fixing and non-N_2 fixing trees used in alley cropping systems, *Plant and Soil*, **167:** 219–226.

Schroth, G., 1999. A review of below-ground interactions in agroforestry, focussing on mechanisms and management options, *Agrofor. Syst.*, **43:** 5–34.

Schroth, G., and Lehmann, J., 1995. Contrasting effects of roots and mulch from three agroforestry tree species on yields of alley-cropped maize, *Agric. Ecosys. Environ.*, **54:** 89–101.

Schroth, G., and Zech, W., 1995. Above- and below-ground biomass dynamics in a sole cropping and alley cropping system with *Gliricidia sepium* in the semi-deciduous rainforest zone of West-Africa, *Agrofor. Syst.*, **31:** 181–198.

Schroth, G., Poidy, N., Morshäuser, T., and Zech, W., 1995. Effects of different methods of soil tillage and biomass application on crop yields and soil properties in agroforestry with high tree competition, *Agric. Ecosys. Environm.*, **52:** 129–140.

Schroth, G., Lehmann, J., Barros, E., Rodrigues, M.R.L., and Macedo, J.L.V., 2001. A review of plant-soil interactions in multi-strata agroforestry systems with perennial crops, *Agrofor. Syst.*, in press.

Smart, J.B., Dilworth, M.J., and Robson, A.D., 1984. Effect of phosphorus supply on phosphate uptake and alkaline phosphatase activity in rhizobia, *Arch. Microbiol.*, **140:** 281–286.

Snoeck, D., 1995. *Interactions entre végétaux fixateurs d'azote et non fixateurs en culture mixte: cas des Leucaena spp. associées à Coffea arabica L. au Burundi*, Ph.D. Thesis, University of Lyon, Lyon, 199 pp.

Sprent, J.I., 1999. Not all nitrogen fixing legumes have a high requirement for phosphorus, *Agrofor. Forum*, **9:** 17–19.

Ssekabembe, C.K., 1985. Perspectives on hedgerow intercropping, *Agrofor. Syst.*, **3:** 339–356.

Sternberg, L.D.L., Green, L., Moreira, M.Z., Nepstad, D., Martinelli, L.A., and Victoria, R., 1998. Root distribution in an Amazonian seasonal forst as derived from delta C-13 profiles, *Plant and Soil*, **205:** 45–50.

Szott, L.T., Palm, C.A., and Buresh, R.J., 1999. Ecosystem fertility and fallow function in the humid and subhumid tropics, *Agrofor. Syst.*, **47:** 163–196.

Teixeira, W.G., and Villani, E.M.A., 1996. Variabilidade espacial de características químicas do solo em um latossolo amarelo, com utilisação agrícola anterior, In: *Sociedade Brasileira de Ciència do Solo, XXII Reunião Brasileira de Fertilidade do Solo e Nutrição de Plantas*, Universidade do Amazonas, Manaus, 132–133 pp.

Tian, G., Brussard, L., and Kang, B.T., 1995. An index for assessing the quality of plant residues and evaluating their effects on soil and crop in the (sub-) humid tropics, *Appl. Soil Ecol.*, **2:** 25–32.

Tiessen, H., Weisbach, C., and Menezes, R.S.C., 1999. Estimating the contribution of trees to available phosphorus, *Agrofor. Forum*, **9**: 57–60.

Timmer, L.A., Kessler, J.J., and Slingerland, M., 1996. Pruning of néré trees (*Parkia biglobosa* (Jacq.) Benth.) on the farmlands of Burkina Faso, West Africa. *Agrofor. Syst.*, **33:** 87–98.

Tuomela, K., Otsamo, A., Kuusipalo, J., Vuokko, R., and Nikles, G., 1996. Effect of provenance variation and singling and pruning on early growth of *Acacia mangium* Willd. plantation on *Imperata cylindrica* (L.) Beauv. dominated grassland. *For. Ecol. Manage*, **84:** 241–249.

Vanlauwe, B., Aman, S., Aihou, K., Tossah, B.K., Adebiyi, V., Sanginga, N., Lyasse, O., Diels, J., and Merckx, R., 1999. Alley cropping in the moist savanna of West-Africa III. Soil organic matter fractionation and soil productivity, *Agrofor. Syst.*, **42:** 245–264, van Wambeke, A., 1992, Soils of the Tropics, McGraw-Hill, New York, 343 pp.

Viera-Vargas, M.S., Souto, C.M., Urquiaga, S., and Boddey, R.M., 1995. Quantification of the contribution of N_2 fixation to tropical forage legumes and transfer to associated grass, *Soil Biol. Biochem.*, **27:** 1193–1200.

Vogt, K.A., Vogt, D.J., and Bloomfield, J., 1991. Input of organic matter to the soil by tree roots, In: *Plant Roots and their Environment*, (McMichael, B.L., and Persson, H., eds.), Elsevier, Amsterdam, 171–190 pp.

Wahid, P.A., Kamalam, N.V., and Antilkumar, K.S., 1993. Coconut, In: *Rooting Patterns of Tropical Crops*, (M.A. Salam, and P.A. Wahid, eds.), Tata McGraw-Hill, New Delhi, 188–203 pp.

Wilson, G.F., Kang, B.T., and Mulongoy, K., 1986. Alley cropping: trees as sources of green manure and mulch in the tropics, *Biol. Agric. Hort.*, **3:** 251–267.

Woomer, P.L., Bekunda, M.A., Karanja, N.K., Moorehouse, T., and Okalebo, J.R., 1998. Agricultural resource management by smallholder farmers in East Africa, *Nature Res.*, **34:** 22–33.

Xu, Z.H., Saffigna, P.G., Myers, R.J.K., and Chapman, A.L., 1993. Nitrogen cycling in leucaena (*Leucaena leucocephala*) alley cropping in semi-arid tropics: II. Response of maize growth to addition of nitrogen fertilizer and plant residues, *Plant and Soil*, **148:** 73–82.

Young, A., 1997. *Agroforestry for Soil Management*, CAB International, Wallingford, 320 pp.

CHAPTER 15

ACCRETION OF CARBON AND NUTRIENTS IN THE FOREST FLOOR AND MINERAL SOIL OF *CASUARINA EQUISETIFOLIA* PLANTATIONS ON THE COASTAL SAND DUNES OF SENEGAL

Daniel Mailly[1] and H.A. Margolis[2]

[1]*Ministère des Ressources naturelles, Direction de la Recherche Forestière, 2700, rue Einstein, Sainte-Foy, Qué., G1P 3W8, Canada,*

[2]Centre de Recherche en Biologie Forestière, Faculté de Foresterie et Géomatique, Université Laval, Sainte-Foy, Qué., G1K 7P4, Canada

ABSTRACT

Casuarina equisetifolia, a nitrogen-fixing tree species, is planted throughout the tropics and is considered to be an excellent source of firewood. In Senegal, it is the preferred tree species for stabilizing shifting sand dunes on the northern coast of the country. Plantations of *C. equisetifolia* generally have a mean annual volume increment of about 4.2 m^3 ha^{-1} $year^{-1}$ on good sites, and are capable of accumulating significant quantities of organic matter on the ground. Forest floor accumulations of up to 200–300 Mg ha^{-1} have been reported in 34 to 37-year-old plantation located on the coastal sand dunes of Senegal. Nitrogen in the forest floor was found to accumulate at a rate of 45 kg ha^{-1} $year^{-1}$ in a chronosequence of *C. equisetifolia* plantations, whereas calcium, aluminum, magnesium, iron, postassium and phosphorus in the forest accumulated at 68, 37, 18, 13, 4.9 and 2.5 kg ha^{-1} $year^{-1}$, respectively. In the mineral soil, total phosphorus, exchangeable calcium, and pH decreased with plantation age. Nitrogen in the mineral soil increased by 30 kg ha^{-1} $year^{-1}$ while organic matter in the mineral soil increased by 62 kg ha^{-1} $year^{-1}$. The pattern of litterfall, forest floor accumulation and nutrient cycling found in these plantations is attributed to a particular combination of species characteristics and ecological conditions present on the coastal sand dunes, such as the almost complete absence of rain during nearly nine months of the year which leads to very dry conditions at the soil surface. All these conditions appear to permit *C. equisetifolia* plantations in the interdune depressions to maintain high levels of productivity and thus substantial levels of litterfall on what would otherwise appear to be very harsh sites.

Key Words: Casuarina plantations, nutrient accretion, coastal sand dunes, nitrogen fixation, plantation management.

Casuarina equisetifolia (Forst. & Forst.) is planted throughout the tropics and is considered to be an excellent source of firewood (Subbarao and Rodriguez-Barrueco, 1995; Boland *et al.*, 1984, National Academy of Sciences, 1980). Its high tolerance to low water and nutrient availability, as well as its capacity to fix atmospheric nitrogen by means of actinorhizal root symbioses, has helped in the success of many plantation programs (National Research Council, 1984). The mean annual increment of *C. equisetifolia* generally varies between 6 and 10 m^3 ha^{-1} $year^{-1}$ (Pandey, 1987), and is about 4.2 m^3 ha^{-1} $year^{-1}$ on good sites in coastal Senegal (Ndiaye *et al.*, 1993). *C. equisetifolia* plantations are also capable of accumulating significant quantities of organic matter on the forest floor. Forest floor accumulations of up to 251 Mg ha^{-1} (ash-free basis) have been reported in a 37-year-old plantation located on the coastal sand dunes of Senegal (Gourbière and Debouzie, 1995). *C. equisetifolia* plantations are also known to produce significant amounts of litterfall, and annual litterfall rates between 3.2 and ~6.5 Mg ha^{-1} have been reported in India (Jamaludheen and Kumar, 1999; Jambulingam and Rangaswamy, 1990; Sugur, 1989). Results from a previous study indicate that pH as well as phosphorus and calcium availability in the mineral soil of *C. equisetifolia* plantations can decline with plantation development (Mailly and Margolis, 1992a).

Because of all these characteristics, it appears that a better understanding of the ecology and growth of *C. equisetifolia* plantations on the coastal sand dunes of Senegal may be useful for management purposes. Our goal in this chapter is to review the effects of *C. equisetifolia* plantations on forest floor and soil nutrient status of these plantations. Much of the results presented in this chapter come from earlier studies of forest floor and mineral soil development in *C. equisetifolia* plantations that took place in the dune area of the Cap-Vert penisula in the southern part of the Niayes region of northwestern Senegal (Dommergues, 1963; Mailly and Margolis, 1992a,b; Gourbière and Debouzie, 1995). The region has been described in detail by Roy-Noël (1974), whereas the *C. equisetifolia* plantations in Senegal are described in Maheut and Dommergues (1959), Andéké-Lengui and Dommergues (1983), and Mailly *et al.* (1994).

CASUARINA EQUISETIFOLIA IN SENEGAL

PHYSICAL ENVIRONMENT

The coastal region of Senegal comprises of shifting live sand dunes which, under the influence of northerly winds during the dry season, advance 10 to 12 metres year^{-1}. The live dunes consist of coarse sands, i.e. regosols (Sanchez, 1976) of marine origin. They are comprised almost exclusively of grains of quartz accompanied by small quantities of magnetite, pyroxene, tourmaline and zircon (Adam, 1971). The remnants of seashells in these soils accounts for their fairly high pH of around 7–8 (Mailly and Margolis, 1992a).

CLIMATE

The coastal region of northern Senegal has the mildest climate in the country due to the presence of maritime trade winds. There are two distinct seasons; a rainy season from mid-July to mid-October with precipitation averaging 500 to 550 mm and a dry season from mid-November to mid-May. The periods from mid-October to mid-November and from mid-May to mid-July are the two transition periods. The mean temperature on the Cap-Vert penisula is around 23°C during the dry season and 28°C during the rainy season (République du Sénégal, 1980).

PLANTATION PROGRAMME

The *C. equisetifolia* plantation programme in the Niayes region of Senegal was started in 1948 with the goal of stabilizing the drifting sand dunes along the northern coast in order to protect adjacent agricultural fields (Maheut and Dommergues, 1959, Andéké-Lengui and Dommergues, 1983, Mailly *et al.*, 1994). Development assistance from the Food and Agriculture Organization of the United Nations, Canada, and the United States has helped the programme to continue and expand.

All the species trials conducted in the Niayes region of Senegal have shown *C. equisetifolia* to be the only tree species sufficiently adapted to the severe environmental conditions occurring on the virgin sand dunes (Maheut and Dommergues, 1959; Mailly *et al.*, 1994). The *C. equisetifolia* plantation programme has occurred during two distinct periods. During the first period (1948 to 1959), 513 ha of plantations were established. The second period began in 1972 and continues to the present. Between 1972 and 1991, 9252 ha of *C. equisetifolia* plantations have been established, for a total of 9765 ha (Table 1).

Table 1 Plantations (ha) of *C. equisetifolia* on the coastal sand dunes of Senegal from 1948 to 1991.

	Zone		
	South	*Center*	*North*
1948	9	–	–
1949	20	–	–
1950	27	–	–
1951	24	–	–
1952	10	–	–
1953	22	–	–
1954	22	–	–
1955	50	–	–
1956	100	–	–
1957	115	–	–
1958	110	–	–
1959	4	–	–
1960–1971	–	–	
1972	30	–	–
1973	40	–	7
1974	50	–	14
1975	27	52	12
1976	107	60	5
1977	105	58	–
1978	85	53	–
1979	40	140	50
1980	–	179	100
1981	600	300	240
1982	1000	310	260
1983	345	60	40
1984	100	110	70
1985	420	100	160
1986	590	300	170
1987	–	250	150
1988	125	250	145
1989	134	300	139
1990	193	300	148
1991	203	300	226
Total	4707	3122	1936

Sources: Maheut and Dommergues, 1959; Mailly *et al.*, 1994

Context of the Study

Preliminary observations in the *C. equisetifolia* plantations on the coastal sand dunes of Senegal revealed that these plantations were capable of accumulating large quantities of organic matter in the forest floor. We hypothesized that this organic layer likely plays an important role in determining rates of nutrient cycling as well as the physical and biological properties of the underlying mineral soil. Thus a knowledge of the pattern of development of the forest floor as well as the rate of depletion or accumulation of nutrients in the mineral soil was thought to be useful in developing management plans for these plantations that would reconcile their potential benefits to the people of Senegal, i.e. (1) stabilizing the coastal sand dunes in order to protect adjacent agricultural areas, (2) improving soil fertility for future rotations of either tree or agricultural crops (e.g. Seck and Lo, 1998), and (3) providing the most productive source of wood fiber in the Sahelian zone of Senegal. Here we report on four plantations representing age classes 6, 8, 17 and 34 years that were sampled on the coastal sand dunes of Senegal (Table 2). For more information about the sampling procedures and details about the stands, please refer to Mailly and Margolis (1992a,b).

Table 2 Plantation location and characteristics of *C. equisetifolia* plantations in Senegal.

Age (years)	*Site*	*Location*	*Survival* rate (*%*)	*Density* (*stems/ha*)	*Mean Diameter* (*cm*)	*Mean Height* (*m*)
6	Notto	15°02 N; 17°03 W	87[1]	1395[18]	6.5[0.1]	8.1[0.4]
8	Kayar	14°54 N; 17°08 W	88[3]	1415[41]	8.6[0.1]	10.9[0.3]
17	Cambérène	14°46 N; 17°27 W	73[2]	1167[30]	9.3[0.3]	9.0[0.3]
34	Lake Retba	14°52 N; 17°12 W	68[3]	1085[50]	15.4[0.7]	16.9[0.8]

Source: Mailly and Margolis, 1992a. Portion of this table are reprinted from Forest Ecology and Management, Vol. 55, Mailly, D., and Margolis, H.A., Forest floor and mineral soil development in *Casuarina equisetifolia* plantations on the coastal sand dunes of Senegal, 259–278, Copyright (1992), with permission from Elsevier Science. Each value represents the mean +/– SE; n=5 for ages 6, 8 and 34, n=3 for age 17.

FOREST FLOOR ACCUMULATION

Forest Floor Accumulation

The thickness of the organic layer in a chronosequence of *C. equisetifolia* stands increased from 4.2 cm to 8.0 cm between age 6 and age 34, an average increase of 0.13 cm^{-1} $year^{-1}$ (Table 3). Accumulation of the forest floor increased from 33.5 Mg ha^{-1} to 227.9 Mg ha^{-1} between age 6 and age 34 (Table 4); an average increase of 6.9 Mg ha^{-1} year^{-1}. Based on these figures, there was no evidence that the accumulation of forest floor biomass in our *C. equisetifolia* plantations reached steady-state. Furthermore, the development of a well-defined humus (>2 cm) layer did not occur until sometime after age 17 (Mailly, 1991). Dommergues (1963) also reported little if any humus formation in a 13-year-old plantation in Senegal.

We are aware of only two other studies on forest floor accumulation in *C. equisetifolia* plantations in Senegal (Table 4). Dommergues (1963) reported a forest floor accumulation of about 21.6 Mg ha^{-1} (not corrected for ash) in a 13-year-old plantation, whereas Gourbière and Debouzie (1995) reported a forest floor accumulation ranging from 20.7 to 250.7 Mg ha^{-1} in a 37-year-old plantation. Large differences in sampling methods between the Gourbière and Debouzie (1995) study and our study may explain the differences in reported values. They sampled a 37-year-old plantation along a 300 m transect perpendicular to the sea shore. It is very likely that their minimum value (20.7 Mg ha^{-1}, Table 4) is the forest floor accumulation closer to the sea shore, and their largest value (250.7 Mg ha^{-1}, Table 4) at the end of the transect. In our study (Mailly and Margolis, 1992a), most of the sampling in the 34-year-old plantation was restricted in the interdune depressions located *ca.* 300-400 m from the seashore, which may explain an average forest floor accumulation of 228 Mg ha^{-1} (Table 4), comparable to the largest value reported by Gourbière and Debouzie (1995). It is also important to note that in the oldest plantation, the F layer was dark and showed important fungal and faunal activities (Gourbière and Debouzie, 1995). There was large amounts of coleoptera fecal pellets (1 × 2 mm) in the H layer, as well as sand particules stuck together with the organic material. When converted to an ash-free basis, the organic matter accumulation in these plantations varied from 23.1 to 120.3 Mg ha^{-1} between age 6 and age 34 (Table 3).

LITTERFALL

We are not aware of any study conducted in Senegal in which litterfall in *C. equisetifolia* plantations was measured directly. We have estimated the rate of litterfall deposition per year by dividing the forest floor accumulation by plantation age in young stands, i.e. by assuming that in younger stands the forest floor had not yet started to decompose and incorporate into the soil layers (cf. also Dommergues, 1963). Estimation of litterfall rates for *C. equisetifolia* plantations in Senegal varies from 1.7 Mg ha^{-1} year^{-1} for a 13-year old plantation (Dommergues, 1963) to about

Table 3 Characteristics of the forest floor in a chronosequence of *C. equisetifolia* plantations on coastal sand dunes in Senegal (mean ± standard deviation in brackets).

Age (years)	*Forest floor thickness* (*cm*)	*Organic matter accretion** (*Mg ha*$^{-1}$)	*N*	*P*	*K*	*Ca*	*Mg* (*kg ha*$^{-1}$)	*Fe*	*Mn*	*Al*	*Na*
6	4.2[0.2]	23.1[2.1]	244.5[25.9]	18.2[2.2]	50.3[5.8]	589.9[57.1]	135.1[11.7]	117.7[10.7]	2.1[0.2]	205.8[18.2]	135.7[28.8]
8	4.4[0.3]	27.2[3.8]	312.0[42.4]	19.0[2.3]	49.0[6.2]	681.2[93.5]	150.7[19.9]	122.3[17.4]	3.0[0.4]	241.4[35.3]	80.5[20.9]
17	4.8[0.2]	43.0[3.6]	547.3[53.1]	33.6[3.2]	92.3[13.5]	916.3[77.2]	368.2[33.0]	269.5[25.9]	8.0[1.0]	654.9[79.6]	233.9[60.5]
34	8.0[0.2]	120.3[6.1]	1567.6[108.9]	108.9[6.2]	188.9[14.0]	2756.5[215.3]	634.7[49.0]	479.2[30.9]	21.6[1.4]	1205.6[81.9]	76.6[10.9]
Average per year**	0.13	3.5	45	2.5	4.9	68	18	13	0.7	37	–

Source : Mailly and Margolis, 1992a,b. Portions of this table are reprinted from Forest Ecology and Management, Vol. 55, Mailly, D. and Margolis, H.A., Forest floor and mineral soil development in *Casuarina equisetifolia* plantations on the coastal sand dunes of Senegal, 259–278, Copyright (1992), with permission from Elsevier Science.

*ash-free basis

**based on the slope of the linear regression of the dependent variable against age

Table 4 Comparison of forest floor weights (Mg ha^{-1}) in different stands of *C. equisetifolia* growing on coastal sand dunes in Senegal.

Plantation age	*Forest floor weight (Mg ha^{-1})**				*Reference*
	Mean	*SD*	*Min*	*Max*	
6	33.5	9.9	16.9	53.3	Mailly, 1991 and unpublished data
8	41.2	22.1	13.2	86.0	Mailly, 1991 and unpublished data
13	21.6	–	–	–	Dommergues, 1963,
17	58.6	29.7	13.5	111.1	Mailly, 1991 and unpublished data
34	227.9	36.2	167.9	313.1	Mailly, 1991 and unpublished data
37	106.2	39.4	20.7	250.7	Gourbière and Debouzie, 1995

*not corrected for ash

5.2 Mg ha^{-1} year^{-1} for an 8-year old plantation (Mailly and Margolis, 1992). These values are within the range of litterfall reported elsewhere in the tropics. Jamaludheen and Kumar (1999) have reported a litterfall rate of approximately 6.5 Mg ha^{-1} year^{-1} in an 8-9 year-old *C. equisetifolia* woodlot in Kerala (India). Jambulingam and Rangaswamy (1990) have reported litterfall rates ranging from 3.7 to 4.6 Mg ha^{-1} year^{-1} in a 4 year-old *C. equisetifolia* plantation in Tamilnadu (India). Sugur (1989) has reported average litter accumulations of 3.2 Mg ha^{-1} year^{-1} for a 10-year-old plantation in India containing 2500 stems ha^{-1} (Sugur, 1989).

Nutrients

The accumulation of macro-nutrients in the organic layer generally followed a linear pattern (Mailly and Margolis, 1992a,b; Table 3). Calcium accumulated at the highest rate of 68 kg ha^{-1} year^{-1}. This was followed by nitrogen (45 kg ha^{-1} year^{-1}), magnesium (18 kg ha^{-1} year^{-1}), potassium (4.9 kg ha^{-1} year^{-1}), and phosphorus (2.5 kg ha^{-1} year^{-1}). Micro-nutrients also accumulated in the organic layer following a linear pattern (Table 5), except for sodium whose accumulation rate was not constant through time. Aluminium accumulated at a rate of 37 kg ha^{-1} year^{-1}. This was followed by iron (13 kg ha^{-1} year^{-1}), and manganese (0.7 kg ha^{-1} year^{-1}).

The nitrogen to phosphorus ratio measured in the forest floor of these plantations varied from 13.4 to 16.5 (Mailly and Margolis, 1992a,b; Table 3). It is interesting to note that forest floors of tropical and mediterranean forests generally have nitrogen to phosphorus ratios of around 30 :1 (Vogt *et al.*, 1986). Alexander (1977) suggested that microbial decomposition readily occurs when the nitrogen to phosphorus ratio is

Table 5 Characteristics of the mineral soil in a chronosequence of *C. equisetifolia* plantations on coastal sand dunes in Senegal (mean ± standard deviation in brackets).

Site	*Age*	*Soil organic matter content*	*pH*	*N*	*P*	*K*	*Ca*	*Mg*
	(*years*)	(*kg* ha^{-1})		(*kg* ha^{-1})	(*kg* ha^{-1})	(*kg* ha^{-1})	(*Mg* ha^{-1})	(*kg* ha^{-1})
Notto	6	787[80]	7.3[0.04]	2194[55]	2410[299]	167.3[9.9]	34.7[2.1]	491.6[29.2]
Kayar	8	1195[151]	7.2[0.13]	2315[98]	2520[237]	65.4[17.0]	35.7[3.4]	308.3[28.5]
Cambérène	17	2226[252]	6.9[0.11]	2567[169]	2430[345]	123.5[35.0]	20.0[4.5]	427.4[74.0]
Lake Retba	34	2618[204]	6.5[0.17]	3058[115]	1940[194]	164.1[22.5]	18.3[3.7]	456.7[40.6]
Average per year*		62	–0.03	30	–19	1.5	-0.6	1.7

Source: Mailly and Margolis, 1992a,b. Portion of this table are reprinted from Forest Ecology and Management, Vol. 55, Mailly, D. And Margolis, H.A., Forest floor and mineral soil development in *Casuarina equisetifolia* plantations on the coastal sand dunes of Senegal, 259–278, Copyright (1992), with permission from Elsevier Science.

*Average accumulation rate based on the slope of the linear regression of the dependent variable against age

10 or lower. The values reported by Mailly and Margolis (1992a,b) clearly highlight the extent to which decomposition processes are likely to be phosphorus-limited in the forest floor of *C. equisetifolia* plantations in Senegal.

MINERAL SOIL

NUTRIENT CONCENTRATIONS

In a previous study, Mailly and Margolis (1992a,b) showed that the concentrations of total phosphorus, exchangeable potassium and exchangeable calcium varied with soil depth but not with plantation age, but that there were significant interactions between age and depth of soil for nitrogen, magnesium, carbon, C/N ratio, iron, aluminium and sodium. In general, concentrations of phosphorus, exchangeable potassium, and exchangeable calcium were much higher in the upper layer (0–2 cm; Mailly and Margolis, 1992a,b). Higher nitrogen and carbon concentrations were found in the 17- and 34-year-old plantations at the 0–2 cm and the 3–20 cm depths (Mailly and Margolis,1992a,b), whereas C/N ratio tended to be greater in the older plantations at all depths.

SOIL ORGANIC MATTER, PH, AND NUTRIENT ACCRETION

The soil organic matter content increased at an average rate of 62 kg ha^{-1} $year^{-1}$ in a chronosequence of *C. equisetifolia* plantations in Senegal (Mailly and Margolis (1992a,b; Table 5), whereas the soil pH declined linearly from 7.3 to 6.5 over the 28-year period separating the 6-year-old and 34-year-old plantation (Mailly and Margolis, 1992a,b). Total nitrogen accumulated linearly with age in the mineral soil to a depth of 100 cm at an average rate of 30 kg ha^{-1} $year^{-1}$ (Table 5). No pattern of potassium and magnesium accretion in the mineral soil was observed. The content of other nutrients such as exchangeable calcium and phosphorus, however, declined with plantation age at rates of 660 and 19 kg ha^{-1} $year^{-1}$, respectively (Table 5). It is likely that the pH of the mineral soil in the two younger plantations may have limited phosphorus availability, since phosphorus availability is maximum at pH levels between 6 and 7 (Bohn *et al.*, 1979). The pH of the 6- and 8-year-old plantations was above 7 but declined to about 6.5 in the oldest plantation (Mailly and Margolis,1992a,b). In general, soil pH tends to decrease following litter decomposition and the liberation of organic acids from the litter accumulated on the forest floor.

NITROGEN-FIXATION AND ACCRETION

N_2-fixation rates published in the literature generally refer to ecosystems that have been subjected to diverse constraints, or to studies that used different techniques to estimate these rates. For example, differences in N_2-fixation rates in *C. equisetifolia* in Senegal have been attributed to variations in soil fertility and proximity to the sea shore (Mariotti *et al.*, 1992). It is therefore not surprising that, for a given tree species, there are large variations in the estimations of the total amount or percentage of N_2-fixed (Dommergues, 1995). In Senegal, values of actual N_2-fixed by *C. equisetifolia* reported in the literature range from 4.3 to 42.4 g N_2-fixed tree^{-1} year^{-1} (Dommergues, 1995). The amounts of N_2 fixed in a constraint-free environment (or N_2-fixing potential, *sensu* Dommergues, 1995) on a hectare basis vary between 15 kg ha^{-1} year^{-1} in a 3-year-old *C. equisetifolia* plantation with a density of 2500 trees ha^{-1} (Mariotti *et al.*, 1992) and 84.8 kg ha^{-1} year^{-1} in a 2-year-old *C. equisetifolia* plantation with a density of 2000 trees ha^{-1} (Sougoufara *et al.*, 1990). Nitrogen accretion rates in *C. equisetifolia* plantations of Senegal as estimated from biomass accumulation measurements range from 58 kg ha^{-1} year^{-1} (Dommergues, 1963) to 75 kg N ha^{-1} year^{-1} (Mailly and Margolis, 1992a). Overall, these figures compare favorably with nitrogen fixation rates reported for tree species planted in the tropics, which typically have input rates of 50 to 150 kg N ha^{-1} year^{-1} (*Casuarina, Albizia, Leucanea* and *Acacia;* Binkley and Giardina, 1997). They are also comparable to similar rates reported for temperate actinorhizal tree species such as *Alnus glutinosa* (16 to 60 kg ha^{-1} year^{-1}; Akkermanns and van Dijk, 1976, Coté and Camiré, 1984), *A. rubra* (40 to 150 kg ha^{-1} year^{-1}; Binkley, 1981) and *A. sinuata* (20 to 150 kg ha^{-1} year^{-1}; Binkley, 1981, Heilman and Ekuan, 1983).

MANAGEMENT OF *C. EQUISETIFOLIA* PLANTATIONS

Results from an earlier study in a chronosequence of *C. equisetifolia* plantations (Mailly and Margolis, 1992a,b) revealed that total phosphorus and calcium availability in the mineral soil declined with plantation age while the accumulation of all essential mineral elements in the forest floor increased. It was also found that the majority of the nutrient capital of the sites was located in the forest floor and upper 2 cm of the mineral soil. In addition, the accumulation of phosphorus and calcium in the forest floor was accompanied by a decrease in the total phosphorus and exchangeable calcium measured in the mineral soil. These results suggest that management and harvesting practices in these plantations should take into account the importance and the fragility of the forest floor and the upper mineral soil if forest managers are to continue building upon the benefits realized during the first rotation.

Organic matter and nitrogen accretion in the forest floor and mineral soil increased from the youngest (6-year-old) to the oldest (34-year-old) plantations and never attained steady-state. Thus from the standpoint of improving the fertility of the sand dunes for future crops, it appears that a minimum of a 34-year rotation would be justified in the interdune depressions. While this may be a reasonable objective on certain sites and under certain conditions, soil improvement is only one aspect of the multiple benefits which can be realized from these plantations. The capacity for wood production is another benefit and thus it is important to take into account the productivity of these plantations. Ndiaye *et al.* (1993) demonstrated that the growth rates of *C. equisetifolia* plantations on the sand dunes of Senegal are a function of microtopography; i.e. growth rates are higher in interdune depressions than they are on the flanks or on the summits of the dunes. Microtopographical position, classified as interdune depressions, flanks, or summits, was found to have a significant effect on height, diameter, and volume growth. Maximum mean annual volume increment was 4.2 m^3 ha^{-1} at age 18 years for the interdune depressions, 2.0 m^3 ha^{-1} at age 29 for flanks, and 1.7 m^3 ha^{-1} at age 30 for summits. Thus, different rotation lengths can be recommended for specific microsites depending on microtopographical position and on whether fibre production or soil amelioration is the dominant management objective.

CONCLUSION

Planted seedlings of *C. equisetifolia* on the coastal sand dunes of Senegal must thrive in a dry environment characterized by an inconsistent rainy season of about three months. Some of the conditions for successful establishment are the proximity to the sea shore in order to benefit from the condensation of fog and seaspray that occurs throughout the year, and the presence of a fresh water table as close as one to two metres below the surface of the dunes. Once established, however, the *C. equisetifolia* plantations located in the interdune depressions have a pattern of forest floor accumulation during stand development significantly different from that normally found in tropical forest plantations. Mailly and Margolis (1992a) attributed this special pattern to a particular combination of species characteristics and ecological conditions present on the coastal sand dunes, such as the almost complete absence of rain during nearly nine months of the year which leads to very dry conditions at the soil surface. All these conditions appear to permit *C. equisetifolia* plantations in the interdune depressions to maintain high levels of productivity (Ndiaye *et al.*, 1993) and thus substantial levels of litterfall on what would otherwise appear to be very harsh sites.

The amount of litter accumulated in a given site is determined by a balance among litter production and deposition, litter destruction and removal (Facelli and Pickett, 1991). Litter production depends primarily on site productivity, although other properties of the environment, as well as chance, may introduce important variation. Climate, rainfall and length of the growing season are clearly important factors controlling litter production, whereas soil fertility, soil water retention, and species composition are important within the same climatic range (Meentemeyer *et al.*, 1982; Facelli and Pickett, 1991). Litter decomposition rates vary greatly among ecosystems: faster rates are generally found in the tropics and slower rates in subapline ecosystems (Olson, 1963). Tropical forests are thus not generally considered to accumulate large quantities of organic matter in their forest floors. Both Ovington (1965) and Rodin and Bazilevich (1967) claimed that there is greater forest floor accumulation with increasing distance from the tropics. Vogt *et al.*, (1986), using a larger data set, reported that any relation between forest floor accumulation and latitude was masked by variability within each forest type. However, they were able to show that forests at greater than 40°N latitude accumulated greater forest floor biomass than forests at less than 40°N latitude. The *C. equisetifolia* plantations on the coastal sand dunes of Senegal are clearly an exception to these generalities.

REFERENCES

Adam, J.G., 1971. Les dunes maritimes de l'Afrique occidentale, la fixation des dunes de Dakar, In: *Colloques Phytosociologiques: I. La végétation des dunes maritimes*, (J.-M. Hehu, ed.), J. Cramer, Vaduz, Liechtenstein, 245–251 pp.

Akkermanns, A.D.L., and van Dijk, C., 1976. The formation and nitrogen fixing activity of the root nodules of *Alnus glutinosa* under field conditions, In: *Symbiotic nitrogen fixation in plants*, (P.S. Nutman, ed.), Cambridge University Press, Cambridge, 511–520 pp.

Alexander, M., 1977. *Introduction to Soil Microbiology*, Second edition, Wiley and Sons, New York, 467 pp.

Andéké-Lengui, M.A., and Dommergues, Y., 1983. Coastal sand dune stabilization in Senegal, In: *Casuarina Ecology, Management and Utilization*, (Midgley, S.J., J.W. Turnbull, and R.D. Johnston, eds.), Proceedings of an International Workshop, 17–21 August 1981, Canberra, Australia, 158–166 pp.

Binkley, D., 1981. Nodule biomass and acetylene reduction rates of red alder and Sitka alder on Vancouver Island, *B.C. Can. J. For. Res.*, **11:** 281–286.

Binkley, D., and Giardina, C., 1997. Nitrogen fixation in tropcial forest plantations, In: *Management of soil, nutrients and water in tropical plantation forests*, (E.K.S. Nambiar, and A.G. Brown, eds.), ACIAR Monograph No. 43, Canberra, Australia, 297–337 pp.

Bohn, H.L., McNeal, B.L., and O'Connor, G.A., 1979. *Soil Chemistry,* Wiley and Sons, New York, 329 pp.

Boland, D.J., Brooker, M.I.H., Chippendale, G.M., Hall, N., Hyland, B.P.M., Johnston, R.D., Kleinig, D.A., and Turner, J.D., 1984. *Forest trees of Australia,* Thomas Nelson Inc. and the Commonwealth Scientific and Industrial Research Organization, Melbourne, Australia, 687 pp.

Coté, B., and Camiré, C., 1984. Growth, nitrogen accumulation, and symbiotic nitrogen fixation in pure and mixed plantings of hybrid poplar and black alder, *Plant Soil,* **78:** 209–220.

Dommergues, Y., 1963. Évaluation du taux de fixation de l'azote dans un sol dunaire reboisé en filao (*Casuarina equisetifolia*). *Agrochimica,* **7**(4): 335–340.

Dommergues, Y.R., 1995. Nitrogen fixation by trees in relation to soil nitrogen economy, *Fertilizer Research,* **42:** 215–230.

Facelli, J.M., and Pickett, T.A., 1991. Plant litter: its dynamics and effects on plant community structure, *Bot. Rev.,* **57**(1): 1–32.

Gourbière, F., and Debouzie, D., 1995. Spatial distribution and estimation of forest floor components in a 37-year-old *Casuarina equisetifolia.* (FORST.) plantation in coastal Senegal, *Soil Biol. Biochem.,* **27**(3): 297–304.

Heilman, P.E., and Ekuan, G., 1983. Nodulation and nitrogen fixation by red alder and Sitka alder on coal mine spoils, *Can. J. For. Res.,* **12:** 992–997.

Jamaludheen, V., and Kumar, B.M., 1999. Litter of multipurpose trees in Kerala, India: variations in the amount, quality, decay rates and release of nutrients, *For. Ecol. Manage.,* **115:** 1–11.

Jambulingam, R., and Rangaswamy, S.R.S., 1990. Litterfall in *Casuarina equisetifolia* in Tamilnadu. Paper presented at the Second International Casuarina workshop, Desert Development Center, A.U.C., Cairo, Egypt, January 15–20, 10 pp.

Maheut, J., and Dommergues, Y., 1959. La fixation par le reboisement des dunes de la presqu'île du Cap-Vert et l'évolution biologique des sols, *Bois et Forêts des Tropiques,* **63:** 3–16.

Mailly, D., 1991. *Accumulation de matière organique et d'éléments nutritifs au sol dans une série de plantations de filaos (Casuarina equisetifolia Forst. & Forst.) le long du littoral nord du Sénégal,* M.Sc. Thesis, Université Laval, Qué., Canada, 76 pp.

Mailly, D., and H.A. Margolis, 1992a. Forest floor and mineral soil development in *Casuarina equisetifolia* plantations on the coastal sand dunes of Senegal, *For. Ecol. Manage.,* **55:** 259–278.

Mailly, D., and H.A. Margolis, 1992b. Corrigendum, *For. Ecol. Manage.,* **65:** 301.

Mailly, D., Ndiaye, P., Margolis, H.A., and Pineau, M., 1994. Fixation des dunes et reboisement avec le filao (*Casuarina equisetifolia*) dans la zone du littoral nord du Sénégal, *For. Chron.,* **70**(3): 282–290.

Mariotti, A., Sougoufara B., and Dommergues, Y.R., 1992. Estimation de la fixation d'azote atmospherique par le traçage isotopique naturel dans une plantation de *Casuarina equisetifolia* (FORST.), *Soil Biol. Biochem.,* **24**(7): 647–653.

Meentemeyer, V., Box, E.O., and Thompson, R., 1982. World patterns and amounts of terrestrial plant litter production, *Bio Science,* **32**(2): 125–128.

National Academy of Sciences, 1980. *Firewood crops: shrub and tree species for energy production,* National Academy Press, Washington, D.C., 237 pp.

National Research Council, 1984. *Casuarinas*: Nitrogen-Fixing trees for adverse sites. Innovations in tropical reforestation, National Academy Press, Washington, D.C., 118 pp.

Ndiaye, P., D. Mailly, Pineau, M., and Margolis, H.A., 1993. Growth and yield of *Casuarina equisetifolia* plantations on the coastal sand dunes of Senegal as a function of microtopography, *For. Ecol. Manage.,* **56:** 13–28.

Olson, J.S., 1963. Energy storage and the balance of production and decomposition in ecological systems, *Ecology*, **44:** 323–331.

Ovington, J.D., 1965. Organic production, turnover and mineral cycling in woodlands, *Biol. Rev.*, **40:** 295–336.

Pandey, D., 1987. Modèle de rendement pour les reboisements tropicaux, *Unasylva*, **157/158:** 74–75.

République du Sénégal, 1980. *Étude de développement régional intégré du littoral-nord (Niayes), Livre blanc*. Annexe 1: le milieu naturel. Ministère du Plan et de la Coopération. 44 pp.

Rodin, L.E., and Bazilevich, N.I., 1967. *Production and Mineral Cycling in Terrestrial Vegetation*, Oliver and Boyd, London, 288 pp.

Roy-Noël, J., 1974. Recherches sur l'écologie des Isoptères de la presqu'île du Cap-Vert (Sénégal): Introduction et première partie: le milieu, *Bulletin de l'Institut Fondamental d'Afrique Noire, Tome*, **26-A**(2): 291–378.

Sanchez, P.A., 1976. *Properties and management of soils in the tropics*, Wiley, New York, 618 pp.

Seck, M.A., and Lo, C., 1998. Two compost types used in the Niayes (Senegal) market garden area: technological innovation and vegetable production, *Acta Horticulturae*, **469:** 325–331.

Sougoufara, B., Danso, S.K.A., Diem, H.G., and Dommergues, Y.R., 1990. Estimating N_2-Fixation and N Derived from Soil by *Casuarina equisetifolia* Using Labelled N_{15} Fertilizer—Some Problems and Solutions, *Soil Biol Biochem.*, **22**(5): 695–701.

Subbarao, N.S., and Rodriguez-Barrueco, C., 1995. *Casuarinas*, Science Publishers, Inc., Lebanon, NH, USA, 240 pp.

Sugur, G.V., 1989. Litter production and nutrient cycling of different species under plantations conditions, *Myforest*, **25**(1): 43–49.

Vogt, K.A., Grier, C.C., and Vogt, D.J., 1986. Production, turnover, and nutrient dynamics of above- and below-ground detritus of world forests, *Adv. Ecol. Res.*, **15:** 303–377.

SUBJECT INDEX

AUTHOR INDEX

ABOUT THE EDITOR

M. Vikram Reddy, a professor of environmental biology in the department of Zoology, teaches applied ecology and environmental science at Kakatiya University, Warangal in Andhra Pradesh, India. He was also qualified for a professorhip in Terrestrial Ecology (soil biology and ecology) in University of Copenhagen, Denmark. Professor Reddy, a founder life-member of Indian Society of Soil Biology and Ecology (UAS, Bangalore), is one of the Vice-Presidents of the society as well as editorial board members of its journal—Journal of Soil Biology and Ecology. He is also a life member of the International Union of Soil Science (New York, USA) in 1996. He worked as a visiting scientist at Kansas State University (USA) with a Fulbright fellowship in 1988. He was awarded the prestigious *Rockfeller Foundation Environmental Research Fellowship* in International Agriculure in 1991, and worked on 'the impact of soil management on beneficial soil arthropods and earthworms in the Semi-Arid Tropics in India', during 1992–1994, at the Natural Resources Management Program of International Crops Research Institute for the Semi-Arid Tropics (ICRISAT) at Hyderabad in India. Last year, he was nominated by INSA—Indian National Science Academy (New Delhi) to JSPS—Japan Society for Promotion of Science (Tokyo) to carry out research at Faculty of Environmental and Information Sciences of Yokkaichi University, Yokkaichi city in central Japan, on 'the effects of land-use on biodiversity of beneficial microarthropods'.

Prof. Reddy has published more than hundred twenty-five research and review papers in reputed international and national journals, as well as in various conferences and congress proceedings. He was edited two books entitled 'Soil Organisms and Litter Decomposition in the Tropics', published by Oxford & IBH Publ. Co. Pvt. Ltd. (New Delhi and co-published by Westview Press (Colorado, San Francisco, Oxford) in 1995 and 'Management of Tropical Agroecosystems and the Beneficial Soil Biota', published by Oxford & IBH publ. Co. Pvt. Ltd. (New Delhi) and Science Publishers, Inc. (U.S.A.) in 1999. More than twelve students have been awarded doctoral degress under his research supervision.